A Mathematician's Journeys

Archimedes

NEW STUDIES IN THE HISTORY AND PHILOSOPHY OF SCIENCE AND TECHNOLOGY

VOLUME 45

Archimedes has three fundamental goals; to further the integration of the histories of science and technology with one another: to investigate the technical, social and practical histories of specific developments in science and technology; and finally, where possible and desirable, to bring the histories of science and technology into closer contact with the philosophy of science. To these ends, each volume will have its own theme and title and will be planned by one or more members of the Advisory Board in consultation with the editor. Although the volumes have specific themes, the series itself will not be limited to one or even to a few particular areas. Its subjects include any of the sciences, ranging from biology through physics, all aspects of technology, broadly construed, as well as historically-engaged philosophy of science or technology. Taken as a whole, *Archimedes* will be of interest to historians, philosophers, and scientists, as well as to those in business and industry who seek to understand how science and industry have come to be so strongly linked.

More information about this series at http://www.springer.com/series/5644

Alexander Jones • Christine Proust
John M. Steele
Editors

A Mathematician's Journeys

Otto Neugebauer and Modern Transformations of Ancient Science

Springer

Editors
Alexander Jones
Institute for the Study of the Ancient World
New York University
New York, NY, USA

Christine Proust
Laboratoire SPHERE
CNRS & Université Paris Diderot
Paris, France

John M. Steele
Department of Egyptology and Assyriology
Brown University
Providence, RI, USA

ISSN 1385-0180 ISSN 2215-0064 (electronic)
Archimedes
ISBN 978-3-319-25863-8 ISBN 978-3-319-25865-2 (eBook)
DOI 10.1007/978-3-319-25865-2

Library of Congress Control Number: 2015959787

Springer Cham Heidelberg New York Dordrecht London

Printed on acid-free paper

Springer International Publishing AG Switzerland is part of Springer Science+Business Media (www.springer.com)

Preface

Otto Neugebauer, more than any other scholar of recent times, shaped the way we perceive premodern science. Through his own hugely productive scholarship and his influence on three generations of colleagues who learned from him as students and collaborators, he inculcated both an approach to historical research on ancient and medieval mathematics and astronomy through precise mathematical and philological study of texts and a vision of these sciences as systems of knowledge and method that spread outward from the ancient Near Eastern civilizations, crossing cultural boundaries and circulating over a tremendous geographical expanse of the Old World from the Atlantic to India. It is impossible for a present-day historian working on these fields not to be constantly conscious of the power, and sometimes the limitations too, of Neugebauer's intellectual legacy.

Neugebauer's career was demarcated by repeated changes in his interests.[1] Born in 1899 in Innsbruck, he was schooled in the Akademisches Gymnasium at Graz, where he showed little enthusiasm for the classical languages but much for technical subjects including mathematics. After serving in the Austrian army in the First World War, he enrolled in turn at the Universities of Graz, Munich, and Göttingen, shifting the focus of his studies from engineering to physics to mathematics. As a doctoral student at the Mathematical Institute at Göttingen, although he studied a broad range of areas of contemporary mathematics, he turned his research entirely to the history of mathematics in antiquity, and the subject of his thesis was ancient Egyptian fractions. Despite his negligible original mathematical research, he earned the respect of his fellow mathematicians for the manifest rigor of his historical work and their indebtedness for his tireless service to the community, in particular as Courant's assistant in the administration of the Mathematical Institute and as the founding editor of the abstracting journal *Zentralblatt für Mathematik und ihre Grenzgebiete*.

[1] For biographical details we direct the reader to N. M. Swerdlow's outstanding article, Swerdlow 1993, as well as to the papers in this volume.

From the late 1920s through the late 1930s, Neugebauer's chief historical project was to study and publish the mathematical texts preserved on cuneiform tablets from ancient Babylonia, culminating in his three volume edition *Mathematische Keilschrift-Texte* (*MKT*, 1935–1937). Following the dismissals of Jewish faculty at Göttingen in 1933, Neugebauer took up a temporary post at the Mathematical Institute in Copenhagen. (Though not Jewish, Neugebauer was politically unacceptable to the Nazis and vice versa.) During his 5-year Danish sojourn, he came to be increasingly preoccupied with the history of ancient astronomy, beginning with late Babylonian tablets and Egyptian papyri. Mathematical astronomy was to be the subject closest to his heart for the rest of his life; among his numerous publications relating to it, the edition *Astronomical Cuneiform Texts* (*ACT*, 1955) and the *History of Ancient Mathematical Astronomy* (*HAMA*, 1975) stand out as indispensible resources for subsequent research.

Neugebauer resigned as editor of the *Zentralblatt* in late 1938 in response to the removal of Levi-Civita from the journal's editorial board and other racially motivated restrictions imposed on its operation. In the meantime, the collapse of the *Zentralblatt* as a genuine international journal led the American Mathematical Society to undertake a new abstracting journal, *Mathematical Reviews*, and Neugebauer was invited to be its editor, simultaneously being offered a position in the Mathematics Department at Brown University. He arrived in the USA in early 1939.

Brown's administration was greatly supportive of Neugebauer's research, and he was enabled to attract younger colleagues who shared interests and possessed complementary areas of expertise, beginning with the Assyriologist Abraham Sachs and the Egyptologist Richard Parker. A special department of History of Mathematics was founded in 1947 for Neugebauer and Sachs, 2 years after the publication of their joint volume *Mathematical Cuneiform Texts*, devoted to tablets, mostly in American collections, that had not been available for inclusion in *MKT*. The department later grew through the appointments of Gerald Toomer in 1965 and David Pingree in 1971; it was closed following Pingree's death in 2005 and the professorship in the history of the exact sciences in antiquity was transferred to the newly created Department of Egyptology and Ancient Western Asian Studies (renamed the Department of Egyptology and Assyriology in 2014). Through a steady flow of visitors and students, it built up a worldwide network of historians who were deeply influenced by Neugebauer. From 1950 until his death in 1990, Neugebauer also was a member of the Institute for Advanced Study at Princeton, where he worked for a part of every year in alternation with Brown.

Younger historians seldom have anything approaching the deep familiarity with older historiography that Neugebauer, through his vast reading, possessed, and it is not uncommon to find Neugebauer treated as the starting point for the fields most strongly represented in his work, especially ancient Near Eastern mathematics and astronomy. The quarter-century since his death gives us enough distance to consider afresh both the researches of others that preceded and laid the foundations for his contributions and the ways in which the study of the ancient exact sciences has taken new directions following his fundamental publications. Moreover, through documentary collections such as the Shelby White and Leon Levy Archives Center of the Institute for Advanced Study (which houses the largest holdings of Neugebauer's papers), new information is available concerning Neugebauer himself and in particular his intellectual formation at Graz, Munich, and Göttingen, helping us to understand his distinctively "mathematician's" approach to the history of science, what Swerdlow calls the "notable tension between the analysis of culturally specific documents… and the continuity and evolution of mathematical methods regardless of ages and cultures."[2]

In 2010, marking the twentieth anniversary of Neugebauer's death, a conference was held at the Institute for the Study of the Ancient World (ISAW), New York University, entitled "A Mathematician's Journeys: Otto Neugebauer between history and practice of the exact sciences."[3] The goal of the conference was to explore facets of Neugebauer's career, his impact on the history and practice of mathematics, and the ways in which his legacy has been preserved or transformed in recent decades, looking ahead to the directions in which the study of the history of science will head in the twenty-first century. This collection of papers includes a large part of the papers presented during the conference, several of them in considerably revised and expanded form. It has three principal focuses: the central interval of Neugebauer's career in the 1920s and 1930s during which he was most closely connected with the mathematical community while making himself, in turn, a mathematician, a historian of mathematics, and a historian of the exact sciences in the broader sense; the historiography of ancient Egyptian and Mesopotamian mathematics centering on Neugebauer's *Grundlagen der ägyptischen Bruchrechnung*, *MKT*, and its sequel *MCT* written in collaboration with Sachs; and the historiography of Babylonian astronomy centering on *ACT*. In the spirit of Neugebauer's own attention to less studied and less regarded texts as a means of better understanding the canonical landmark works of science, we hope that these papers will contribute to a more exact appraisal of the nature of Neugebauer's achievement and our relation to it.

---

[2] Swerdlow 1993, 141.

[3] The conference was organized by the editors of this collection together with John Britton, who, alas, died unexpectedly 5 months before it took place; we deeply missed his presence while editing the volume. The sponsors included New York University's Institute for the Study of the Ancient World and Courant Institute of Mathematical Sciences (New York), the Department of Egyptology and Ancient Western Asian Studies (now Egyptology and Assyriology) of Brown University (Providence), the Institute for Advanced Study (Princeton), the Center for International Research in the Humanities and Social Sciences (a joint research center between New York University and the Centre National de la Recherche Scientifique, France), and the CNRS research group Recherches Epistémologiques et Historiques sur les Sciences Exactes et les Institutions Scientifiques—now part of the CNRS research group SPHERE (Paris).

New York, NY, USA — Alexander Jones
Paris CX 05, France — Christine Proust
Providence, RI, USA — John M. Steele

Reference

Swerdlow, N. M. 1993. Otto E. Neugebauer (26 May 1899–19 February 1990). *Proceedings of the American Philosophical Society* 137.1: 138–165.

Otto Neugebauer (1899–1990), unidentified photographer, not later than 1936 (Courtesy of the Mathematisches Institut, Georg-August-Universität Göttingen). The photograph was published in *National Mathematics Magazine* 11, 1936, 16

Contents

Contributors

Béatrice André-Salvini Musée du Louvre, Paris, France

Lis Brack-Bernsen History of Science, University of Regensburg, Regensburg, Germany

Teije de Jong Astronomical Institute 'Anton Pannekoek', University of Amsterdam, Amsterdam, The Netherlands

Jens Høyrup Section for Philosophy and Science Studies, Roskilde University, Roskilde, Denmark

Duncan J. Melville Department of Mathematics, St. Lawrence University, Canton, NY, USA

Mathieu Ossendrijver Excellence Cluster TOPOI/Institute of Philosophy, Humboldt University, Berlin, Germany

Jim Ritter Institut de mathématiques de Jussieu-Paris Rive Gauche, Paris, France

UMR 7586, Sorbonne Universités, UPMC-Univ Paris 06, Paris, France

Univ Paris Diderot, Sorbonne Paris Cité, CNRS, Paris, France

David E. Rowe Johannes Gutenberg Universität, Mainz, Germany

Reinhard Siegmund-Schultze Faculty of Engineering and Science, University of Agder, Kristiansand S, Norway

From Graz to Göttingen: Neugebauer's Early Intellectual Journey

David E. Rowe

Otto Neugebauer's early academic career was marked by a series of transitions. As he moved from Graz to Munich and then on to Göttingen, his interests shifted from physics to mathematics, and then, quite suddenly, to the history of mathematics and the exact sciences, the field in which he would stake his formidable reputation. Still, the training he received during his formative years before he became a historian left a deep and lasting mark on his later work. Indeed, as I will argue here, Neugebauer remained throughout his career a leading representative of a distinctive Göttingen mathematical culture, an influence he felt even before he arrived there in the spring of 1922. Moreover, Neugebauer's close association with Göttingen had much to do with the personal friendship he established there with Richard Courant, himself a Göttingen product from the pre-war era when he came under the influence of Felix Klein and David Hilbert. Both Klein and Hilbert placed great hopes in Courant, a talented mathematician who proved to be a gifted organizer even as he served in the German army during the Great War. Afterward, they successfully engineered Courant's appointment to the chair formerly held by Klein, and in 1922 he became Director of the newly created Mathematical Institute in Göttingen. Despite striking temperamental differences, Neugebauer and Courant became lifelong collaborators, allies, and friends, bound from the outset by deep mutual trust and respect. Owing to these circumstances, any adequate picture of Neugebauer's early career must likewise take due account of Courant's own remarkable story, including various legends about mathematics in Göttingen that he helped to create and sustain.

Like many who had studied in Göttingen, Courant loved to tell stories about Hilbert, Klein, and other local heroes. Various versions of these tales became part of the lore found in scattered secondary sources, but especially in Constance Reid's biographies of Hilbert and Courant (Reid 1970, 1976). One easily recognizes that much of the information in these books was based on oral interviews. Indeed, her

---

D.E. Rowe (✉)
Johannes Gutenberg Universität, Mainz, Germany
e-mail: rowe@mathematik.uni-mainz.de

A. Jones et al. (eds.), *A Mathematician's Journeys*, Archimedes 45,
DOI 10.1007/978-3-319-25865-2_1

portraits of Göttingen mathematics drew heavily on eyewitness accounts, belated recollections that were often vivid, but not always accurate. In the case of Courant's life story, her version of certain important events in his life sometimes flies in the face of hard documentary evidence. A striking example concerns the events that led up to his appointment as Erich Hecke's successor (Hecke held Klein's former chair) in Göttingen in July 1920. By clarifying the record we can better appreciate how Courant benefitted from a set of unusual circumstances shortly before Neugebauer's arrival on the scene. Moreover, contextualizing these events sheds light on the vulnerability of Courant's enterprise during the early years of the Weimar Republic, a period when economic and political turmoil deeply affected relations in German mathematics. At the heart of this story also lies the pervasive issue of the "Jewish question" in German academia, particularly as this relates to the emergence of liberal/leftist elites who identified strongly with the ideals of the new political order.

Reid's biography tends to view Richard Courant through the prism of his later triumphs. Undoubtedly, he came to see himself through that very same glass, a perspective already apparent in her biography of Hilbert, a project he helped her to write. After its completion in 1970, Courant's close associate, K. O. Friedrichs, wrote Reid to ask if she would be willing to work on Courant's life story, in some ways a more delicate task. By the time she began, he was nearing death and could offer her little assistance. As an old man, living mainly in the past, no one would have begrudged him the privilege of basking in his fame, a success story that had taken many astonishing turns. Reid ended up relying heavily on Friedrichs for advice, but she also conducted interviews with many others who had been close with Courant, including Otto Neugebauer. Owing to these circumstances, Friedrichs naturally enters often in her biography of Richard Courant, subtitled "The Story of an Improbable Mathematician." Neugebauer's role in that story, on the other hand, was downplayed for the most part. Still, those familiar with mathematical life in Göttingen during the Weimar era knew very well that it was he, together with the institute secretary Hilde Pick, who managed routine daily affairs at Courant's institute (Schappacher 1987). As Courant's right-hand man, Neugebauer oversaw all aspects of the enterprise, reporting problems to his boss, while quietly implementing his policies, but also helping him pursue his larger schemes. There were many other key players as well, but without Neugebauer it would seem hard to imagine how Courant could have realized his vision for Göttingen mathematics.[1]

Filling Klein's Chair

In retrospect, Courant liked to view his life as a vindication of the ideals championed by his mentors, Hilbert and Klein. If we turn back the clock, though, and imagine him returning to Göttingen after the war, our picture of his place in the

[1] For a brief account of Courant's deep identification with Göttingen mathematics, see (Rowe 2015).

mathematical world of that day looks very different indeed. Surely no one then would have thought him a likely candidate for the chair once held by Felix Klein, a position that symbolized Göttingen's pre-eminent place in the world of mathematics. Within Germany, two universities had traditionally dominated the mathematical scene, Berlin and Göttingen. To be offered a full professorship from either of these two institutions represented the highest possible distinction a German mathematician could hope to attain. Courant was young and ambitious, but no one would have compared him with Hilbert's brilliant pupil, Hermann Weyl. Still, much had changed during the protracted period of the Great War that ended so disastrously for Imperial Germany. One consequence, even if little noticed amid the mayhem, was a marked loss in allure for its two leading universities, Berlin and Göttingen. Within mathematics, this became apparent when these two institutions sought to fill empty academic chairs.

After Klein's formal retirement in 1913—he never let go of the administrative reins completely until 1922—his professorship was occupied by Constantin Carathéodory, a Göttingen product who had taken his doctorate under Minkowski. Just before the war ended, however, Carathéodory was offered the chair in Berlin vacated by the death of the algebraist Ferdinand Georg Frobenius. Thereafter, Klein's former chair was briefly taken by Erich Hecke, one of Hilbert's most gifted protégés. Hecke, however, did not remain long, departing for Hamburg after the summer of 1919. Around this same time, Carathéodory was in the process of leaving Berlin to accept a position at the new University of Smyrna. This meant there were now professorial vacancies in mathematics at the two leading centers in Germany, a situation that had last arisen in 1892 when the death of Kronecker and the retirement of Weierstrass set off a chain reaction of appointments (Biermann 1973). But back then the domestic situation within Germany was tranquil, nothing like the atmosphere in 1919, when the country was nearly overwhelmed by political chaos, runaway inflation, and sporadic outbursts of violence. As before, the traditional rivalry between Germany's two leading mathematical centers manifest itself in their pursuit of the most distinguished candidates available, but with a very different outcome than in years past. The post-war climate in Germany thus decisively influenced the academic appointments that would eventually be made in Göttingen as well as in Berlin.

Among Hilbert's many distinguished students, Hermann Weyl stood in a special category all his own. His personal relationship with Hilbert, on the other hand, was highly ambivalent, in part because Weyl, unlike Hilbert, strongly preferred research over teaching. After joining the faculty at the ETH in Zürich in 1913, he afterward turned down a series of attractive offers from leading German universities, preferring to remain in Switzerland until 1930 (Frei and Stammbach 1992). In that year he was offered Hilbert's chair, an honor even Weyl could not refuse. Courant, of course, knew Hermann and Hella Weyl very well from their student days in Göttingen. In recalling his earlier life, however, he may well have forgotten that his own career owed much to Weyl, in particular the latter's reluctance to leave the beautiful surroundings of Zürich for the buzz-saw of mathematical activity in Göttingen. In this respect, Weyl was the polar opposite of Courant, who loved to be at the center of the storm.

In Reid's biographies of Hilbert and Courant, she tells a curious fable about a new professorship that Courant "apparently negotiated" in 1922; this was supposedly offered to Weyl, who then declined (Reid 1976, 90). One might naturally wonder how the Prussian Ministry of Education could have funded a new professorship at this time given the scarcity of financial resources available; moreover, if Weyl had in fact turned this position down, why was it not then offered to someone else? But, in fact, the true situation can easily be clarified and corrected: the year was 1920 and the position was Klein's former chair, the professorship Courant ultimately obtained. Moreover, contrary to what one reads further on in Reid's book, Courant's call to Göttingen came about not through some carefully calculated plan hatched by Klein and Hilbert, but rather as the result of a complicated series of events that no one could have foreseen at the time. The actual course of negotiations in both Göttingen and Berlin can, in fact, be reconstructed from extant ministerial and faculty records, sources we can assume to be far more reliable than human memory. These documents not only clarify the chain of events that led to Courant's appointment but, even more, they throw fresh light on the surrounding circumstances as well as the truly abysmal living conditions in Germany at this time.

Given the prestige attached to these two vacant professorships, the faculties in Göttingen and Berlin naturally set their sights on the most accomplished mathematicians of the day. Both universities focused on three outstanding candidates: the Dutch topologist L.E.J. Brouwer, Leipzig's Gustav Herglotz, and Hermann Weyl. In Göttingen, these three were nominated in just that order, whereas the Berlin faculty placed Weyl after Brouwer, but ahead of Herglotz. Clearly, a strong consensus of opinion had been reached about these three men, but then something happened that would have been unthinkable in earlier times: all three candidates turned down *both offers*, preferring to remain in Amsterdam, Leipzig, and Zurich, respectively. In view of the ongoing political unrest in Berlin, which culminated with the unsuccessful Kapp Putsch in March 1920, one can easily understand their reluctance to reside in the Prussian capital. Weyl, for one, quickly dismissed this possibility, but not the idea of leaving Switzerland for Göttingen. It took him a good six months before he finally declined, thereby opening the way for Courant's dark horse candidacy.

In the meantime, Courant's personal ties to Göttingen had become stronger than ever (Rowe 2015). Immediately after the war he was eking out a living as *Assistent* to Carl Runge, Göttingen's Professor of Applied Mathematics. His relations with the Runge family grew even closer when in January 1919 he married their daughter, Nina. Housing being scarce, the newlyweds resided with her parents, and early the next year Nina gave birth to their son Ernst. Not long afterward, Courant was offered a professorship in Münster, the chair formerly occupied by Wilhelm Killing. He accepted, despite the drudgery of travelling back and forth from Göttingen. At this time he had no idea that he might be offered Klein's former position, though he probably knew that Hilbert and Klein were agitated over Weyl's inability to reach a decision. Still, there was no immediate plan to recruit Courant from Münster, contrary to Reid's version of the ensuing events. In fact, the extant documentary evidence suggests a very different picture.

When Hecke left for Hamburg in the fall of 1919, Arthur Schoenflies wrote to Hilbert, offering him advice about potential candidates.[2] As a former protégé of Klein and Hilbert, Schoenflies was well aware of their general views regarding academic appointments. He thus left Courant's name off his list on the assumption that he could not be promoted from a mere titular professor, a status he acquired in 1918, to an *Ordinarius*. Even more to the point, Schoenflies explicitly noted that such a nomination would contravene the principle prohibiting *Hausberufungen* (in-house appointments) since Courant had never held a position outside Göttingen. Schoenflies thus understood very well that Courant had strong support, but he also knew that his candidacy would have encountered great resistance, if only on purely formal grounds.[3]

Schoenflies did not need to raise another inevitable hurdle, one that he, as a Jew, knew all too well. The philosophical faculty in Göttingen had long been open to accepting Jewish colleagues with the understanding that there should never be more than one in a given field (Rowe 1986). Thus when Minkowski suddenly died in 1909, he was succeeded by Landau, one of three Jews nominated for the position (the others were Otto Blumenthal and Adolf Hurwitz). It seems this chair was informally reserved for Jewish candidates, which clearly posed an obstacle for Courant's appointment to one of the other three chairs in mathematics. Thus, to gauge what was at stake here in 1920 one must take into account the larger issue of the "Jewish question" as this relates to career opportunities in mathematics (Siegmund-Schultze 2008).

After the turn of the century, German universities had gradually drawn large numbers of talented Jewish students, many of whom excelled in fields like mathematics and theoretical physics. Institutions of higher education, however, were none too eager to employ them, particularly recent arrivals from the east. By the end of the war, several distinguished Jewish mathematicians still awaited their first regular academic appointment. A few had been passed over on several occasions in favor of Christian candidates; they were either stymied by resistance at the faculty level or occasionally at the higher level of the state ministries.

Max Born, who had studied alongside Courant in Göttingen, managed to gain an appointment in Berlin during the war as an associate professor of theoretical physics. This was a fairly new professional field of research, spawned over the course of the preceding 30 years at several German universities. During the Weimar era, this field came to symbolize the ascendency of Jews in German science, spearheaded by figures like Born (Jungnickel and McCormmach 1990). As a side benefit of his appointment, Born got to strike up a useful friendship with Albert Einstein: both were determined to do what they could to promote the careers of their kinsmen in mathematics and physics. Thus, in 1919 Einstein wrote to Felix Klein urging him to take steps with the Prussian Ministry of Education so that the brilliant female mathematician, Emmy Noether, would finally be given the title of *Privatdozent* in

[2] Schoenflies to Hilbert, 1919, Hilbert Nachlass 355, Niedersächsische Staats- und Universitätsbibliothek Göttingen.

[3] Courant's appointment in Münster thus helped pave the way for his return to Göttingen in 1920.

Göttingen.[4] Her lecture courses on modern algebra soon thereafter attracted a throng of enthusiastic young talent.

Born lacked Einstein's influence, but he nevertheless did his part to promote the cause of Jewish academics. In 1919 he wrote to Carl Heinrich Becker in the Prussian Ministry that the time had come to level the playing field, pointing to Courant's case as an example of past injustices.[5] That same year a chair in mathematics came open in Halle. Hilbert was contacted by the physicist, Gustav Mie, who sought advice about prospective candidates. Hilbert named Isaai Schur, Paul Koebe, and Courant in that order. Nevertheless, the chair instead went to Heinrich Wilhelm Jung, who had been full professor in Kiel since 1913.[6] In 1917 the Berlin faculty had already placed Schur *aequo loco* with Carathéodory, hoping that the Ministry would appoint both. Instead the Greek was chosen over the Jew; as a sign of the times, Isaai Schur, a Russian-born Jew, was nominated no fewer than nine times to various chairs at German universities before he was finally appointed to a full professorship (Biermann 1973).

A particularly striking instance illustrating the tensions aroused by this backlog of talented Jewish mathematicians can be seen from the private correspondence of Otto Toeplitz, who contacted several leading Göttingen mathematicians in early 1920 about a vacancy in Kiel. Like Courant and Born, Toeplitz came to Göttingen from Breslau, where his father taught mathematics at a local Gymnasium. He and Ernst Hellinger, both Jews from Silesia, became leading experts on Hilbert's theory of integral equations. In 1913 Toeplitz took a post as associate professor in Kiel, where he was promoted to full professor after the war. When his colleague, H. W. Jung, decided to accept the call to Halle, Toeplitz found himself in the unenviable position of serving on a commission charged with nominating candidates for this vacant chair in mathematics. As was surely expected, he began by seeking the advice of senior colleagues in Göttingen, including Felix Klein. Toeplitz wished to learn what Klein thought about five particular individuals—Isaai Schur, Ernst Steinitz, Leon Lichtenstein, Ernst Hellinger, and Felix Bernstein—all of whom happened to be Jewish. In responding, Klein added some remarks questioning the wisdom of this approach, particularly in the present political climate: "… on the one hand we have not only the enormous advance of Jews as a result of their peculiar abilities but also through the rise of Jewish solidarity (where Jews seek in the first instance to help and support their clansmen in every way). On the other hand, in reaction to this, we have rigid anti-Semitism. This is a general problem in which Germany plays only a secondary role, if we leave the new immigration from the East out of consideration. No one can say how things will develop." (Siegmund-Schultze 2008, 26). Klein, who was well aware of past injustices, commented fur-

[4] Einstein to Klein, 27 December 1918, *Collected Papers of Albert Einstein*, vol. 8B (Princeton: Princeton University Press, 1998), pp. 975–976.

[5] Born to C. H. Becker, 1919, Geheimes Staatsarchiv Preußischer Kulturbesitz, Berlin, I.HA. Rep.92. C. H. Becker.7919.

[6] It is possible, though, that Schur may have been offered this position; in 1919 he attained a long-sought promotion to full professor in Berlin, where he had been the star pupil of Frobenius.

ther: "One could also almost argue that the flourishing of anti-Semitism at all universities has given Christian candidates such an advantage that only Jewish candidates are now available. But I ask you, please, to think about it again. We are potentially entering into conflicts that could become disastrous for our situation as a whole." (*ibid.*)

Toeplitz was more than a little surprised by the frankness of Klein's letter, but he responded in kind by explaining some of the special circumstances in Kiel. In the meantime Toeplitz had received a letter from Hilbert, who wrote in praise of Steinitz, also naming Felix Hausdorff, Ludwig Bieberbach, and Lichtenstein as worthy of a second or third place on the list. Hilbert seems to have given little weight to the issue of racial background, though Bieberbach was an ethnic German, of course.[7] His colleague, Edmund Landau, saw this as a potential problem, however, and thus advised Toeplitz to add one or two non-Jewish names, even if they were not likely to accept an offer from Kiel. Landau, who would later suffer the indignity of having his lectures boycotted by young Nazis, no doubt shared some of Klein's misgivings (Schappacher 1987). Soon thereafter, Steinitz received the call to Kiel. These background events at a provincial university would hardly be worth describing in such detail were they not symptomatic of much larger issues clearly reflected in the exchanges cited above. Ethnic and religious factors had always played a major role in academic appointments at the German universities, but in this new political climate the "Jewish question" took on a special urgency that strongly shaped and influenced concurrent deliberations over suitable candidates including the two positions that remained to be filled in Göttingen and Berlin.

In the meantime, the situation in Göttingen had become quite complicated due to the departure of the Dutch theoretical physicist Peter Debye, who chose to accept an attractive offer from the University of Zürich. Debye had worked quite closely with Hilbert, who was intent on finding a suitable successor. His first choice was Max Born, who formerly studied in Göttingen and was now teaching in Frankfurt. Thus, in mid-February Hilbert wrote to Einstein, asking him to send a letter assessing Born's abilities as well as his suitability for the position in Göttingen. Einstein was happy to sing the praises of his friend, whom he once regarded as primarily a mathematical talent. Einstein now thought, however, that Born's more recent work showed a strong sense for physical reality.[8] This letter, written the very day the Philosophical Faculty convened, may well have given Hilbert the ammunition he needed. In any case, Born's name appeared second on the faculty's list, behind Arnold Sommerfeld's. No one imagined the latter would be tempted to leave Munich, as proved to be the case, so Born quickly emerged as the candidate of first choice.

Born wrote Einstein for advice and then plunged into a series of complex negotiations with the Berlin Ministry as well as the Göttingen faculty (Greenspan 2005,

---

[7] Hilbert to Toeplitz, 8 February 1920. Universitäts- und Landesbibliothek Bonn, Toeplitz B: Dokument 47. Hilbert mistakenly thought that Hausdorff, too, was of non-Jewish background.

[8] Einstein to Hilbert, 21 February 1920, *Collected Papers of Albert Einstein*, vol. 9, (Princeton: Princeton University Press, 2004), p. 440.

96–102). Hilbert had already signaled to Born that he would have the opportunity to recommend an experimental physicist to fill another vacancy, so he already had a bargaining chip in hand. He played it forcefully by making plain that he would not leave Frankfurt unless the Göttingen faculty agreed to a double appointment; furthermore, he insisted that the second chair in experimental physics had to be offered to his friend James Franck, then director of the physics division at the Kaiser Wilhelm Institute for Physical Chemistry in Berlin. Franck found this plan highly amenable, but various complications quickly ensued. Since both men were of Jewish background, this bold venture was bound to encounter resistance within the Philosophical Faculty, more than Born bargained with. As it turned out, the negotiations dragged on for several months. Some years later, Hilbert recalled how Born's appointment proved to be "the most ruthless and hardest fight [he] ever had to endure in the faculty."[9] There had been many such fights, in fact (Rowe 1986). Hilbert had a well-deserved reputation as a fearless warrior for liberalism when it came to academic politics, a prime reason why he was much admired by those in that same camp and so loathed by his conservative colleagues.[10]

This particular battle had not yet ended when, in early July, Weyl's letter finally arrived; after much soul-searching he decided to reject the Göttingen offer (Frei and Stammbach 1992; Weyl 1932). Now that the original list of candidates had been exhausted, the idea of calling Courant from Münster could at last come into play. Klein decided to lay all his cards on the table. He composed a letter to Courant, which he read in Hilbert's presence, setting forth the mutual understanding he assumed all three of them shared. This began: "As you may have heard from other sources, I intend to advocate your appointment in Göttingen. It would be extremely helpful for me if you would confirm explicitly in writing that you are willing to promote energetically tasks which, in my opinion, have long been unduly neglected in our educational system as well as new demands which I can foresee as coming up." (Reid 1976, 83) He then proceeded to enumerate which reforms he had in mind, and summed up by saying he was sure that none of these points would come as any surprise. Klein thereby obtained the proper assurances from Courant, who surely realized he would be assuming an awesome responsibility.

Klein and Hilbert now took their case to the faculty, but there they encountered a potential roadblock: Edmund Landau was not to be persuaded. Landau saw no reason to doubt Courant's abilities, but he expressed strong reservations with regard to what he perceived as an unhealthy trend in Göttingen, one that was creating an imbalance between pure and applied mathematics. As a number theorist, Landau

[9] Hilbert to Hermann Wagner, 1926, Cod. Ms. H. Wagner 27, Niedersächsische Staats- und Universitätsbibliothek Göttingen.

[10] One of his admirers, Albert Einstein, wrote on the occasion of Hilbert's 60th birthday: "Nur ein Zipfel Ihres gewaltigen Lebenswerkes kann ich Beschränkter (und Fauler) überschauen, aber gerade genug, um das Format Ihres schaffenden Geistes zu ahnen. Dazu den Humor und den sicheren, selbständigen Blick in alle Dinge und—einen harten Schädel wie kein zweiter nebst zwei starken Armen, um von Zeit zu Zeit den Fakultätsstall auszumisten." (Einstein to Hilbert, 18 February 1922, *Collected Papers of Albert Einstein*, vol. 13, (Princeton: Princeton University Press, 2012), p. 92).

had long felt isolated in a community where analysis, mathematical physics, and applied mathematics dominated the scene, so he saw no reason to appoint yet another applied type like Courant. Instead he pushed for a pure mathematician, nominating Berlin's Isaai Schur in a strongly supportive letter. This went out to the Ministry on 12 July (just 4 days after Weyl had declined the offer) together with the counter-proposal, signed by Klein and Hilbert, with very different arguments in favor of Courant (including his bravery during the war). Even now, no one could have been sure that the Ministry would agree to either of these two candidates, though soon thereafter Courant received the good news.

What transpired afterward in Berlin would also eventually have profound consequences for mathematics in Germany. Following the initial failure to fill Carathéodory's chair in Berlin, the Prussian Ministry opened negotiations with Hamburg's Erich Hecke. However, he too declined, forcing the Berlin faculty to reconvene in order to start the search process all over again. It took until the end of 1920 before they could agree on a new list. This time they named the Austrian geometer, Wilhelm Blaschke, Frankfurt's Ludwig Bieberbach, and the geometer Gerhard Hessenberg, who taught in Tübingen. After Blaschke declined the position, Bieberbach agreed to accept the post, one that accorded with his ambitions and inflated self-esteem. Nevertheless, his opposition to the Göttingen faction only gradually emerged during the late 1920s.

Courant's sense of loyalty to Klein, Hilbert, and Runge ran very deep. No doubt his sincerity and sense of belonging was fully appreciated when they chose him. Small and soft-spoken, Richard Courant must have appeared as the unlikeliest imaginable successor to Felix Klein, and yet he promoted the legacies of both Klein and Hilbert brilliantly (Rowe 2015). As a pupil of Hilbert, he took up classical analysis—variational methods, Dirichlet's principle and conformal mapping—a program that kept him busy all his life. His *Doktorvater* had, in fact, already formulated this research agenda a few years before Courant arrived on the Göttingen scene in 1907. Still, these research interests represented only one strand within the Hilbertian legacy, as Courant and those close to him well knew. Kurt Friedrichs, one of the most distinguished members in the Courant circle, thought his mentor was intellectually far closer to Klein than to Hilbert (Reid 1976, 241). To many, Courant seems to have been an anomaly—at once a daring innovator, but at the same time a conservative with a deep belief in the vitality of older traditions. What he accomplished in Göttingen was largely to build on the shoulders of Klein and Hilbert, the giants who dominated the scene during the pre-war years. With Courant's return, followed by the double appointment of Born and Franck, Göttingen suddenly acquired an impressive trio of talent; they were not only gifted but, just as importantly, all three got along with each other exceptionally well. That they all happened to be secular Jews did not escape notice either; each got to know firsthand about various forms of local anti-Semitism.

Two years later, during Neugebauer's first semester in Göttingen, Niels Bohr came to deliver seven famous lectures on quantum physics, an event that came to be known as the *Bohr Festspiele*. For Courant and Born, this brought back memories of the good old days when Poincaré, Lorentz, Planck, and other stellar figures came to speak about current research in physics. Indeed, the *Bohr Festspiele* represented a revival of the pre-war Wolfskehl lectures funded by the Göttingen Academy—the last such lectures, in fact, owing to the collapse of the Mark—while marking one of the highlights in Bohr's singularly successful career (Mehra and Rechenberg 1982, 345–368). Beyond that, his very presence served to reinforce the close personal ties that developed between the Göttingen trio and their friends in Copenhagen, above all Niels and Harald Bohr. International relations between German and Danish scientists had never been particularly problematic, but Courant's special friendships with the Bohr brothers would later help pave the way to support from other foreign sources, in particular the Rockefeller Foundation. Thus seen, the events of 1920 that led to the appointments of Courant, Born, and Franck gave Göttingen the impetus needed to re-emerge as a vibrant scientific community with plentiful interactions between younger mathematicians and physicists.

Physics in Graz

Little is known about Otto Neugebauer's early life growing up in Graz, where he attended school.[11] As a teenager, he probably found the curriculum at the classical Gymnasium rather dry and somewhat stifling: heavy loads of Latin and Greek, relatively little mathematics, and almost nothing of relevance pertaining to the exact sciences. He graduated early, at age 17, so that he could join the Austrian army. Throughout the latter half of the war, Neugebauer served as a forward observer for artillery stationed on the Italian front. Throughout this time he kept a detailed notebook of his numerous travels, apparently in order to have a record of what he had seen and done.[12] True to his character, it contains nothing at all about what he thought or felt, even though he must have been in dangerous situations quite often. Toward the end of the war, Neugebauer's company surrendered to the Italians, who took him into custody as a prisoner of war. If Neugebauer had any plans for his life after the war ended, he seems to have kept these to himself.

Soon after his release, he returned to Graz where he began studying theoretical physics (Table 1). This period in his life also remains quite obscure, but physics clearly remained his dominant interest up until he arrived in Göttingen. His principal instructors during these two years in Graz were the theoretical physicist, Michael Radaković, and the mathematician, Roland Weitzenböck. Radaković, an expert in ballistics, had studied under Helmholtz and Kirchhoff in Berlin; his younger colleague, Weitzenböck, was known as a leading authority on differential invariants. Both were well abreast of recent developments in theoretical physics, but more

[11] For much of the information relating to Neugebauer's training and educational activities in Graz, Munich, and Göttingen I have drawn heavily on sources that can be found in the Otto Neugebauer Papers, The Shelby White and Leon Levy Archives Center, Institute for Advanced Study, Princeton, NJ, USA, cited hereafter as "Neugebauer Papers, IAS".

[12] Neugebauer Papers, IAS, Box 13, Tagebuch, 1917–1919.

Table 1 Courses attended by Otto Neugebauer as a student in Graz

WS 1919–20	Radaković	Theory of Electricity
SS 1920	Radaković	Theory of Electricity
	Weitzenböck	Differential Equations
		Seminar Theoretical Physics
WS 1920–21	Radaković	Theory of Special Relativity, Thermodynamics
	Weitzenböck	Mathematical Foundations of Relativity
		Seminar Theoretical Physics
SS 1921	Radaković	Theory of Special Relativity, Theory of Radiation
	Weitzenböck	Mathematical Foundations of Relativity
		Seminar Theoretical Physics

importantly for Neugebauer their courses gave him a solid grounding in the theory of relativity, including the mathematical machinery required for understanding Einstein's approach to gravitation, the general theory of relativity.

Alongside these physics courses, Neugebauer also studied mechanical engineering, at least initially. He left some notes from a course that dealt with the efficiency of machines, for which he made several drawings showing various designs and linkages together with calculations of pressure based on various data. If nothing else, these reveal his talent as a draftsman, a skill that would serve him well in Göttingen and also later when he turned to the history of astronomy. Thankfully, Neugebauer preserved detailed notes from the physics courses he attended during his years as a student in Graz. He copied these into neatly written notebooks, carefully organized in a manner that already bore the stamp of his personality.[13] Neugebauer worked through the material from these lecture courses; no doubt these notebooks were based on rough notes which he then cast aside. The finished products were thus *Ausarbeitungen* in the classical style: carefully elaborated reconstructions of the subject matter covered in these lectures; some of Neugebauer's notebooks are of nearly publishable quality. One recognizes immediately the work of a diligent hand and sharp mind, a writer who abhors sloppiness.

In July 1921, Neugebauer ended his studies in Graz by delivering two lectures in Weitzenböck's seminar on the mathematical foundations of relativity. Again, we can only form an impression of these oral presentations from his carefully written text, a document that shows all the marks of a brilliant, budding scholar.[14] Clearly, Neugebauer had learned a great deal from his two teachers, yet these two lectures reveal a bold, over-arching vision of the history of geometry from Euclid to Riemann and beyond, a viewpoint that stresses the rich interplay between mathematical and physical conceptions. The text resonates with the grandeur of a dramatic story, though not the more familiar one in which Einstein plays the dominant role. Neugebauer only alludes to Einsteinian *Leitmotifs*—Mach's principle and the ear-

[13] Neugebauer Papers, IAS, Box 7.

[14] Neugebauer Papers, IAS, Box 13, Vortrag über A.R.T. und (Weyl 1921).

lier principle of equivalence—while noting their importance for general relativity. His main focus remains throughout the geometrization of physics, culminating with recent efforts to unite gravity with electromagnetism in hopes of creating a new paradigm for field physics, a goal that would eventually become Einstein's dominant obsession. The first bold step in this direction, however, was taken by Hermann Weyl, who conceived of a new geometrical framework that went well beyond Einstein's general theory of relativity. Neugebauer's two lectures thus dealt not only with Einstein's theory of gravitation, but also with Weyl's even more recent unified field theory based on a generalization of Riemannian geometry.

Given that he was only in his fourth semester of studies, his knowledge of these complex matters, both technically and historically, can only be described as truly remarkable. When we consider this text in the light of Neugebauer's subsequent career, however, its significance would seem greater still. For here we see him writing as a 22-year-old neophyte historian of science, without any awareness of his future calling. In setting the stage for understanding Weyl's theory, he begins with a brief picture of the history of geometry, understood as the oldest of all exact sciences, a field of inquiry that had accompanied theorizing about the natural world from Euclid to Einstein. Clearly, Neugebauer's technical understanding of some of the very most recent developments in mathematical physics had been accompanied by a deep interest in the historical background. Indeed, his presentation alludes to sweeping vistas from the past, a hallmark of some of his best-known later work as an historian of mathematics and astronomy.

His larger vision in these two lectures thus concerns the dialectical interplay between geometry and physics, understood not merely in conjunction with modern field physics but rather as an enduring theme that profoundly shaped the historical development of both disciplines from antiquity to the present. Modern physics, he notes at the outset, treats the phenomena under discussion by means of idealized concepts drawn from complex observable events. Early mankind felt a deep urge to understand the natural world, even when the only possibility for doing so was by explaining its workings by means of myths, gods, or demons. Only one aspect of nature proved amenable to a scientific theory, namely geometry, which began as the study of idealized objects according to their various forms.

When at the very beginning of the *Elements* Euclid defines entities such as points and lines, he was merely describing familiar properties of idealized figures with well understood counterparts in the physical world. Geometry, so understood, forms part of the larger field of physical inquiry; it arises as the earliest and most elementary branch of the science of physics. From the very outset, Neugebauer set forth a larger philosophical theme that he uses as a framework for these lectures. His central claim, somewhat akin to Thomas Kuhn's much later notion of scientific paradigms, is that the natural historical tendency within every exact science—since *all such sciences* necessarily draw their own boundaries by means of idealization—will follow a pattern of development illustrated by the case of so-called Euclidean geometry. Thus, once a scientific theory reaches a certain stage of maturity and completeness, it will tend to raise new questions or problems that point beyond the rather arbitrary boundaries that were originally constructed by the discipline. Only at the end of the two lectures, when he returns to this theme, do we find a hint of at least one source of inspiration for this grandiose vision: there he cites Hilbert's views on the axiomatization of physics.[15]

With these general thoughts in mind, Neugebauer proceeds to describe briefly some of the key ideas that eventually led to the realization that Euclid's parallel postulate was merely one formulation of a global property that held within a special type of geometry. For 2000 years mathematicians had sought in vain to prove what ancient Greek geometers (possibly even Euclid himself) regarded as a fundamental postulate for establishing the theorems of plane geometry. These various efforts culminated in the eighteenth century with the probing investigations of Saccheri and Lambert (here Neugebauer drew on the standard source book on the pre-history of non-Euclidean geometry (Engel and Stäckel 1895)). Then, during the early nineteenth century, Lobachevsky and Bolyai made the essential breakthrough, even though few took note of what they had achieved. Gauss, who anticipated both, refused to publish anything on the topic, preferring to remain silent rather than enter thorny public discussions with regard to the geometry of physical space. Nevertheless, he did publish fundamental results on the differential geometry of surfaces which led to a key insight, namely that the localized measurements of geometrical figures depend merely on a suitable metric, hence they do not depend on any special theory of parallels or, as Neugebauer puts it, "relations at infinity."

Gauss thus opened the way to studying the intrinsic geometry of surfaces by means of their curvature properties. A Euclidean plane, being everywhere flat, has vanishing Gaussian curvature, but spheres have constant positive curvature, and by the 1860s Beltrami introduced surfaces of constant negative curvature as a means for visualizing the geometries of Lobaschevsky and Bolyai. Soon afterward, mathematicians suddenly became aware of Riemann's daring generalization of Gauss's surface theory, which not only opened the doors to higher-dimensional differential geometry but also placed age-old questions about the geometry of physical space in an entirely new light. Riemannian geometry went hand in hand with field forces, which act locally on objects in space, as opposed to the theories modeled on Newtonian gravitation, where objects act on one another at a distance.

Neugebauer sketched this background, including the rise of field physics, starting with the electromagnetic theories of Faraday and Maxwell, passing to their generalization by means of the Hamiltonian formulation of Mie's field theory of matter, and finally taking up Einstein's field theory of gravitation. Einstein viewed gravitational and inertial forces as induced by the effects of space-time curvature. This meant passing from a flat space-time geometry, introduced by Minkowski for special relativity, to a Riemannian geometry, thereby generalizing the theory of relativity to non-inertial frames of reference. Neugebauer clearly recognized that these were subtle matters, which had led Levi-Civita, Weyl, and many other mathematicians to formulate new analytic theories appropriate for space-time physics. Thus he emphasized the central importance of Weyl's concept of a guiding field, which

[15] Presumably Neugebauer had read Hilbert's 1917 lecture on "Axiomatisches Denken" (Hilbert 1917), which suggests a very similar viewpoint.

accounts for force-free motion in a curved space-time, thereby underscoring the central role of differential geometry in Einstein's theory. To clarify these connections, he presented a flow chart showing the interlocking relationships between key mathematical formalisms in the general theory of relativity. He then ended this first lecture by alluding to Weyl's bold new approach to geometrizing field physics by uniting gravity with electromagnetism. Weyl went beyond Einstein's original framework, but in order to do so he had to create a purely infinitesimal geometry that generalized the metrics of Riemannian geometry (Scholz 2001).

For his second lecture, Neugebauer adopted a parallel form of presentation. This time, however, he described a different line of geometrical ideas, beginning with Klein's classification of geometrical properties by means of appropriate transformation groups. Klein had already sketched this approach in his Erlangen Program of 1872, though it was not until the 1890s that this group-theoretic understanding of geometry gained wider appreciation. By 1910 Hermann Minkowski's geometrization of special relativity provided a natural link between space-time physics and group theory. In this way, Einstein's original theory came to be interpreted as the invariant theory of the Lorentz group: its transformations leave Maxwell's equations unchanged. Einstein's general theory of relativity was built using more general space-time manifolds, those with an invariant semi-Riemannian metric. Weyl's geometries, on the other hand, were more general still. These are manifolds equipped with local linear transformations that allow for the infinitesimal parallel displacement of vectors: manifolds with a so-called affine connection. Such purely infinitesimal geometries suffice for the introduction of a guiding field (*Führungsfeld*), a structure that serves to generalize the inertial frames of special relativity.

Neugebauer contrasted Einstein's theory with Weyl's by noting that, whereas in the former the laws of physics (such as Einstein's gravitational field equations) are invariant under arbitrary coordinate transformations, Weyl's transformations also require that an associated linear differential form remain invariant under gauge transformations. Weyl identified this form with the electromagnetic four-potential. After delving into some of the conceptual notions underlying Weyl's theory, Neugebauer discussed its striking formal features, again by making use of an elaborate flow diagram. Taking as axiomatic the existence of an invariant action integral under both coordinate and gauge transformations, he shows how Maxwell's equations for electromagnetism and Einstein's gravitational field equations fall out as special cases. Following these remarks, he wrote that no other theory has succeeded in uniting so many fundamental laws of physics. Still, he expresses caution: "We are dealing here with a pure *continuum theory*—derived from Riemann's introduction of a *continuous* manifold. How this will fare when it comes to the conceptions of quantum theory is entirely unknown. Perhaps these questions are connected with a great difficulty in Weyl's theory, namely its failure to distinguish between positive and negative charges." He then added that Weyl hoped to evade this problem by investigating the structure of the singularities in electromagnetic fields, which he interpreted as the seat of matter in this theory.

In summing up, Neugebauer ended this remarkable *tour de force* with these reflections: "The conclusion that we had already reached at the end of the first part of our survey now appears through Weyl's train of thought even stronger than before. What would seem to be two entirely *foreign* fields—geometry, or the theory of extended magnitudes, and physics—can be brought under the spell *of one* idea. This entirety is captured by a single great formalism: the *action principle*. Here we might recall an idea once expressed by Hilbert: if a science reaches a certain height it becomes amenable to the axiomatic method and thereby, indirectly, to mathematics. And now we touch again on much that was said at the start of our entire survey: a completed science leads beyond itself—or returns to idealization. Shall that not be a sign for the limits of our knowledge?" (Neugebauer Papers, IAS, Box 13, Vortrag über A.R.T. und (Weyl 1921), my translation).

Although he had just turned 22, Neugebauer was already writing about profound matters with the clarity that marks all his later work as an historian. Moreover, his intellectual trajectory pointed in a clear direction; one has merely to note again the names mentioned above: Gauss, Riemann, Klein, Hilbert, and Weyl. Thus, well before he ever set foot in Göttingen, Neugebauer was already steeped in its rich mathematical tradition.

From Munich to Göttingen

When Otto Neugebauer left Graz after the summer of 1921 to enter the Ludwig Maximilians Universität in Munich, he presumably did so knowing there was no better place in the world to study theoretical physics. Arnold Sommerfeld, a former protégé of Felix Klein, was now drawing some of the era's most brilliant minds to Munich. Wolfgang Pauli had just taken his doctorate when Neugebauer arrived; he would spend the academic year 1921–22 working with Max Born in Göttingen. In the meantime, Sommerfeld still had another young bull in his stall: Werner Heisenberg, a budding genius destined like Pauli to put his stamp on the new *Knabenphysik*.

During his semester in Munich (Table 2), Neugebauer took two courses in physics. The first, taught by Sommerfeld, dealt with mathematical techniques used to solve boundary value problems in Maxwell's theory. In the second, Neugebauer learned about quantum theory from Karl F. Herzfeld, a *Privatdozent*. Herzfeld focused on Bohr's quantum theory, which sought to account for spectroscopic results by introducing small perturbations in the classical dynamical systems of Hamiltonian theory. This theory was elaborated in Sommerfeld's recently published monograph, *Atombau und Spektrallinien*, but it remains unclear whether Neugebauer

Table 2 Courses attended by Otto Neugebauer as a student in Munich

WS 1921–22	Sommerfeld	Boundary Value Problems in Maxwell's Theory
	Herzfeld	Quantum Mechanics of Atomic Models
	Rosenthal	Theory of functions of a real variable

Table 3 Courses attended by Neugebauer in Göttingen, 1922–27

SS 1922	Courant	Partial Differential Equations, Seminar: Algebraic Functions
	Hilbert	Statistical Methods, especially for Physics
	Born	Kinetic Theory of Gases
WS 1922–23	Landau	Transcendental Functions, Analytic Number Theory
SS 1923	Landau	Analytic Number Theory
SS 1925	Noether	Algebraic Functions
WS 1926–27	Herglotz	Celestial Mechanics

studied this celebrated text. In fact, he took only 40 pages of notes for this course, far less than what one finds in his notebooks from the courses he took in Graz. It seems likely that he found this whole new subject rather murky and unappealing. At any rate, Herzfeld's course seems to have been his only direct encounter with quantum physics.

However, another course taken at this time did prove of lasting value for him. In Graz he had imbibed large doses of mathematical physics, but he still lacked a solid grounding in pure analysis. This he gained in a course on functions of a real variable taught by Arthur Rosenthal, from whom he learned both classical and modern methods as exemplified by the integration theories of Riemann and Lebesgue.

Neugebauer surely planned to stay longer when he first arrived in Munich, but soon thereafter he changed his mind. Evidently he felt little sympathy for the methods of quantum theory, whereas Sommerfeld and Rosenthal both thought his talents lay more in the direction of mathematics than physics. He might have considered leaving for Heidelberg, since Rosenthal had just accepted a new position there, but instead he opted for Göttingen. In all likelihood, Sommerfeld helped pave this transition, for shortly after Neugebauer arrived, Richard Courant invited him to participate in his seminar. He also decided to attend Courant's course on partial differential equations, while at the same time deepening his knowledge of mathematical methods in physics by taking courses taught by Hilbert and Born (Table 3).

Courant was hardly a brilliant lecturer, but he did have the ability to spark interest in students by escorting them to the frontiers of research in analysis. In his course on partial differential equations, he stressed two sharply opposed types of literature: works that expound general theory, on the one hand, and those that pursue special problems and methods, on the other.[16] To the first category, he counted standard textbooks in French, English, and German by Goursat, Jordan, Forsyth, and Lie & Scheffers, a highly eclectic list. For the second type of literature, Courant's background and personal preferences came to the fore. Here he recommended the *Ausarbeitung* of Riemann's lectures on partial differential equations prepared by Karl Hattendorf together with Heinrich Weber's subsequent volume on Riemann's theory of PDEs in mathematical physics. Courant also drew attention to the text-

[16] Neugebauer Papers, IAS, Box 7.

book by Friedrich Pockels, which had grown out of a lecture course offered by Klein, as well as Sommerfeld's article on boundary-value problems in the *Encyklopädie der mathematischen Wissenschaften*. All these works were deeply rooted in the Göttingen tradition, whereas the last was, of course, closely linked with material Neugebauer already knew, having attended Sommerfeld's course the previous semester.

Courant's plan that semester was highly ambitious, as he hoped to deal with no fewer than five different topics. To cover this material, he decided to leave aside all the fine points of the theory, including the proofs for the required existence theorems. He began with a general introduction, motivated by simple examples, before proceeding to the theory of first-order PDEs. For his third topic, Courant chose various second-order PDEs of special important for physics, after which he would turn to related boundary-value problems. Finally, he planned to wrap up the course with a discussion of relevant problems in the calculus of variations. Neugebauer's notes leave the impression that Courant must have had other things on his mind that semester besides teaching. He took down only around 50 pages, and these contain almost nothing on boundary-value problems and not a word about the calculus of variations.

Even so, he seems to have fared even less well in his other two courses. By this time, Hilbert had also begun to plunge into Bohr's recent work on quantum physics. His doctoral student, Hellmuth Kneser, had just published his dissertation (Kneser 1921) on the mathematical foundations of quantum theory, and his new colleagues in physics, Born and Franck, took a deep interest in these breaking developments. Neugebauer, on the other hand, attended only the first part of Hilbert's two-semester course, which was devoted to classical statistical mechanics. Thus, he bypassed the chance to enroll in Hilbert's course on the Mathematical Foundations of Quantum Theory (Hilbert 2009, 507–601), opting instead to study number theory with Edmund Landau. This choice surely marks an important personal transition; here he had the chance to gain a deep knowledge of the mathematical methods needed in quantum theory—a topic Hilbert had in mind when he quipped that "physics had become too difficult for the physicists"—but decided in favor of pure mathematics. Hilbert's lecture course was worked out by his physics assistant, Lothar Nordheim, who went on to become a pioneer in the application of quantum mechanics to solids.

Neugebauer also seems to have found Born's course on thermodynamics—or more probably the subject itself—not really to his liking. His notes break off abruptly when Born came to a famous paradox, first debated by Zermelo and Boltzmann in the 1890s, concerning the proper interpretation of Poincaré's recurrence theorem (Zermelo insisted *contra* Boltzmann that Poincaré's theorem for dynamical systems stood in contradiction to the second law of thermodynamics). Neugebauer wrote a brief marginal note that indicated he found Born's explanation less than convincing ("Hier ist etwas sicher nicht in Ordnung! Mengenlehre!").[17]

[17] Neugebauer Papers, IAS, Box 7. Neugebauer once made reference to Poincaré's theory of lunar motion to refute standard teleological views of astronomers in the sixteenth century: "The investigations of Hill and Poincaré have demonstrated that only slightly different initial conditions would

Apparently he never attended the course afterward, a clear sign that he was drifting away from physics.

Neugebauer's transition to pure mathematics was most definitely hastened by his experiences in Courant's seminar on algebraic functions. Countless stories have been told about episodes that took place in Göttingen seminars, many of them reflecting their highly charged, competitive atmosphere. All participants surely knew that their performances would be scrutinized carefully, not only by the professors but even more so by their peers. One needed a thick skin in Göttingen, especially in the seminars, which served as true testing grounds for young mathematical talent. Neugebauer suddenly found himself amidst a stellar gathering of ambitious young men that included Courant's *Assistent*, Hellmuth Kneser, two brilliant young post-docs, Carl Ludwig Siegel and Emil Artin, and another newcomer, Kurt Otto Friedrichs from Düsseldorf. All would go on to become world-class researchers.

Decades later, Friedrichs still had vivid memories of the special atmosphere in that particular seminar. Courant had a way of attracting talent and then coaxing them to take part in seminars, so: "they all had to come and participate. Such a group of people, who knew everything about everything—it was very exciting to me" (Reid 1976, 91). Friedrichs also recalled how shocked he was at the level of difficulty. In Freiburg he had taken a course based on the first two chapters of the same text, whereas Courant simply skipped that material and started with chapter three. The book Friedrichs referred to in this conversation with Constance Reid was the classic volume by Hensel and Landsberg, *Theorie der algebraischen Funktionen einer Variabeln und ihre Anwendung auf algebraische Kurven und abelsche Integrale*, first published in 1902.

Neugebauer probably had even less exposure to this subject than Friedrichs; the course he took with Rosenthal in Munich was not likely to have touched on this abstract arithmetical approach to algebraic curves and Riemann surfaces. Nevertheless, he jumped head first into the fray with presentations on three separate occasions.[18] His first talk on May 11th dealt with the notion of a basis for a field of algebraic functions; he spoke a second time on 22 June, presenting an overview of the foundations as well as some important problems in the theory of algebraic functions; finally, on 20 July he gave a talk that dealt with Weierstrass's *Lückensatz*. Courant, himself no expert on this subject, was duly impressed; he obviously had found another bright young talent for his budding group of researchers. Neugebauer, too, found this type of communal learning environment much to his liking, and he surely realized that Courant was a mathematician of unusual breadth.

---

have caused the moon to travel around the earth in a curve [with nodal loops and] ... with a speed exceedingly low in the outermost quadratures as compared with the motion at new and full moon. Nobody would have had the idea that the moon could rotate on a circle around the earth and all philosophers would have declared it as a logical necessity that a moon shows six half moons between two full moons. And what could have happened with our concepts of time if we were members of a double-star system (perhaps with some uneven distribution of mass in our little satellite) is something that may be left to the imagination." He even drew a figure to illustrate this (Neugebauer 1969, 152–153).

[18] Neugebauer Papers, IAS, Box 7, Drei Vorträge aus dem Gebiet der algebraischen Funktionen.

No doubt Courant had chosen this topic because he wanted to learn something about the arithmetical approach to algebraic functions, a theory that was still quite unfamiliar in Göttingen. Hilbert had earlier been drawn to explore the deep connections between fields of algebraic functions and algebraic number fields, an analogy explored by Kronecker and Weierstrass, and then elaborated by Dedekind and Weber in a famous paper from 1882. Hilbert's own work in this area, however, predated Courant's arrival in Göttingen. After 1900, when Hilbert attracted a huge flock of talented doctoral students, the master focused on analysis and mathematical physics. Practically all of his students did their dissertation work on topics in these areas. Courant certainly had broader interests than most, but these were still confined to pure and applied analysis.

On the other hand, Emmy Noether, who came to Göttingen during the war years, was a mathematician of comparable breath and even greater originality. As an algebraist, she gave Göttingen the perfect complement to Courant's strength in analysis. While mainly remembered today as the "mother of modern algebra," Noether was, like her father Max Noether, a true mathematical scholar steeped in the classical literature of the nineteenth century. In 1919 she wrote a major expository article in the *Jahresbericht der Deutschen Mathematiker-Vereinigung* summarizing research in the arithmetical tradition of Dedekind and Weber, one major signpost being the book by Hensel and Landsberg (Noether 1919). Courant surely knew that report, and if he read it carefully he would have realized that this tradition had spawned a central field of research in modern mathematics, a field that was still rapidly unfolding.

So with people like Artin and Siegel at hand, what better way to catch up with these developments than to offer a seminar on algebraic functions and algebraic number fields, then twist some arms to make sure plenty of bright people got involved? The arm-twisting was perhaps new, but not the main impulse: Courant had experienced this over and again as a student in Hilbert's seminars. Indeed, he knew perhaps better than anyone how this spirit had animated nearly every memorable mathematics seminar in Göttingen since his days as a student.

In subsequent semesters, Neugebauer worked closely with Courant, but he also began to gravitate more and more toward pure mathematics. Thus, during his second and third semesters in Göttingen he took courses with Edmund Landau, probably the world's leading authority on analytic number theory. Unlike Courant, Landau cultivated a distinctly rigorous approach in his books and lectures. This did not suit everyone's taste, to be sure, but Neugebauer was clearly drawn to it. Neugebauer also had an excellent opportunity to acquaint himself with a decidedly different style of thought that semester when Niels Bohr came to Göttingen to deliver ten lectures on quantum theory, one of the highlights of his still young career.

This "Bohr Festspiele" took place in mid June and drew roughly 150 listeners, including Hilbert, Runge, and Courant, their colleagues in physics, Born, Franck, and Robert Pohl, and a large entourage of advanced students. Plenty of distinguished guests turned up, too: Sommerfeld and Heisenberg came from Munich; Lenz and Pauli from Hamburg; Landé, Gerlach, and Madelung from Frankfurt; and Ehrenfest from Leyden. Niels Bohr did not arrive alone either; he was accompanied by two

Swedish physicists, Oskar Klein and Wilhelm Oseen. Later that year, they would honor him again when the Royal Swedish Academy of Sciences announced that Bohr would be the next recipient of the Nobel Prize in Physics.

If Otto Neugebauer attended these lectures, which is more than likely, the experience could well have clinched his decision to give up further study of contemporary physics. Like all the other younger people, he would have been forced to sit toward the back of the lecture hall.[19] Not only was it hard to hear from that vantage point, one also had to cope with the fact that Bohr tended to mumble. Of course, the real problem was what he seemed to be saying; even those who had already read Sommerfeld's *Atombau und Spektrallinien* had great difficulty understanding what it was all about. Neugebauer might have been able to follow some of it, but since he also prized clarity his close encounters with the latest quantum physics surely helped propel him in a very different direction. By the spring of 1924, he made the reverse journey from Göttingen to Copenhagen, not to study physics with Niels Bohr, but to do mathematics with his brother, Harald.

Courant and the Springer Connection

Like all the German universities after the war, Göttingen fell on hard times. When Neugebauer arrived for the summer semester of 1922, Courant was struggling to keep his new Mathematics Institute afloat. Nearly everyone in Germany faced serious problems as the country struggled to recover from the war, and Neugebauer witnessed many of these difficulties first hand. His arrival coincided with the period of mounting hyperinflation throughout the Weimar Republic. Housing and food shortages continued to plague the country, while political chaos undermined stability, particularly after the French decided to take over the Ruhr region in January 1923. The paper Mark, introduced to replace the gold Mark during the war, had been steadily losing value after the fighting ended. Over the three-year period that preceded French occupation its buying power shrank by a factor of a thousand, a tenfold loss per year. Yet the German government took no effective measures to protect the currency, allowing private banks to issue credit to speculators, who exploited the steady deterioration of the Mark on a daily basis.

Years later, Neugebauer liked to tell humorous stories about Courant's Göttingen during the "good old days", glossing over all the hardships with tales of ingenuity and cunning. One such story took place shortly after Neugebauer arrived, so in all likelihood the original version stems from Courant himself. Neugebauer told the following version to the audience that attended the celebration of Courant's 75th birthday. After describing the famous *Lesezimmer* in the old *Auditorienhaus*, he went on to speak of Courant's "first revolutionary step" as an empire builder in Göttingen:

[19] Young Friedrich Hund described this atmosphere many years later, see (Mehra and Rechenberg 1982, 345).

> He applied to the Ministry of Education in Berlin for permission to change the simple heading "Universität Göttingen" on the stationery to "Mathematisches Institut der Universität Göttingen." When, after due delay, permission was granted, he made the prophetic remark: "They don't know how much that will cost them." Indeed the total library budget at that time was some 500 Marks (about $120) per year. (Neugebauer 1963, 2)

Money was indeed a major concern at Courant's institute, though there were plenty of others as well. Neugebauer went on to describe how Courant managed to cope with the horrendous inflation that ruined so many lives during these terrible times. Salaries and prices had to be re-calculated every week, immediately after the government released a new (and rapidly increasing) numerical coefficient for the recalibration of monetary value. Thus, each week the university administration was faced with the task of carrying out these calculations for the new salaries, which ran to very high figures. Courant, on the other hand, found a way to help out. His new Institute had recently purchased an electronic desk computer with a range of 19 digits, just what was needed to perform these calculations. So he agreed to loan this machine to the business office, and in return they agreed to inform him *immediately* once they obtained the new multiplicative coefficient from Berlin, which is to say a few hours before that number was made public. "This simple device," Neugebauer recalled, "greatly increased our purchasing capacity. In particular, it became possible to buy books on a great scale and to organize widespread international exchanges by means of which the war gaps in the library were filled" (Neugebauer 1963, 4). Courant's later reputation as a "tricky operator" surely harkens back to his early exploits in Göttingen, but this particular use of inside information in order to upgrade the *Lesezimmer* was merely a clever tactic. Courant's real talents were as a bold-thinking scientific entrepreneur.

In the period from 1919 to 1925 mathematics publishing took on a vital new importance for Germany, both scientifically and economically. In an era of growing international contacts, German mathematicians and scientists were generally barred from attending congresses and meetings held in the countries of their wartime enemies. Many thought of German science as the last bastion of power within which the country still stood supreme, but it, too, was clearly vulnerable, particularly if the products of German intellectual activity never found their way to the marketplace. Engineering and the applied sciences were hard pressed, but in the case of an ivory-tower field like mathematics the situation was particularly acute given the adverse political climate. A more aggressive approach to marketing the products of German mathematicians and scientists was needed, an approach embodied in the business practices of the firm of Julius Springer. Taking advantage of the vacuum created when B. G. Teubner pulled back from the mathematics market after the war, Springer soon emerged as a bold new player in this small niche within the publishing industry, promoting a surge in productivity that gave "mathematics made in Germany" an enduring allure (Remmert and Schneider 2010).[20]

Courant had already met Ferdinand Springer during the war; Courant was then temporarily stationed in Ilsenburg, a village in the Harz Mountains, working on

[20] The famous "yellow series" founded by Courant in 1920 continues to occupy a central niche in Springer's publishing program, though its character changed quite dramatically after 1945 when English became the dominant language for international publications in mathematics.

terrestrial telegraphy (Sarkowski 1996, 262). This meeting, which took place on 28 September 1917, was facilitated by the editor of *Die Naturwissenschaften*, Arnold Berliner, whom Courant had known growing up in Breslau. By the following year, plans for Courant's *Grundlehren der mathematischen Wissenschaften*—better known as the "yellow series" or the "yellow peril" ("die gelbe Gefahr")—were already underway. Courant not only lined up Hilbert's support for the plan, he also persuaded Hamburg's Wilhelm Blaschke and his father-in-law, Carl Runge, to join him as associate editors. By 1921 the first volume, Blaschke's *Vorlesungen über Differentialgeometrie, I*, was already in print with several more due to follow. That same year Springer opted to put Courant on his payroll as a consultant; he was paid the generous sum of 1500 marks (ca. 450 gold marks) quarterly (Sarkowski 1996, 264).

Courant's yellow series had just been launched when Neugebauer showed up in Göttingen. Not surprisingly, he soon became an integral part of this local publishing project. Neugebauer was still only a student without a doctorate when Courant took him under his wing. Yet beginning already in the winter semester of 1923–24, he began to assume various administrative duties at the institute, while helping Courant to write some of his books. Years later, he offered a vivid account of a typical scene during the end phase of this production process:

> A long table in Runge's old office was the battleground on which took place what Courant's assistants used to call the "Proof-Reading-Festivals" ("Korrekturfeste"). ... During this period Courant wrote his first group of famous books, the second edition of the "Hurwitz-Courant," the first volume of the "Courant-Hilbert," and the "Calculus." All of his assistants during these years participated at one or the other time in the preparation of the manuscripts: [Kurt] Friedrichs,[Hans] Levy, [Willy] Feller, [Franz] Rellich, [B. L.] van der Waerden, and others; red ink, glue, and personal temperament were available in abundance. Courant had certainly no easy time in defending his position and reaching a generally accepted solution under the impact of simultaneously uttered and often widely divergent individual opinions about proofs, style, formulations, figures, and many other details. At the end of such a meeting he had to stuff into his briefcase galleys (or even page proofs) which can only be described as Riemann surfaces of high genus and it needed completely unshakeable faith in the correctness of the uniformisation theorems to believe that these proofs would ever be mapped on *schlicht* pages. (Neugebauer 1963, 6–7)

One particular case deserves special attention here: Volume III by Hurwitz-Courant mentioned above. Its full title already hints at an unusual undertaking: *Vorlesungen über allgemeine Funktionentheorie und elliptische Funktionen* von Adolf Hurwitz, herausgegeben und ergänzt durch einen Abschnitt über *Geometrische Funktionentheorie* von Richard Courant. As Courant explained in the introduction to the first edition, Hurwitz had planned to publish these lectures before his death in 1919, so little by way of editing was actually needed. The contents of this first part of Hurwitz-Courant drew much of their inspiration from Weierstrass' lecture courses, offered during the late 1870s and early 1880s, which Hurwitz himself had attended.[21] Young Richard Courant had the opportunity to hear Hurwitz lecture on

[21] Hurwitz's original *Ausarbeitungen* from that time can still be found among his scientific papers: they are numbers 112, 113, and 115 in the Hurwitz Nachlass, ETH Bibliothek, Zürich.

function theory before he came to Göttingen. So did another student from Breslau, Max Born, who called these "perhaps the most perfect [lectures] I have ever heard." (Born 1978, 72).[22] Other *Ausarbeitungen* of Weierstrass' lectures were in circulation, of course, but it is surely ironic that Hurwitz, who had been Felix Klein's star pupil, proved to be such an influential conduit for the ideas of the once revered Berlin *Meister*.[23]

Courant, however, was not content to publish a volume that contained nothing but Hurwitz's version of Weierstrass' theory. As the new standard bearer for Göttingen mathematics, he felt compelled to add a dose of Riemannian function theory into the mix. In the preface to the first edition he wrote: "The viewpoint of the Weierstrassian theory can today no longer alone satisfy the student, despite the inner consistency with which it is erected."[24] Courant's supplementary text, however, did not meet with the same high critical acclaim as did Hurwitz's lectures. In fact, another Hilbert pupil, the American Oliver Kellogg, found it quite lacking in rigor. "It gives the impression," he wrote, "of being the work of a mind endowed with fine intuitive faculties, but lacking the self-discipline and critical sense which beget confidence. ... The proofs offered often leave the reader unconvinced as to their validity and, at times, uncertain even as to whether they can be made valid" (Kellogg 1923, 416). In view of this criticism, Courant engaged Neugebauer to help him rewrite the Riemannian portion of the book, which came out in 1925 as the second edition. Several more editions of Hurwitz-Courant appeared after this, and the book grew thicker and thicker each time.

Courant's motivation in producing this work was thoroughly Kleinian; he was guided by the notion that geometric function theory contains vital ideas that keep on giving life, whereas Weierstrassian complex analysis, while beautiful, was already complete and hence lifeless. Tributes to Klein abound in the yellow series, beginning with the very first volume written by Blaschke. Courant prepared new editions of Klein's *Elementarmathematik vom höheren Standpunkte aus* (Bände XIV–XVI); he had Neugebauer and Stephan Cohn-Vossen edit Klein's wartime lectures on the mathematics of the nineteenth century (Bände XXIV–XXV); and he published authorized editions of several of the lecture courses that Klein had earlier circulated locally through mimeographed copies. All of this had a strikingly conservative, not to say nationalistic, tendency. Throughout his life, Courant saw himself as the great protector and defender of the Göttingen legacy associated with Klein and Hilbert, both of whom had far more mathematical breadth than did he. During the Weimar years Hilbert's star continued to shine on brightly, in no little part due to the reverence Courant held for him. Indeed, Hilbert's name and fame continued to grow long after his heyday in mathematical research had passed. Thanks to his assistants, he

---

[22] Born also related that he gave Courant his notebook for use in preparing the Hurwitz-Courant volume.

[23] Another influential figure for the reception of Weierstrass' theory in Italy was Salvatore Pincherle.

[24] Bei aller inneren Konsequenz des so errichteten Gebäudes kann der Lernende sich heute mit den Gesichtspunkten der Weierstraßschen Theorie allein nicht mehr begnügen. (Hurwitz and Courant 1922, v).

continued to pursue his research program in foundations of mathematics throughout the Weimar period. The legendary old man, who became increasingly eccentric with the years, remained a living symbol of past glory even after the demise of Göttingen as a world-class center in 1933.

Yet Courant was hardly a hidebound traditionalist, even if his mathematical tastes ran toward classicism. The single most famous volume in the yellow series, his Courant-Hilbert, *Mathematische Methoden der Physik*, attests to a vision that went far beyond the legacies of his teachers. In the preface to the first edition, Courant decried the tendency among analysts to focus undue attention on "refining their methods and finalizing their concepts" at the cost of forgetting that analysis has its roots in physical problems. At the same time, he emphasized that theoretical physicists had begun to lose touch with the mathematical techniques most relevant to their own research. As a result, two new disciplinary cultures had developed, each with its own language and methods, neither able to communicate in a meaningful way with the other. Courant, writing in February 1924 just after the country had nearly succumbed to runaway inflation, saw this not just as an unfruitful use of resources, to him this represented a familiar danger that both Klein and Hilbert had earlier tried to counteract: "Without doubt this tendency represents a threat to science itself; the stream of scientific development is in danger more and more of branching out, seeping away, and drying up."[25] In preparing this volume, Courant relied on Hilbert's publications and *Vorlesungen* from the period 1902–1912. He also leaned heavily on the support of his own school of *Mitarbeiter*. These young men remained anonymous in 1924, but in the preface to the revised second edition from 1930 he gave credit to Kurt Friedrichs, Franz Rellich, Rudolf Lüneburg, among others. He also alluded to the mathematical difficulties that had caused him to delay the publication of Courant-Hilbert II, which finally appeared in 1937.[26]

If mathematical physics was Courant's domain, one should not overlook Max Born's role as associate editor of the *Grundlehren* series. Born was not only a close friend of Einstein's, he was also at the vanguard of the relativity movement from the time he became Minkowski's assistant in Göttingen. Like Courant, he too grew up in Breslau, where he befriended Arnold Berliner.[27] After November 1919, when the British announced the confirmation of Einstein's prediction for the bending of light in the vicinity of the sun's gravitational field, Born delivered a series of well attended popular lectures on relativity in Frankfurt am Main. Springer published Born's own *Ausarbeitung* the following year under the title *Die Relativitätstheorie Einsteins*, which eventually went through several revised editions. In the meantime, Berliner

[25] "Ohne Zweifel liegt in dieser Tendenz eine Bedrohung für die Wissenschaft überhaupt; der Strom der wissenschaftlichen Entwicklung ist in Gefahr, sich weiter und weiter zu verästeln, zu versickern und auszutrocknen." (Courant and Hilbert 1924, vi).

[26] Courant-Hilbert II was not listed in the bibliography of the Deutsche Bücherei. It was still listed in the Springer catalogues, however, in 1940. The Sicherheitsamt of the Reichsführer of the SS established a liaison office in the Deutsche Bücherei in 1934 to oversee the listing of books by Jewish authors. See (Sarkowski 1996, 353).

[27] See (Born 1978, 79–80).

had begun a new series as a spinoff from his periodical, *Die Naturwissenschaften*, so Born's book originally came out as volume III in *Naturwissenschaftliche Monographien und Lehrbücher*. Interestingly enough, volume I in this series was a book entitled *Allgemeine Erkenntnislehre* by Moritz Schlick, the founder of logical positivism, a reminder that physics was about to have a major impact on philosophy.

All this shows that even before he became professor of theoretical physics in Göttingen, Born had forged a working alliance with Courant in cooperation with Berliner and Springer. Just as Hilbert and Klein had seized on Einstein's theory during the war years as an ideal arena at the interface between mathematics and physics, so after the war could Springer exploit this new-found convergence of interests between mathematicians, physicists, and philosophers while filling the vacuum left after Teubner's departure. Evidence of Springer's role in promoting these developments is not hard to find. One need look no further than the masthead of *Mathematische Annalen*, which became a Springer journal in 1920 (joining the firm's new journal, *Mathematische Zeitschrift*). Alongside the familiar names in the *Hauptredaktion* (Klein, Hilbert, Otto Blumenthal) there appear two others: Constantin Carathéodory and Albert Einstein; and among the associate editors we find Born, Courant, and Sommerfeld.

Klein and Hilbert had been deeply immersed in mathematical physics, and yet Courant went further than they, in part because he was an analyst by training, temperament, and talent. Klein approached analysis from his standpoint as a visionary geometer, whereas Hilbert's contributions to modern analysis bore the mark of his background as an algebraist. Courant admired Klein's approach to Riemannian function theory and he absorbed all the new techniques for solving differential equations that Hilbert introduced. His own style was a kind of fusion of what he learned from Klein and Hilbert. Courant may not have been one of the more brilliant mathematicians of his era, but no one embodied the older Göttingen tradition better than he, not even Weyl.

As editor of Springer's "yellow series" Courant turned local oral knowledge—in the form of the edited lectures of famous mathematicians like Hurwitz, Klein, Hilbert, et al.—into internationally accessible knowledge in print form. The scope of this undertaking eventually went far beyond the intellectual confines of the Göttingen tradition, and while its range was truly encyclopedic, Courant's brainchild exerted a far deeper and more lasting influence than Klein's massive *Encyklopädie der mathematischen Wissenschaften*. The latter was a reference work comprised of lengthy scholarly reports filled with footnotes that pointed to the vast specialized literature; it reflected Klein's penchant for detail rather than the needs of working mathematicians. The best volumes in the yellow series, on the other hand, were living mathematics of a kind that a younger generation of mathematicians could not only learn from but also build upon. That was precisely what Courant and his *Mitarbeiter* showed in producing the various new editions of Hurwitz-Courant and Courant-Hilbert, books that drew on research traditions with a long and rich history.

Courant's success as Springer's front man in the world of mathematical publishing had much to do with an uncanny ability to motivate others, but it also had much to do with his own unusual background. His life had been one long, hard, uphill battle, but he had never fought alone. He could empathize with others because of the gratitude he felt toward those who had helped him along the way. Courant often straddled the line: an innovator, who welcomed change, he also held on to a deep belief in the vitality of older traditions. His yellow series looked backward as well as forward; in fact, surprisingly few of its volumes betray his commitment to what came to be identified as modern, abstract mathematics. Far more evident was the way in which Courant and his co-editors built on the tradition of Klein and Hilbert.

Neugebauer, a far more direct, even blunt personality, came to share these same values. Indeed, he would ultimately devote himself to the study of the very same nexus of mathematical sciences, but within the realm of ancient cultures. Once again, the Springer connection paved the way: its short-lived *Quellen und Studien* series, launched in 1929 and edited by Neugebauer, Julius Stenzel, and Otto Toeplitz, set a new standard for studies in the history of the ancient exact sciences. Like Courant, Neugebauer was a visionary, but neither man could have foreseen the explosion of interest in ancient as well as modern mathematics that would make this difficult decade a remarkably productive time for scholarly publications in Germany. Along the way to becoming a historian, Neugebauer gained an ever deeper respect for the unity of mathematical knowledge; much of that came through his interactions with Göttingen mathematicians.

From Modern to Ancient Things

Having gravitated into the midst of a singularly rich mathematical environment, Neugebauer still had no clear ideas about his own future as a researcher. Now that he had given up on physics, he seemed to have two main options. Certainly he had plenty of opportunities to immerse himself in Courant's brand of analysis, but he also sensed that his tastes ran closer to Landau's purist style. Yet along the way to becoming an historian, he also began to take a deep interest in other matters that were hotly debated at this time. Göttingen had long stood as a bastion of support for the radically new ideas of Georg Cantor, the founder of modern set theory. For over two decades, Hilbert had openly championed Cantor's *Mengenlehre*, often in the face of strong opposition from other quarters. During the early years of the Weimar Republic—an era when open clashes between revolutionaries and reactionaries had become commonplace—this longstanding conflict between Hilbert and his antagonists flared up again, this time with plenty of political overtones.

Once again, Neugebauer was in the right place at the right time. Indeed, the year 1922 marks the beginning of a new phase in the so-called foundations crisis, which pitted Hilbert against the Dutch topologist L. E. J. Brouwer. Neugebauer, who was little inclined to enter such philosophical discussions in later years, nevertheless showed a keen interest during the 1920s in Hilbert's ideas and their relevance for the history of mathematics. Although Brouwer had long expressed doubts about the validity of Hilbert's formalistic views, Hilbert took little notice of these stirrings until after the war ended. His star pupil, Hermann Weyl, had similar misgivings, however, so when he openly sided with Brouwer this long simmering conflict heated up rapidly. Even Neugebauer would have been forced to admit that, seen retrospectively, Weyl's earlier reluctance to accept Klein's former chair in Göttingen was surely an omen of what was to come. One year later, Weyl published a long essay entitled "Über die neue Grundlagenkrise der Mathematik" (Weyl 1921), a text full of rhetoric appropriate to the times. This began with a barely veiled attack on Hilbert and other Cantorians, accusing them of intellectual dishonesty:

> The antinomies of set theory are usually considered as border conflicts that concern only the remotest provinces of the mathematical empire and that can in no way imperil the inner solidity and security of the empire itself or of its genuine central areas. Almost all the explanations of these disturbances which were given by qualified sources (with the intention to deny them or to smooth them out), however, lack the character of a clear, self-evident conviction, born of a totally transparent evidence, but belong to that way of half to three-quarters attempts at self-deception that one so frequently comes across in political and philosophical thought. Indeed, every earnest and honest reaction must lead to the insight that the troubles in the borderland of mathematics must be judged as symptoms, in which what lies hidden at the center of the superficially glittering and smooth activity comes to light, namely the inner instability of the foundations, upon which the structure of the empire rests. (Weyl 1921, 39)

Weyl's brilliant use of political metaphor was most effective when it came to his critique of Hilbert's use of the logical principle of the excluded middle, which enabled him to prove far-reaching existence theorems in mathematics. Brouwer had long objected to this methodology, claiming that mathematical truths were fundamentally constructive in nature. Now Weyl brought this point home by comparing existence theorems with paper money, soon to become virtually valueless in the Weimar Republic. Thus, mathematics built on "general and existential propositions" should be seen as nothing more than:

> a monstrous "paper economy". Real value, comparable to food products in the national economy, is only manifest in the singular; everything general, and all the existential statements, participate only indirectly. And yet we mathematicians seldom think of cashing in this "paper money"! The existence theorem is not the valuable thing, but rather the construction carried out in the proof. Mathematics is, as Brouwer has occasionally asserted, more an activity than a theory (mehr ein Tun als eine Lehre). (Weyl 1921, 55)

Weyl thus famously bemoaned the "threatening disintegration of the statehood of analysis" and openly sided with the opposition: "Brouwer—das ist die Revolution!" (Weyl 1921, 40)

Hilbert struck back the following year in a speech delivered in Hamburg:

> [Mathematicians] have pursued arguments based on the concept of number sets to the utmost extent without any trace of inconsistency resulting anywhere: if Weyl notes an "inner untenability of the foundations upon which the empire rests" and worries about "the threatening disintegration of the statehood of analysis," then he sees ghosts. ... What Weyl and Brouwer are doing amounts, in principle, to a walk along the same path that Kronecker once followed: they are attempting to establish the foundations of mathematics by throwing

> everything overboard that appears uncomfortable to them and erecting a dictatorship à la Kronecker. This amounts to dismembering our science, which runs the risk of losing a large part of our most valuable possessions. ... no: Brouwer is not, as Weyl contends, the revolution but rather only the repetition of a Putsch attempt with old means. If earlier it was carried out more sharply and still completely lost out, now, with the state so well armed and protected through Frege, Dedekind, and Cantor, it is doomed to failure from the outset. (Hilbert 1922, 159–160)

Hilbert's last remark here was sheer bluff: he knew all too well that the "state of analysis" was by no means well armed and that the theories of Cantor, Dedekind, and Frege ultimately provided no real protection.[28] Indeed, he himself had long promoted the purely axiomatic method as the only viable approach that could secure the foundations of analysis as well as Cantor's theory of infinite sets. That goal, however, proved far easier to announce than to attain, but after a lapse of nearly two decades he decided to try again by launching a more sophisticated proof theory, supported by his new assistant Paul Bernays.[29]

When Hermann Weyl spoke of a "new foundations crisis" in mathematics (Weyl 1921), he also had an older crisis in mind. Presumably he was alluding to the ancient myth that the discovery of incommensurable magnitudes had caused a crisis among mathematicians who held to the Pythagorean doctrine that "all is number". The notion that this famous breakthrough led to an "ancient foundations crisis" soon emerged as a parallel theme among historians of mathematics, who tried to draw a picture of subsequent developments up to the time of Euclid (Christianidis 2004, 233–256). Neugebauer, who was always skeptical about the influence of older Pythagorean thought on Greek mathematics, would later draw a different parallel, linking Hilbert's proof theory to the Egyptian number system (see below).

In the meantime, however, he immersed himself in a new field: analytic number theory, the specialty of Edmund Landau. During the winter semester of 1922–23, Neugebauer attended his course on this subject along with another on entire transcendental functions. Many found it hard to take notes in Landau's lectures, and not only because of the intrinsic difficulties of the mathematics. He spoke quickly and wrote even faster; it was what people in Göttingen called the "Landau style": lean, precise, formal, and at times pedantic—theorem, lemma, proof, corollary. Dirk Struik, who attended one of his courses a few years later, recalled how Landau sometimes offered teasers to his audience: "Occasionally he would present a well-known theorem in the usual way, and then while we sat there wondering what it was all about, he pontificated: 'But it is *false—ist aber falsch'*—and, indeed, there would be some kind of flaw in the conventional formulation" (Rowe 1989, 20). Landau's blackboard technique was also famous and simply superb: while lecturing he would fill several boards with carefully numbered formulas. He used these as back-references while unfolding his arguments so that he would have everything at hand

[28] In the same text, he later went on to admit that his own theory starts from precisely the opposite standpoint as that taken by Frege and Dedekind (Hilbert 1922, 163).

[29] For a detailed account of Hilbert's work on foundations during the final decades of his productive career, see (Hilbert 2013).

in order to prove a theorem. These were masterful performances, but not at all easy to follow. So students generally adopted one of two main strategies: some tried to take down careful notes without really pausing to think (unless they could think about one thing, while writing about another: not easy); others took only sporadic notes while listening carefully and trying to follow the main arguments. The latter approach, which Neugebauer evidently followed, was probably chosen by most listeners. After all, most of what Landau had to say could be found in his books, and he was a prolific writer.

Since Neugebauer liked to read books, this probably explains why he did not bother to take reams of notes in Landau's courses.[30] For Landau's lectures on transcendental functions, however, he compiled a glossary of technical terms used by Landau, setting these down next to comparable lists taken from two other standard texts, one written by Emile Borel, the other by Ludwig Bieberbach. Such a systematic working method—aimed at a comparative analysis of the technical vocabulary employed by three different authors—was surely most unusual for a young mathematician. It betrays a meticulous mind and the passion of a true scholar: clearly Neugebauer was on his way to becoming a learned mathematician. Still, the question remained: did he have the creative talent needed for original research?[31]

In his course on analytic number theory, Landau mainly lectured on the properties of the Riemann zeta-function, which plays a major role when studying distributions of prime numbers. This was a famous and notoriously difficult subject; in fact, the Riemann conjecture on the zeroes of the zeta-function remains to this day perhaps the outstanding unsolved problem in mathematics (Hilbert made it his eighth Paris problem in 1900 and a century later it was chosen as one of the seven Clay millennial problems). Landau had published a great number of papers on the zeta-function, including a few with his Danish collaborator Harald Bohr. Their partnership went back to the pre-war years, when Bohr often visited Göttingen, a place he thought of as his second home. During this period he also struck up a warm friendship with Richard Courant. They, too, wrote a paper together that dealt with the Riemann zeta-function, a topic Neugebauer came to know well. Whether he thought about doing a dissertation on a related problem in analytic number theory remains unclear, however.

In any event, Landau shifted to a quite different theme the following semester when he lectured on the theory of lattice points. This theory has its roots in a theorem of Gauss, but it only became an active field of research with the publication of Minkowski's *Geometrie der Zahlen* in 1896. Landau's interest in this topic, however, seems to have been even more recent still. It was aroused in 1920 when the Dutch mathematician, Johannes van der Corput, came to Göttingen to study with Landau. Van der Corput had only recently taken his doctorate in Leyden with a dissertation on lattice points in number theory. Three years later, during the summer semester of 1923, Landau made this topic the central theme of his course on number

[30] Neugebauer Papers, IAS, Box 7.

[31] See Neugebauer's letter to Erich Bessel-Hagen, 21 April, 1924, cited in Siegmund-Schultze's paper for this volume.

theory. Oddly enough, he seems to have passed over Minkowski's work entirely, starting with results of Sierpinski in order to move quickly to van der Corput's recent findings (Van der Corput 1920). This, at any rate, is the impression left by Neugebauer's rather sketchy notes, which suggest that he had no burning interest in this subject. After completing this second course with Landau, he seems to have entered the final phase of his educational journey as a budding mathematician.

At this point he apparently stopped taking courses and began to search in earnest for a suitable dissertation topic. Unfortunately, Neugebauer's extant papers in Princeton offer no real clues about his activities over the course of the next year. Perhaps he felt stuck trying to solve a difficult problem? Maybe he needed to get away from Göttingen so that he could clear his mind again? Whatever the state of his mind may have been, the following spring Neugebauer traveled to Copenhagen in order to work with Harald Bohr. Probably he knew that this was a golden opportunity since Bohr had just cracked open an important new branch of analysis: the theory of almost-periodic functions. In all likelihood, Courant arranged this trip, but Landau may well have been a party to the arrangements as well, given his close relationship with the congenial Dane. At any rate, this new venture did help to stir Neugebauer's imagination; it even led to a brief collaboration with one jointly written paper on a special problem in Bohr's theory (Bohr and Neugebauer 1926). Yet strangely enough, this trip to Copenhagen sparked a new interest that would prove decisive for Neugebauer's future career as an historian of mathematics.

Quite by chance, Bohr invited him to write a review of T. Eric Peet's edition of the Rhind Papyrus for the Danish journal *Matematisk Tidsskrift*. He surely knew of Neugebauer's interest in this subject, but Bohr could never have imagined how much time his Austrian guest would spend in preparing this review (Neugebauer 1925).[32] Puzzling over the mysteries of representations by sums of unit fractions, he got hooked on a special problem that has continued to fascinate scholars ever since Neugebauer first decided to tackle it (Gillings 1982; Knorr 1982).[33] Returning to Göttingen, he no longer thought about a career as a research mathematician: he was all afire to become an expert on ancient mathematics. Had Neugebauer been a student of Hilbert or Landau, his chances of completing his studies with a dissertation on Egyptian unit fractions would have been very scant, indeed. Luckily, however, he had Courant on his side, someone who valued human character above formal requirements.[34]

Courant did not mind "breaking the rules" in order to promote a good cause; moreover, he was deeply convinced that Neugebauer's talents should be utilized rather than wasted. Early on he had given him various duties to perform at the hub of operations, which were then located on the third floor of the *Auditorienhaus*. There Neugebauer was put in charge of the famous *Lesezimmer*, created by Felix

[32] A copy of (Neugebauer 1925) can be found in Neugebauer Papers, IAS, Box 14, item 1.

[33] This fascination actually began with J. J. Sylvester, who called attention to the mysteries of unit fractions soon after publication of the Rhind papyrus.

[34] Neugebauer ended his tribute to Courant on his 75th birthday with some moving remarks about how he had made it possible for him to pursue his chosen career (Neugebauer 1963, 9).

Klein, with its impressive collection of books and off-prints. He also looked after the magnificent collection of mathematical models that lined the corridors. These had a long prehistory that also reached its climax under Klein, the great champion of *anschauliche Geometrie*. By now Klein was an infirm old man who rarely left his home, which overlooked the botanical garden immediately behind the *Auditorienhaus*. Nevertheless, he still kept very busy with a tightly organized schedule that included regular reports from visitors like Richard Courant. It was probably he who first brought Neugebauer's new interest in Egyptian mathematics to Klein's attention. The "Great Felix" also heard a complaint that Neugebauer had stuffed all the books on mathematics education tightly together on a high shelf, making them nearly inaccessible. Neugebauer remembered how one day Klein called him over to his home for a gentle scolding. When he arrived, Klein greeted him by saying: "there came a new Moses into Egypt and he knew not Pharaoh!" (Reid 1976, 100), (a play on: "Now there arose up a new king over Egypt, which knew not Joseph," Exodus I.8).

Courant could not, of course, pass judgment on a dissertation in Egyptology, even if its contents were mathematically intelligible to him. He could, however, recruit an expert with the requisite competence. Courant's former colleague, Kurt Sethe, author of an authoritative article entitled "Von Zahlen und Zahlworten bei den alten Ägyptern" (1916), expressed his willingness to cooperate. Since Sethe was now Professor of Egyptology in Berlin, his successor in Göttingen, Hermann Kees, had to be contacted as well. In the end, all three wrote glowing reports on Neugebauer's novel work, *Die Grundlagen der ägyptischen Bruchrechnung* (Neugebauer 1926), published by Springer in the same year he took his degree. This study focused on the famous *2/n* table in the Rhind Papyrus, which reveals how the Egyptians represented these numbers as sums of unit fractions. Just one year later, Neugebauer's findings drew fresh attention owing to the publication of new source material. This was made available when the British Museum finally published the contents of the Egyptian Mathematical Leather Roll, which like the Rhind Papyrus had been purchased by Henry Rhind back in 1858. When it appeared, Neugebauer's analysis of these texts was considered highly controversial among Egyptologists.

By the time he finished his dissertation, Neugebauer enjoyed considerable support and respect among key members of the Göttingen faculty. So, as was to be expected, his *rigorosum* went smoothly.[35] In these oral exams, taken after completion of the dissertation, doctoral candidates underwent questioning in three subject areas—in Neugebauer's case these were geometry, analysis, and theoretical physics—an ordeal that could last up to 2 h. Neugebauer's final exercise as a student began late in the afternoon on 21 April 1926, when he faced a formidable battery of examiners: Hilbert, Courant, and Born. Hilbert's questions dealt with curves and surfaces, covering both topological and algebraic properties. He asked about Plücker's formulas for the singularities of plane curves and their duals, then he turned to the curvature properties of surfaces and their associated geodesics. Clearly,

[35] Neugebauer's *Rigorosum* took place on 21 April, 1926. Promotionsverfahren Otto Neugebauer, Universitätsarchiv Göttingen, Math. Nat. Prom. Spec. N.I.

Neugebauer knew these things very well: Hilbert recorded that his answers displayed not just very good knowledge but a solid understanding ("bestes Verständnis").

After this came Courant's exam in "analysis." Apparently wanting to put his candidate's aptitude for higher mathematics on display, Neugebauer's thesis advisor asked him questions that would hardly have been expected in a standard oral in this subject. Courant started with algebraic number fields, distribution of primes, and related properties of the Riemann zeta-function. He then went on to Minkowski's geometric theory of numbers, Riemannian function theory, and the theory of Abelian groups. Neugebauer was familiar with the last of these topics thanks to a course he took the previous year with Emmy Noether. Most of the others he knew from Landau's courses, so only the questions on function theory reflected Courant's influence. Ironically enough, Neugebauer had mainly learned about this when he helped Courant rewrite the Hurwitz-Courant volume for the yellow series! Indeed, this part of his final oral exams confirms that, mathematically speaking, Neugebauer probably stood closer to Landau than to his mentor, Richard Courant. The latter's verdict went even beyond Hilbert's "sehr gut," normally the highest grade given; in analysis he received an "excellent" ("ausgezeichnet").

Such a high mark was not to be expected in theoretical physics, especially since he had done virtually no course work in this area since coming to Göttingen. Max Born began by asking about Coulomb's law, electrostatics, and the potential theory of electromagnetic fields, leading up to Maxwell's theory and equations. He then moved on to van der Waal's theory of ideal gases, the Joule-Thomson effect, and the second law of thermodynamics. Neugebauer may well have found it amusing that Born ended with this last topic, since he surely remembered Born's unconvincing explanation when he had lectured on Zermelo's paradox in thermodynamics. At any rate, Born found Neugebauer's responses convincing enough to deserve the grade of "good." In the end, his final oral grade was "sehr gut", his grade for the dissertation even better: "ausgezeichnet".

As a motto for this work, Neugebauer chose a quotation from Hermann Hankel's inaugural lecture at Tübingen in 1869: "Es ist eben Mathematik auch eine Wissenschaft, die von Menschen betrieben wird, und jede Zeit, sowie jedes Volk hat nur einen Geist" (Hankel 1869). Undoubtedly, he first learned about this rather obscure text by reading Felix Klein's war-time lectures on the mathematics of the nineteenth century. Courant had a longstanding interest in these lectures; in fact, just before the war broke out he had been briefly involved in helping Klein prepare them. He was attending an informal seminar, in which Klein asked him to do some background research on the arithmetization of analysis; Hilbert's *Assistent*, Alfréd Haar, also attended and presented a brief history of mathematical astronomy.[36] Such communal work would later become a hallmark of the Courant-Neugebauer partnership.

Years later, Klein was still hoping to edit his manuscripts for publication, but poor health prevented him from doing so. After his death in 1925, Courant enlisted

[36]These documents are located in Cod. Ms. F. Klein 21 F, Niedersächsische Staats- und Universitätsbibliothek Göttingen.

Neugebauer to complete that task, for which he was assisted by Erich Bessel-Hagen.[37] Others helped out as well, including the Dutch differential geometer Dirk Struik, then a Rockefeller fellow in Göttingen. The project moved quickly: volume one came out already in 1926 (Klein 1926); volume two, prepared by Stephan Cohn-Vossen, a year later. Struik later made ample use of Klein's lectures when he wrote his popular *Concise History of Mathematics* (Struik 1948). Neugebauer also came away from this undertaking deeply impressed by Klein's highly personal style as well as his overall approach to history.

Clear evidence of this can be found in an early paper (Neugebauer 1927c), written around the time he completed his *Habilitation* (Neugebauer 1927a). Here Neugebauer discussed the larger purpose of studies in the history of mathematics, both as new type of scholarly discipline, but especially as an antidote to specialization in mathematical research, about which he wrote:

> It is clear that the rigorous grounding of the newly formed sciences can be accomplished only by the greatest division of labor in careful individual investigations. As a consequence of this, however, there has arisen not only a sharp separation of the sciences from each other but also a crumbling of disciplines into subdivisions that are scarcely intelligible to one other and lack any common interests. No doubt a serious reaction must set in against this condition, and in part this has already taken place in a quite perceptible manner. Questions about the past and future course of the sciences, about their place in the broader sphere of our entire civilization, are being asked more and more frequently. In all fields we observe that only in the synthesis of modem research methods with the less hampered perspectives of a deeper intellectual substance can we hope to guarantee the restoration of unity among all the sciences.
>
> Like no other work, Felix Klein's *Lectures on the Development of Mathematics in the Nineteenth Century* shows what a historical view in this sense can mean for mathematics. Truly historical thinking combined with an intimate familiarity with research activity speaks to us here, beckoning us to see and to understand our own research inclinations as bound within a greater historical process. It will not be vouchsafed to many to write the history of a science in this sense. (Neugebauer 1927c, 44–45)[38]

As this passage shows, the young Otto Neugebauer saw himself as a proselytizer for a new approach to the history of mathematics as an integral part of what came to be called the exact sciences. As seen here, he stressed the underlying unity of the mathematical sciences, long a watchword for Klein, Hilbert, and of course Courant. Neugebauer identified with this Göttingen tradition completely; at the same time he was harshly critical of older approaches to the historiography of science and mathematics.

In late 1926, Otto Toeplitz invited Neugebauer to speak to the mathematics colloquium in Kiel about the results of his dissertation research.[39] He took this opportunity to describe his views on modern studies in ancient mathematics by emphasizing the need for an interdisciplinary approach that drew together all related

[37] On Neugebauer's relationship with Bessel-Hagen, see Reinhard Siegmund-Schultze's contribution to this volume.

[38] A copy of this early essay can be found in Neugebauer Papers, IAS, Box 14, item 9.

[39] Lecture in Kiel, "Über die Mathematik im alten Ägypten," 11 December 1926, in Neugebauer Papers, IAS, Box 14, item 5.

knowledge of the cultures under study. This he regarded as the only truly scientific approach to historical studies, and he expressed his confidence that it would soon be recognized as such. He thus looked forward to the day when the history of ancient mathematics would no longer be dominated by writers who preferred bold speculations over firmly grounded evidence.[40]

Only a year after completing his doctorate on *Stammbrüche*, Neugebauer completed his *Habilitationsschrift* (Neugebauer 1927a), in which he investigated the roots of the Babylonian sexagesimal system in the earlier number systems of Sumerian culture. He had already begun to study Assyriology in Göttingen when, in February 1927, he applied for funding from the Göttingen Academy of Sciences to support further research under Anton Deimel in Rome. For this purpose, he outlined the goals of his study while criticizing previous theories that had been forwarded as explanations for the evolution of the sexagesimal system. These theories, in Neugebauer's view, all suffered from two main defects. First, they tended to regard the historical process as one that led to a single unified system that, once established, simply persisted afterward. Second, their tendency was always to explain the base 60 under the assumption that it was derived from astronomical and theological speculations in Babylonia. Neugebauer's research, on the other hand, aimed to refute all such claims.

Three years earlier, Deimel had published a large collection of source material on early number signs (Deimel 1924), which already provided a good basis for a fresh new study. Neugebauer's working hypothesis was that the Babylonian number system had been preceded by a far older one that arose through the simplification of a chaotic system of still older Sumerian number signs. These numbers, in turn, had a direct metrological importance, as they were used for such weights and measures as those that played a key role in daily economic life. Thus, his research in Rome aimed to study the rich new artifacts that had been excavated by the German Oriental Society. Soon after he returned to Göttingen, he submitted the results of his investigations, "Zur Entstehung des Sexagesimalsystems" (Neugebauer 1927a), as his *Habilitationsschrift*. This research enabled him to become a member of the faculty later that year.

Neugebauer may well have been the first *Privatdozent* to be awarded a *venia legendi* for the history of mathematics at a German university. As part of the customary ritual, he was required to submit three different titles for his *Habilitationsvortrag* in his formal application, which he submitted on 27 July 1927. For this purpose he chose: (1) ancient Oriental mathematics and its relation to Greek mathematics: (2) On the history of mathematics: its problems and methods; and (3) The Moscow Mathematical Papyrus (unpublished) and our knowledge of Egyptian

[40] "Diese Arbeitsrichtung, die seit einigen Jahren immer mehr zur Geltung kommt, bedeutet hoffentlich das Ende eines Zustandes in der Geschichte der Mathematik den man – jedenfalls was die Geschichte der vorgriechischen Mathematik anlangt – manchmal nur mit dem Faustrecht vergleichen kann, wo es mehr auf Kühnheit des Behauptens als auf Gründlichkeit der Untersuchung ankommt." (S. 1). As a representative of the earlier tradition, he no doubt thought of Hermann Hilprecht, who speculated about the connection between certain numbers found on cuneiform tablets and Plato's number mysticism, about which see (Neugebauer 1957, 27).

mathematics. The faculty chose the first of these for Neugebauer's *Probevortrag*, which he delivered on 17 December 1927 in the *Auditorienhaus*. In the meantime, however, he had also prepared the text for the second theme on his list, so he arranged to have that published the same year (Neugebauer 1927b). He probably felt some regret that he had not been able to present this text as his official "tryout lecture," knowing that Hilbert was likely to be seated in the front row as he spoke. For what he had to say about problems and methods in the history of mathematics surely would have caused the grand old man to sit up and take notice.

In fact, in that text Neugebauer began with a direct reference to Hilbert's lecture "Über das Unendliche" (Hilbert 1925), which he had delivered at a meeting in Münster held in honor of Weierstrass. This address had already become famous within mathematical circles, particularly in view of the circumstances—namely, the clash with Brouwer and the intuitionists, soon to reach its climax. Hilbert's text was rhetorically brilliant; young André Weil soon thereafter decided to translate it into French for publication in *Acta Mathematica* (Hilbert 1926). Neugebauer's unlikely appeal to it was even more telling, for he took up its historical motifs in order to ascend to an even higher level.[41] Hilbert began with some laudatory remarks describing Weierstrass' contribution in resolving longstanding difficulties in the foundations of analysis, problems that stemmed from the impossibility of grasping infinitesimally small quantities directly. Cantor's theory, on the other hand, offered a bold new theory that enabled mathematicians to embrace the infinitely large, even though many had balked at doing so. Cantor's ideas had sparked much controversy, particularly in view of various paradoxes that had arisen in set theory. Hilbert now claimed that his proof theory would soon supply the necessary tools to place Cantor's theory once and for all on firm logical foundations. If he were right, then this promised to open a new era in the history of mathematics.

In response, Neugebauer took up this lead by drawing a daring parallel between Hilbert's proof theory and the earliest number systems in human history, thereby leaping across 5000 years. He began, however, by noting the larger significance of Hilbert's program for anchoring the foundations of mathematics, which he described as part of the long struggle to find adequate methods for reasoning about the infinite in mathematics. He then cited a *Leitmotif* from Hilbert's earlier lecture in Hamburg (Hilbert 1922). On that occasion Hilbert had introduced the intuitive aspect in his theory by invoking biblical language: "In the beginning is the sign" ("Am Anfang ist das Zeichen") (Hilbert 1922, 163). Within the context of proof theory, signs or strokes function as primitive objects, or immediately knowable things.[42]

Neugebauer now gave this a slight twist—"Am Anfang *war* das Zeichen"—and with that temporal twist he whisks us back to the time of the Pharaohs. This might

---

[41] In view of this, one might wonder why this text does not appear in the three volumes of Hilbert's *Gesammelte Abhandlungen*. Presumably, he and the editors chose to leave it out because the latter part of the text sketches a faulty proof of Cantor's continuum hypothesis.

[42] In (Hilbert 1925) he explicitly identified with Kant's epistemology, while noting that Kant's position already spelled doom for the logicist theories of Frege and Dedekind. This brought him closer to Brouwer's intuitionism, but it was left to Brouwer to point this out in (Brouwer 1928).

Table 4 Courses taught by Neugebauer as a Privatdozent in Göttingen

SS 1926	Analytic Geometry
SS 1929	History of Mathematics: Geometry up till Euclid
WS 1929–30	Descriptive Geometry
SS 1931	Analytic Geometry I
WS 1931–32	Analytic Geometry II
SS 1932	History of Mathematics after Euclid
WS 1932–33	Special Topics in the History of Mathematics

seem beyond daring, perhaps even bizarre, until one considers the precise function of such signs in Hilbert's meta-mathematical philosophy as well as in Egyptian arithmetic. In both cases, signs lose all earlier symbolic associations; they become pure signifying instruments whose only function is to represent some mathematical object, whether an infinite set or a concrete number, starting with a special symbol for the number "one." So the analogy Neugebauer had in mind almost surely would have resonated with Hilbert, who had a penchant for similar global reflections.

Although he was officially qualified to offer courses in the history of mathematics beginning in 1927, it would seem that Neugebauer only rarely exercised that right (Table 4). On the other hand, he taught courses in analytic and descriptive geometry on a fairly regular basis, which makes it likely that his teaching activities reflected the demands for courses in a given semester. If his extant notebooks give an accurate overall impression, then he only had the opportunity to teach history of mathematics on three different occasions. That he chose each time to teach topics in ancient mathematics comes, of course, as no surprise; a glance at the topics he covered, on the other hand, suggests that his aim was mainly to learn and impart standard material rather than enter deeply into his own particular research interests.

The notes from his first course on "Geometry up till Euclid"[43] are broken down into three periods: pre-Greek developments, Greek geometry up to the founding of Plato's Academy, and geometry from the period of the Academy up until Euclid's time. The idea for the course, he wrote, was motivated by his reading of an older book by C. A. Bretschneider, first published in 1870. One might have anticipated that his aim would be to tie certain topics from the first period to developments in the following two. Yet his notes contain only 25 pages for the first two periods, followed by 50 pages for the third, which mainly dealt with famous problems found in standard sources. So this course, which he taught during the summer semester of 1929, apparently covered very traditional material, nearly all of it centered on Greek geometry.

[43] Neugebauer Papers, IAS, Box 1, Geschichte der Geometrie bis Euclid 1929.

Three years later he offered what, from the title, might look like a sequel to the earlier course.[44] Presumably, he thought there was little point in teaching a course on the geometry in Euclid's *Elements*, perhaps on the assumption that much would be common knowledge for educated mathematicians. So he jumped to the *Conica* of Apollonius, covering the first four books in a 2-h course. His notes are full of beautiful drawings that bear a close resemblance to some of those found in his article "Apollonius-Studien" (Neugebauer 1932). Thus, unlike the earlier course, here he had a real research objective in mind, namely to show how the various geometrical techniques for transforming areas in the *Conica* could be understood as a kind of disguised algebra, much as Zeuthen had argued earlier in (Zeuthen 1896).

The following semester Neugebauer offered a very different type of course.[45] This time the approach was far broader, but he still concentrated mainly on topics in Greek mathematics. The overall structure was in two parts, neither showing much concern for chronology. Part I—headed "Geometric and Algebraic Problems"—began with an overview of the history of conic sections. Neugebauer then took up some special problems that required higher curves. From here he passed to the theme of "geometric algebra," describing both the early history and its role in Euclid's *Elements*. This part of the course then ended with a brief discussion of algebra in pre-Greek mathematics. Part II—"The Quadrature Problem and other Foundational Questions"—consisted of five topics: Archimedes' work on quadratures, earlier methods and results, irrationals and related proportion theories, number theory from Euclid to Diophantus, and pre-Greek number systems and counting methods. As a supplement, he added one final topic: trigonometry and spherics in Ptolemy's *Almagest*.

These brief synopses provide at least a vague idea of the types of courses Neugebauer taught during his last years in Göttingen. Much of the material he covered was quite standard, but one can still discern a clear trajectory of interests—the confluence of "geometric and algebraic problems," early Greek "geometric algebra," and its highly elaborated form as found in the *Conica* of Apollonius—preoccupations that would converge into a new vision of ancient mathematics a few years later, though in Copenhagen rather than Göttingen. That particular physical transition followed the shattering of the Göttingen community, a painful period that witnessed the dismantling of the Mathematics Institute Courant had founded and that Neugebauer had helped to build.

Targeting Courant's Institute

Richard Courant assumed an awesome responsibility in July 1920 when he accepted the professorship formerly occupied by Felix Klein. For him, winning the services of young Otto Neugebauer proved a godsend, especially given Courant's plans for

[44] Neugebauer Papers, IAS, Box 1, Geschichte der Geometrie nach Euclid 1932.

[45] Neugebauer Papers, IAS, Box 1, Ausgewählte Kapitel der Geschichte der antiken Mathematik.

modernizing and expanding the Mathematical Institute. He badly needed someone who could manage its daily routine affairs; Neugebauer not only performed such tasks, he did much else besides, thereby freeing Courant to pursue his larger plans. In many respects, the blueprints for modernizing the Göttingen Institute were already in place thanks to Klein's efforts.

During the 1880s and 1890s Klein was at the height of his fame as a mathematician with many students and admirers from the United States. He, on the other hand, was deeply impressed by the generosity of America's elite when it came to funding private universities. In 1893 he had the opportunity to visit the University of Chicago, which opened just one year earlier, its neo-gothic architecture no doubt reflecting the tastes of the man who paid for it: John D. Rockefeller. Klein toured the East coast as well, visiting a number of campuses with impressive modern facilities that were often financed by wealthy individuals and other private donors. Nothing remotely like this existed in Germany, so when he returned to Göttingen Klein began building up a network of contacts that eventually brought entrepreneurs, scientists, and industrialists together. These efforts led to the founding in 1898 of the Göttingen Association for the Promotion of Applied Physics, which helped fund a whole new complex of buildings and laboratories for scientific research in Göttingen.[46]

In keeping with Klein's original vision, this innovative plan would culminate with a new building exclusively for mathematics. The Göttingen Association bought the property for the site and allocated funds for the structure. Blueprints were drawn up for the building along with a timetable for its realization, but then the war broke out and all plans had to put on hold. After four long years that ended in defeat, Klein saw no way forward. With the Weimar Republic plagued by a constant series of financial crises, the project had to be shelved, though it was not entirely forgotten. Courant, who was long aware of Klein's plans, eventually found a way to finance it, indeed, to build a mathematics institute in Göttingen even bigger than the one Klein had dreamed of (Reid 1976, 122–125).

Soon after Klein's death in 1925, he took up negotiations with the Rockefeller Foundation, which had already played an active role in promoting international research through grants provided by its International Education Board.[47] Encouraged by his Danish friends, Harald and Niels Bohr, Courant turned to representatives of the IEB to make the case for a new building. In a letter from 2 October, 1926, he adopted a line of argument that echoed Klein's language, appealing to the singular character of the Göttingen mathematical tradition:

> The close association of mathematics and physics has at all times been a characteristic feature—and the strength—of the Göttingen tradition, in our special sphere. I need only recall the names of Gauss, Weber, Dirichlet, Riemann, H. Minkowski, Felix Klein. The last named entertained for decades the project of establishing a fixed home for mathematics and physics where both sciences would be cared for on the broadest possible basis, and in inti-

[46] For insight into Klein's interests in technology and his impact on Göttingen and beyond, see (Eckert 2013, 67–193).

[47] For a detailed analysis of the scope and impact of the IEB, see (Siegmund-Schultze 2001).

mate mutual conjunction. In this way a series of new buildings and establishments has come to the front [leading to a] concentration of all university activities connected with our special domain.... (Siegmund-Schultze 2001, 146)

Working behind the scenes, Neugebauer helped draw up new blueprints for a T-shaped building designed with an eye toward functionality and comfort. This design called for two large lecture halls, special rooms for drawing and applied mathematics, a colloquium room for a more intimate atmosphere, and a spacious *Lesezimmer* that would house the institute's large and rapidly growing library. Neugebauer realized that the previous facilities had become hopelessly inadequate; even before the war the *Lesezimmer* felt cramped and the hallways were cluttered with glass cases for models that left no room to spare. He was also relieved that the earlier plans for an institute building went unrealized (Neugebauer 1930). That design would have been inadequate, especially due to the explosion in new mathematical literature. In the new quarters, the spacious upstairs foyer offered an ideal setting for the institute's impressive collection of mathematical models and instruments. These had long been a trademark of Klein's regime, but it was Neugebauer who gave them a whole new aesthetic effect by placing some 400 mathematical artefacts in 40 glass cases, most of them displayed in a spacious open hall that came into view as one ascended the front stairs.

Construction went on for two and one-half years before the new quarters could open in November 1929. Neugebauer offered readers of Arnold Berliner's *Die Naturwissenschaften* a vivid description not only of the interior of the building but also the various administrative measures adopted to ease the use of the library facilities, improve the learning environment, and provide for a more effective curriculum (Neugebauer 1930). He also presented statistics to show how dramatically the number of users of the *Lesezimmer* had risen since Klein first founded it in 1886. Back then, there were barely 50 registered members per year, growing to just over 100 by 1900. In 1929, the year Neugebauer was writing it had risen to 450, a trend that would have led to chaos had he not devised a whole new administrative system for filing and ordering books, etc. These were major changes, necessary for modernization, but Neugebauer stressed that the new institute was constructed for its users, where they could meet, talk, and learn mathematics together.

Courant and Neugebauer skilfully promoted their new institute as the realization of Klein's old dream, but they were not alone. In January 1930, the Göttingen community celebrated the opening of the new Institute, on which occasion Hilbert gave a short speech to those gathered. Not surprisingly, he heaped lavish praise on Courant:

If I may say just one thing at the outset: there is one man without whom nothing of what you see here would have come to be, a man who set in motion the mechanism for the creation of this institute and who maintained it to its successful end. It is our dear friend and colleague Richard Courant. The idea of this institute, cherished and nurtured by Felix Klein, lived once upon a time and was charming and lovely, like Sleeping Beauty, and we who have been here a while took proud pleasure from it; but the Wicked Witch of inflation put this Sleeping Beauty into a sleep so deep that all forgot about her until Prince Charming Courant awakened her to new life ... Everyone should know that the institute is, in reality,

> his work. And that if he was able to achieve this great success—and here we have the one and only reason for this accomplishment—it was because he put his wholehearted energy in the work from the beginning. And for this reason we want to give him our wholehearted thanks. (Bergmann et al. 2012, 78)

Neugebauer stood very much in the background on such public occasions, and one may fairly doubt that Hilbert had any idea of how Courant's Institute operated, who worked there, or how daily business was conducted. In this hierarchical world, Neugebauer expected to be overlooked; he was, after all, a mere *Oberassistent*, a title that accurately described his principal occupation. He was something akin to a business manager for Richard Courant, which meant he ran the floor operation. And Neugebauer ran a tight ship, which probably not everyone appreciated. But he also shared a distinctive vision with his boss that was evident in everything they achieved together. Just three years later, their collective efforts were washed away in a storm (Schappacher 1987).

On 7 April, 1933 Hitler's government passed the "Law to Restore the Professional Civil Service," the first of several measures that targeted Jews and other "undesirable" individuals (Becker et al. 1987). Those, who like Courant, had served their country in the First World War, were exempted from the provisions of §3, its "Aryan paragraph." According to §4, however, persons could also be dismissed if it was determined that their former political activities did not give assurance that they would firmly support the national state. All German universities were immediately affected by this statute, and the Prussian Ministry of Education acted to enforce it by first collecting the necessary data through a questionnaire which every state employee was required to fill out and to submit.

Courant consulted with Neugebauer as well as his two physicist colleagues, James Franck and Max Born. Although the latter two were Jewish, only Born seemed to be affected by the law since he had not been involved in combat during the Great War. Franck, however, refused to accept the special exemption he enjoyed as a war veteran. Once the first dismissals were announced, he resigned his professorship, publicly proclaiming that "we Germans of Jewish descent are being treated as aliens and enemies of our homeland."[48] His resignation prompted 42 members of the Göttingen faculty to issue a statement condemning Franck for lending support to anti-German propaganda in the foreign press. This reaction closely paralleled another case that arose in the Prussian Academy of Sciences following Einstein's resignation: some of its members vented their anger by accusing him of public agitation and atrocity mongering (Rowe and Schulmann 2007, 269–278).

Courant was one of the many German Jews who disliked these open protests, which to his mind only undermined the position of those who had proven their loyalty to the country. Ten days after Franck's voluntary resignation, however, he learned that his name was among those Göttingen academics to be put on leave. Among others on the Nazis' dismissal list were Max Born, Emmy Noether, and Felix Bernstein. Courant got this unexpected news either by way of a telegram sent

[48] https://www.uni-goettingen.de/de/brief-von-james-franck-an-den-rektor-der-georg-august-universitaet-vom-17-april-1933/85743.html.

by the Ministry, or possibly by reading an article in one of the local newspapers. He was devastated; immediately thereafter, he wrote to his friend Harald Bohr: "I feel so close to my work here, to the surrounding countryside, to so many people and to Germany as a whole that this 'elimination' hits me with an almost unbearable force." (Reid 1976, 143).

Neugebauer took this letter with him to the Danish consulate in Hamburg in order to ensure that it reached Bohr's hands. He and Friedrichs then mounted a campaign to gain signatures for a petition to the Prussian Ministry of Education aimed at reinstating Courant as director of the Göttingen Institute of Mathematics. After much effort, including many refusals, they came up with 28 distinguished "Aryan names", including Artin, Blaschke, Carathéodory, Hasse, Heisenberg, Herglotz, Hilbert, von Laue, Mie, Planck, Prandtl, Schrödinger, Sommerfeld, van der Waerden, and Weyl (Reid 1976, 151–152). Predictably, this action elicited no response from the Ministry.

Courant's Göttingen Institute had long been a focus for right-wing agitation. In 1926 he came under personal attack by anti-Semites after he was accused of sexually intimidating a female student.[49] After a Göttingen court refused to hear the case, it was taken up by a Nazi newspaper, the "Niedersächsischer Beobachter," in an article entitled "Schändung deutscher Frauen und Mädchen durch Juden."[50] Now that the Nazis had gained power, however, the Berlin Ministry clearly did not want to risk losing control of the local situation. This being so, these dismissals may well have been designed as part of a preemptive action aimed at preventing a grassroots firestorm (Schappacher and Kneser 1990). Neugebauer stepped in as director, but had to resign after just one day since he was deemed unacceptable by the students, many of whom were ardent Nazis. Hermann Weyl then assumed the directorship, but only for a few months since his wife was Jewish. Rather than waiting for the next blow to fall, he stepped down as director to accept a position on the faculty of the newly-founded Institute for Advanced Study. Neugebauer, suffering from nervous exhaustion, requested vacation leave so that he could recover. Almost overnight, the mighty fortress of Göttingen mathematics and physics had been obliterated. Looking backward, though, its vulnerability would seem all too clear.

Courant no doubt had many more enemies than he realized. Still, he must have known that he stood on shaky political ground, despite the fact that he had served as an infantry officer during the war and had been seriously wounded in battle. After recovering, he had rejoined the war effort as a technician in charge of underground telegraphy, again proving his leadership abilities. Nor did Courant shy away from politics, like so many German academics. After the collapse of the Western front, he was elected to head the Council of Soldiers and Workers in Ilsenburg, a small town in the middle of the Harz. Such Soviet-style councils died out quickly, but Courant continued to be politically active when he returned to Göttingen after the war. He

[49] Details about this case, which was dismissed in court, can be found in Universitätsarchiv Göttingen, UAG.Kur.PA.Courant, Richard; Bd.1.

[50] A copy of this newspaper article can be found in the Universitätsarchiv Göttingen, UAG. Sek.299.e.

later claimed he was only following the advice of Felix Klein when he got involved in politics with the local Social Democratic Party; both apparently saw the moderate wing of the SPD as Germany's only hope during these desperate times (Reid 1976, 72–74). Yet what would have been seen in 1920 as the political activism of a German patriot now looked in 1933 like the typical scheming of an ambitious Jew, one who sought to exploit his privileged position within the Göttingen mathematical community.

It should not be overlooked that when Courant founded the Institute of Mathematics in 1922, he took this action in the wake of the dissolution of the Philosophical Faculty (Dahms 1999, 401–402), an event of great significance that reflects the tense political atmosphere in Göttingen over many years.[51] Up until this split, the faculty had been composed of two *Sparten*—one for mathematics and the natural sciences, the other for social sciences and humanities—though both voted together on matters that affected the faculty as a whole. During the war, however, a series of conflicts arose that ultimately made this working arrangement untenable.

One of the more dramatic conflicts involved Emmy Noether, who had come to Göttingen from Erlangen in 1915 in order to habilitate (Tollmien 1990). Her petition soon turned into a test case to determine whether women could be allowed to assume the duties of a *Dozent* at a German university. Noether, a Jew with decidedly left-wing political views, clearly could expect no sympathy from the many conservatives on the faculty, but these factors were largely irrelevant; the key problem was her sex. Klein, Hilbert, and Landau all vouched for her abilities, noting that she was better qualified than most male candidates. Together with their colleagues, Runge and Caratheodory, they presented a memorandum to the Ministry supporting Noether's candidacy for the *venia legendi*. Their proposal, however, was repudiated by nearly all their colleagues in the humanities, who filed a separate report that effectively stymied her case. This blocking action led to an acrimonious dispute in the faculty during which Hilbert personally insulted a colleague (he later agreed to issue an official apology). Two years later the same group of mathematicians, together with the physicists Woldemar Voigt and Peter Debye, petitioned the Ministry directly only to receive a definitive response, namely that neither Noether nor any other woman would be allowed to teach at a Prussian university. Only with the fall of the Reich two years later did this situation change. Emmy Noether was then promptly made a *Privatdozent*, though she was never promoted beyond the level of an unofficial professor (Dick 1981).[52]

Beyond this particular conflict within the faculty there were a number of other open disputes involving the admission and promotion of foreign students, general requirements for German students, and the qualifications of candidates under consideration for chairs in philosophy. According to a spokesman for the humanities, writing in 1918, the two *Sparten* had "completely different principles and viewpoints" regarding these and other matters, which would make it impossible to avoid

[51] Dekanatsakten, "Spaltung der Fakultat," II Ph/lk, Uuniversitätsarchiv Göttingen.

[52] Fakultatsakten IIPh Nr. 4e, Universitätsarchiv Göttingen.

friction in the future.[53] Hilbert was singled out as the ringleader within the scientific faction, which might seem out of character given that he has often been portrayed as either naive or eccentric, for example in (Reid 1970). No doubt he was flamboyant, but when it came to academic affairs he could be a tenacious, hard-headed negotiator whose views carried considerable weight, not only among friendly colleagues but also within the Prussian Ministry as well. In 1915, for example, not long before his unsuccessful campaign on behalf of Emmy Noether, he singlehandedly prevented Johannes Stark from being called to the chair in experimental physics at Gottingen. At that time, Stark was running torpedo tests in Kiel for German U-boats. Hs candidacy was strongly backed by two prominent right-wing physicists, Willy Wien and Philipp Lenard; even Hilbert had to admit that Stark had excellent scientific credentials. Nevertheless, he insisted that Stark, as a *völkisch* nationalist and outspoken anti-Semite, was simply unacceptable for Gottingen, a view his colleague, Peter Debye, came to accept as well.[54]

Hilbert's most spectacular fight with the humanists in Göttingen involved a longstanding feud over the philosopher Leonard Nelson. The *Geisteswissenschaftler* loathed Nelson, not only because of his unorthodox philosophical views, but also because he was a radical fanatic who agitated for causes like pacifism and socialism. Added to this, he also happened to be of Jewish extraction. His sole support within the faculty came from mathematicians and scientists. Hilbert and Klein were the only members to vote in his favor in 1906 when he made his first, unsuccessful attempt to habilitate (he was approved three years later). Nevertheless, Nelson had unusual talent as a teacher, so much so that he eventually became a cult figure in Göttingen. He and his devotees had numerous run-ins with fraternal organizations and other nationalist-oriented student groups, scenes that became more frequent and intense after the war broke out. Nelson's fights with the Gottingen faculty were mostly of a petty nature, but his notoriety spread after the *Göttinger Tageblatt* falsely reported that he had evaded military service by feigning illness. He took the paper's editor to court, and the ensuing trial, which lasted over a year, was reported in considerable detail by the *Tageblatt* (Hieronimus 1964, 97, 101–107).

So while Courant inherited Klein's role as administrator in 1920, he also carried the burden imposed by these fairly recent conflicts, which literally tore the Philosophical Faculty apart. He could not possibly undo the damage or make people forget Hilbert's ugly fights with colleagues during the war years.[55] Einstein had been

---

[53] Memorandum of 10 Aug. 1918, sent by the historische-philologische Abteilung to the Ministry, Rep. 76 Va Sekt. 6, Tit. IV, 1, Vol. XXV, Bi. 400–402, Geheimes Staatsarchiv Preußischer Kulturbesitz.

[54] For Hilbert's campaign against the appointment of Stark, see the documentation in Rep. 76 Va Sekt. 6, Tit. IV, 1, Vol. XXIV, Bi. 341–376, Geheimes Staatsarchiv Preußischer Kulturbesitz. This contains a letter from Stark, dated 1 Feb. 1915, from which it is clear that he would have accepted the call had he received it.

[55] After the war, Hilbert pointedly attacked the conservative Germanist Eduard Schröder for his role in the persecution of pacifists on the faculty during wartime. Hilbert refused to attend meetings of the Göttingen Academy so long as Schröder presided (see Hilbert to Carl Runge, Cod. Ms. D. Hilbert 457: 13).

well aware of this situation after he visited Göttingen in the summer of 1915. When in March 1920 Born asked for his advice about whether he should leave Frankfurt to accept the Göttingen offer, Einstein answered that he would choose to stay. "For me it would be unbearable," he explained, "to be dependent so entirely on such a small circle of vain scholars, most of closed hearts and minds (no other social interactions). Think of what Hilbert has had to put up with from this society" (Born 1979, 47). One might add here that Hilbert had a thick skin and a massive ego as protection. Courant had neither, but he and Neugebauer did share Hilbert's sense of humor, a blend of sarcasm and delight over the follies of life. So they told humorous stories that helped capture the absurdity of a particular situation. They had lived, after all, during the most trying of times when Germany stood on the brink of collapse. How then should a sane person view the petty-minded disputes of a handful of provincial academics? By telling jokes, of course. So when these earlier conflicts led to the final breakup of the faculty in 1922—a split that Klein had long tried to prevent—Courant took that "first revolutionary step" of ordering new stationary.

Born, too, spoke of the revolutionary atmosphere that enabled him to push through the double appointment with Franck, but he was hardly joking. In fact, he indirectly confirmed what Hilbert later recalled as "the most ruthless and hardest fight" he had ever endured. Two years afterward, when their colleague Ludwig Prandtl was offered an attractive position in Munich (which he ultimately declined), Born and Courant contemplated the prospect of bringing Theodor von Kármán back to Göttingen. Kármán, a Hungarian Jew and close friend of Born, had studied under Prandtl before taking a professorship at the *Technische Hochschule* in Aachen. Born had the greatest respect for Kármán's talents, but he could not bring himself to fight for his candidacy in the faculty after the tremendous turmoil that ensued when he insisted on an appointment together with Franck. Guilt stricken, he wrote Kármán a long letter, hoping to ease his own state of mind: "I had to decide if I wanted to carry the fight for you against the enemies of Israel. I felt sick about it. I simply did not have the strength in body and soul to take on this goal myself unequivocally. I had had enough of the molestations of the Lenard people ... I wanted my peace" (Greenspan 2005, 115–116).

The overall atmosphere in the Göttingen mathematical community reflected many typical elements of Weimar era counter-culture with its intermingling of Jews, foreigners, and even some women (Rowe 1986, 444–449). Nothing like these open internationalist tendencies surfaced within the Berlin mathematical community, as reflected by the fact that Erhard Schmidt, Richard von Mises, and Ludwig Bieberbach all supported the call for a boycott of the Bologna International Congress. Thus by the late 1920s, the traditional rivalry between Göttingen and Berlin had taken on clear political overtones. A deep rift separated the nationalist-oriented mathematicians in Berlin, led by Bieberbach, and the outspoken internationalists in Göttingen.

After this boycott effort backfired, Bieberbach engaged in another losing power struggle with the Göttingen mathematicians over control of review journals. This began when he engaged the Prussian Academy to undertake a series of measures to bolster the Berlin-based *Jahrbuch über die Fortschritte der Mathematik*, originally founded in 1869. Bieberbach then reacted in dismay when Courant began making plans with Springer in 1930 to launch the *Zentralblatt* (Siegmund-Schultze 1993). The latter undertaking barely got off the ground before both Courant and the *Zentralblatt's* main workhorse, Otto Neugebauer, were compelled to flee Nazi Germany.[56]

Shortly after the Nazis assumed power, Bieberbach quickly tried to promote his position within the German mathematical community (Mehrtens 1987). Exploiting his connections with prominent Nazis, he encouraged the decimation of the "Courant clique" in Göttingen, a group often likened with supposed "Jewish conspiracies" that had weakened Germany during the Weimar years. Courant's Jewish background and longstanding ties with Göttingen's power elite made him a natural target, though Bieberbach's foremost enemy was the distinguished number theorist, Edmund Landau. Proud and defiant, Landau was ready to test the new waters when the winter semester opened in October 1933. On the way to his lecture hall, he found SA-guards posted at the doorway and *one* student inside. As a consequence, he "voluntarily" agreed to enter early retirement (he died in 1938). The boycott of Landau's class had been led by none other than Oswald Teichmüller, the one truly brilliant mathematician among the Nazi rabble (Schappacher 1991).

Bieberbach had been late to jump onto the Nazi bandwagon, but only months after the destruction of the Mathematics Institute in Göttingen he made a bold move to politicize German mathematics. In a lecture that purported to clarify differences in mathematical styles by appealing to racial and national stereotypes, he seized on Landau as a prime illustration of the "Jewish type" and praised the Göttingen students who boycotted his classes for their "manly action" (*mannhaftes Auftreten*) (Schappacher and Kneser 1990, 58). When Harald Bohr criticized him in a Danish newspaper article, Bieberbach, as *Schriftenleiter* for the *Jahresbericht der Deutschen Mathematiker-Vereinigung*, published an "Open Letter to Harald Bohr" calling him a "pest on all international cooperation" (Bieberbach 1934, 3).

Since Bieberbach had published the letter against the expressed wishes of the only two co-editors who knew about it beforehand (Helmut Hasse and Konrad Knopp), this action set the stage for a dramatic confrontation that culminated in September 1934 at the annual DMV meeting held in Bad Pyrmont. There, flanked by a throng of right-wing students whom he had invited, Bieberbach vied to implement the "Führerprinzip" so dear to Nazi ideologues, by entering a motion in which he nominated Erhard Tornier, a second-rate mathematician with a first-rate Nazi pedigree, for the post of *Führer* of the DMV. When this coup attempt failed, he later tried to intimidate his rivals in the DMV by exploiting his connections with Theodor Vahlen. But having survived the showdown in Bad Pyrmont, Oskar Perron and Konrad Knopp were now prepared to call his bluff, and in early 1935 they forced Bieberbach out of the executive committee of the DMV (Schappacher and Kneser 1990, 62–71).

After losing his position in Göttingen, Courant spent the following academic year in Cambridge, England (Siegmund-Schultze 2009). From there he remained in steady contact with Oswald Veblen, now a member of the new faculty at Princeton's

---

[56] On Neugebauer's role with both *Zentralblatt* and *Mathematical Reviews*, see Siegmund-Schultze's paper in this volume.

Institute for Advanced Study, which opened in 1932. Veblen tried to assure him, writing that "your friends in America are trying to find a worthy position for you" (Reid 1976, 156). The best they could come up with, however, was a two-year appointment at New York University. Courant had never heard of this place, so he asked Veblen a simple question: "Who are the mathematicians there?" Veblen knew only one: a topologist by the name of Donald Flanders.

In the meantime, Courant had been corresponding with Abraham Flexner, Director of the Institute for Advanced Study, who was contemplating the possibility of founding a school for studies of science and culture. George Sarton, a pioneering figure for such studies in the United States, thought that Neugebauer was just the man for such an enterprise. In a letter to Flexner from September 1933, he made this point by humbly contrasting their styles of research: "As compared with Neugebauer I am only a dilettante. He works in the *front trenches* while I amuse myself way back in the rear—praising the ones, blaming the others; saying this ought to be done, etc.—& doing very little myself. What Neugebauer does is fundamental, what I do, secondary" (Pyenson 1995, 268).

Sarton's hopes that Neugebauer could be brought to Princeton came to naught, however, mainly because this would have posed the problem of coping with Springer's *Zentralblatt* operation. Soon afterward, a better option emerged when Harald Bohr arranged a three-year appointment as professor at Copenhagen beginning in January 1934. From this new outpost Neugebauer continued editing Springer's *Zentralblatt* until 1938, when he resigned in protest of Nazi racial policies that led to the removal of Jewish colleagues from its board. These events then paved the way for the founding of *Mathematical Reviews*, which Neugebauer co-managed beginning in 1940, after his arrival at Brown University. Courant, too, had severed his publishing connections with Springer. Ten days after the devastating blow to Jewish property and life during the *Kristallnacht* (Night of Broken Glass), he wrote to Ferdinand Springer informing him that he wished to resign as editor of the *Grundlehren* series (Reid 1976, 208–209). In the United States during the Second World War he soon found ways to promote his own special style of applied mathematics. All the while, his publishing projects reflected the sense of communal enterprise he identified with the Göttingen atmosphere he had helped mold.

Neugebauer as Visionary

Neugebauer was able to get most of his property out of Germany, though he had to abandon a house with a partially paid mortgage (Swerdlow 1993). While in Copenhagen, his research was supported in part by the Rockefeller Foundation. Almost immediately he began preparing a series of lectures on Egyptian and Babylonian mathematics that he would publish in Courant's yellow series as *Vorgriechische Mathematik* (Neugebauer 1934). According to Noel Swerdlow this volume was "as much a cultural as a technical history of mathematics" and represents "Neugebauer's most thorough and successful union of the two interpretations"

(Swerdlow 1993, 145) More striking still is the unfinished character of this work, which represents the first and final volume in a projected trilogy that would remain incomplete. Neugebauer had planned to tackle Greek mathematics proper in the second volume, whereas the third would have dealt with mathematical astronomy, both in the Greek tradition culminating with Ptolemy as well as the largely unknown work of late Babylonian astronomers. Thus, his original aim, as spelled out in the foreword to the first volume, was to achieve a first overview of the ancient mathematical sciences in their entirety, something that had never before been attempted.

Swerdlow has offered compelling reasons to explain why Neugebauer dropped this project, one being that he simply found the rich textual sources for Mesopotamian mathematical astronomy far more important than anything he could ever have written about Greek mathematics. Another even more significant reason had to do with Neugebauer's shifting opinion with regard to the status of Babylonian scientific achievements. As Swerdlow put it, "by the time he had published MKT ["Mathematische Keilschrift-Texte" (Neugebauer 1935/37)] and was deeply engrossed in Babylonian astronomy, his respect for Babylonian mathematical science was far too great for him to treat it as preparatory to anything. .... Never again was Neugebauer to subsume Babylonian mathematics and astronomy under the title *vorgriechische*, and to the best of my knowledge the corresponding term pre-Greek never occurs in his English publications" (Swerdlow 1993, 148). Thus, the mid-1930s saw another major transition in Neugebauer's intellectual journey: from this time forward his primary focus shifted from ancient mathematics per se to the history of mathematical astronomy.

Nevertheless, we can trace a fairly clear picture of the line of argument Neugebauer originally had in mind by examining the summary remarks at the conclusion of his *Vorgriechische Mathematik* as well as some of his other publications from the 1930s. Neugebauer's writings from the 1920s contain few hints that his understanding of ancient mathematics was fundamentally opposed to older views. By the early 1930s, however, his analyses of Babylonian texts led him to a new conception, namely that the Greek penchant for geometrization represented a retrograde step in the natural development of the exact sciences. This did not mean, of course, that he held a low opinion of Euclid's *Elements*; he simply thought that historians and philosophers had distorted its true place in the history of mathematics. Thus, he once imagined how scholars in some future civilization might easily form a deceptive picture of mathematical knowledge *circa* 1900 if the only important text that happened to survive was Hilbert's *Grundlagen der Geometrie* (Neugebauer 1931, 132).

In the course of this transition, Neugebauer's assertions about the character of ancient mathematics often took on a strident tone. Particularly striking in this connection is the essay entitled "Zur geometrischen Algebra," published in 1936 in *Quellen und Studien* (Neugebauer 1936). There, Neugebauer took as his motto a famous fragment from the late Pythagorean Archytas of Tarentum, which reads: "It seems that logistic far excels the other arts in regard to wisdom, and in particular in treating more clearly what it wishes than geometry. And where geometry fails, logistic brings about proofs" (Neugebauer 1936, 245).

Much has been written about this passage, in particular about what might be meant by the term "logistic." This notion pops up in Platonic dialogues and quite clearly it has more to do with ancient arithmetic than it does with logic. The whole matter was discussed at great length by Jakob Klein in his study "Die griechische Logistik und die Entstehung der Algebra" (Klein 1936), which appeared alongside Neugebauer's article (it was later translated into English by Eva Brann (Klein 1968)). In fact, both scholars were chasing after the same elusive goal, though there the similarity ends. Klein was a classical philologist who later became a master teacher of the "Great Books" curriculum at St. Johns College in Annapolis Maryland. Not surprisingly, he was intent on squeezing as much out of Plato as he possibly could. Thus he distinguished carefully between practical and theoretical logistic, offering a new interpretation of Diophantus' *Arithmetica* that placed it within the latter tradition. Neugebauer had no patience for the nuances of meaning classicists liked to pull out of their texts. Indeed, he had an entirely different agenda. His point was that rigorous axiomatic reasoning in the style of Euclid arose rather late, and that Archytas, a contemporary of Plato, was bearing witness to the primacy of algebraic content over the geometrical form in which the Greeks dressed their mathematics. With that, we can move a little closer to understanding Neugebauer's visionary approach to the history of mathematics.

Decades earlier, the Danish historian of mathematics, H. G. Zeuthen, already advanced the idea that the Greeks had found it necessary to geometrize their purely algebraic results after the discovery of incommensurable magnitudes (Zeuthen 1896). Neugebauer took up this by then standard interpretation, adopted by Heath and nearly everyone else, but he now went much further, arguing that the algebraic content—found not only in Book II of Euclid but throughout the entire corpus of Apollonius' *Conica*—could be traced back to results and methods of the Babylonians:

> The answer to the question what were the origins of the fundamental problem in all of geometrical algebra [meaning the application of areas, as given by Euclid's propositions II.44 and VI.27-29] can today be given completely: they lie, on the one hand, in the demands of the Greeks to secure the general validity of their mathematics in the wake of the emergence of irrational magnitudes, on the other, in the resulting necessity *to translate the results of the pre-Greek "algebraic" algebra as well.* Once one has formulated the problem in this way, everything else is completely trivial [!] and provides *the smooth connection between Babylonian algebra and the formulations of Euclid.* (Neugebauer 1936, 250, my translation, his italics)

The mathematical concepts underlying this argument are by no means difficult. It should be emphasized, however, that what may seem mathematically trivial (*i.e.* obvious) should hardly be thought of as historically self-evident. Since Zeuthen's time, it had been customary to interpret Greek problem-solving methods as manipulations closely related to techniques like "completing the square", used to solve quadratic equations. These Greek methods, called applications of areas, occupy a prominent place in Euclid's *Elements* as well as in his *Data*, a kind of handbook for problem solving. Neugebauer was struck by the parallelism between certain standard Babylonian problems and the Greek methods for solving very similar problems geometrically (Neugebauer 1957, 40–41, 149–150).

A typical algebra problem found in several cuneiform tablets from the Old Babylonian period requires that one find two numbers whose sum (or difference) and product are both given (Neugebauer called this the "normal form" leading to a single quadratic equation). This pair of problems, depending on whether the sum or difference is given, can also be found as Propositions 84 and 85 in Euclid's *Data*. Moreover, according to the neo-Platonic commentator Proclus—on the authority of Aristotle's student, Eudemus, author of a lost *History of Geometry* written just before Euclid's time—the three types of applications of areas (later used by Apollonius to distinguish the three types of conic sections: ellipse, parabola, and hyperbola) were discovered long before Euclid: "These things, says Eudemus, are ancient and are discoveries of the Muse of the Pythagoreans" (Heath 1956, 343).

Neugebauer would have been the last to argue that the Pythagoreans had anything to do with this ancient knowledge; for him, the key fact was merely that the original ideas were old, hence likely to have roots in still older cultures from which the Greeks borrowed freely. Having established that the mathematical content of the Babylonian texts was fundamentally algebraic, he now claimed that Mesopotamia was the original source of the algebra underlying the "geometric algebra" uncovered by Zeuthen at the end of the nineteenth century. Neugebauer was fully aware, of course, that his interpretation required a really bold leap of the historical imagination, since making a claim for the transmission of such knowledge over such a vast span of time meant accepting that this mathematical linkage sufficed to fill a gap devoid of any substantive documentary evidence. Summarizing his position, he offered these remarks: "Every attempt to connect Greek thought with the pre-Greek meets with intense resistance. The possibility of having to modify the usual picture of the Greeks is always undesirable, despite all shifts of view, ... [and yet] the Greeks stand in the middle and no longer at the beginning" (Neugebauer 1936, 259).

This study was, in fact, the third in a series of articles in *Quellen und Studien* subtitled "Studien zur Geschichte der antiken Algebra." One need only read the introductory remarks in his first two papers to gain a clear impression of Neugebauer's grandiose vision for radically revising our standard picture of the history of ancient mathematics. In the first of these, he emphasizes that what he means by "ancient algebra" goes far beyond the conventional usage of this term. For him, ancient algebraic methods can be found as an essential component of mathematical thought practically everywhere and at all times, including within the Greek geometrical tradition. Moreover, owing to the sparse source material available for the study of Greek mathematics, he saw no other way to obtain a deeper understanding of the Greek achievements than to broaden the scope of ancient studies to other ancient cultures. Taking this wider range of mathematical activity into account can open a scholar's eyes to important new perspectives on the works of leading Greek authorities. This, he thought, was an important first step toward appreciating the underlying role of "algebraic" methods in their geometrical work.

In Neugebauer's view, historians of mathematics and science needed to recognize that their discipline could only mature by emancipating itself from traditional humanistic studies. This meant, in particular, that modern practitioners should cast aside all the misleading categories and periodizations borrowed from political and cultural history.[57] What insights can possibly be gained by describing Euler as a mathematician of the baroque era? For him, the history of the exact sciences should be seen as consisting of just two periods: ancient and modern. The same then applied to the history of algebra, the modern period beginning with Viète, Fermat, and Descartes, the founders of *symbolic* algebra. All prior forms of algebraic practice he thought of as ancient, which essentially meant that earlier mathematicians should be seen as part of this older tradition whenever their underlying methods reflected techniques of an algebraic nature. What he meant by this can best be seen from the second article in this series, his Apollonius-Studien (Neugebauer 1932), which was devoted to an analysis of key examples of such algebraic methods in the first four books of Apollonius' *Conica.*

As noted earlier, this paper grew out of the course he taught in Göttingen on "Greek geometry after Euclid." Neugebauer began with a lengthy explanation of the methodology behind this study, realizing of course that an ordinary reader would have to wonder why a technical study of various propositions in the theory of conic sections belonged in a series of articles on the history of "ancient algebra." Not wanting to get tied up by the word "algebra", he offered discomfited readers another alternative: he had in mind something like "algorithmic" methods, by which he meant techniques that were carried out according to a set of procedures that required technical skill, but no really innovative thought. What he emphasized, however, was the need to go beyond the formal language—in this case manipulations of plane geometric figures—in order to extract the deeper mathematical content behind it. Reading this introduction to his "Apollonius-Studien"—a paper that clearly left a deep impression on van der Waerden with its bold vision of a broadly based algebraic tradition—one is struck by the resemblance to Hilbert's general outlook. Nor do the similarities end there: Neugebauer, like Hilbert, hungered for unified knowledge, and both famously scorned narrow specialization. In a certain sense, Neugebauer was, indeed, a disciple of the great Pied Piper of Göttingen (Rowe 2012).

During these early years, Neugebauer clearly wanted to create a new discipline, one that could point historical studies of the exact sciences in a direction relevant to the interests of research mathematicians. That larger goal proved chimerical, but he did succeed in creating a small niche for the kind of technical studies he had in mind. For Neugebauer, historians of the exact sciences should have the ability to identify core content in ancient texts. He had no doubts or reservations that this

---

[57] This viewpoint was no mere rhetorical stance; in *The Exact Sciences in Antiquity* (Neugebauer 1957) he illustrated what he meant by discussing the number systems that appear in the famous *Book of Hours* of the Duc de Berry in order to show that "[f]or the history of mathematics and astronomy the traditional division of political history into Antiquity and Middle Ages is of no significance" (Neugebauer 1957, 3).

could be achieved precisely because the content was mathematical and, hence, independent of the ephemeral forms in which the knowledge happened to be couched. This conception of the exact sciences was an elemental part of his scholarship from the beginning.

Neugebauer's research represented part of a large-scale intrusion by mathematicians into a field formerly dominated by classicists, namely the history of Greek mathematics, which was traditionally seen as strongly linked with the works and influence of philosophers, especially Plato and Aristotle. Ever the anti-philosopher, Neugebauer wanted to undermine this standard German fascination with Greek philosophy, most particularly the Platonic tradition (Rowe 2013). In this respect, his work stood poles apart from that of Oskar Becker, or for that matter, Otto Toeplitz, both of whom, like Neugebauer, published regularly in *Quellen und Studien*. These two older contemporaries combined finely tuned mathematical arguments with careful philological analyses of classical Greek texts. Neugebauer, on the other hand, showed very little interest in studies of this kind. Furthermore, he had an entirely different agenda than they: he aimed to overthrow the standard historiography that made mathematics look like the handmaiden of Greek philosophy.

Neugebauer's original vision thus entailed a radical rewriting of the history of ancient mathematics and exact sciences, a vast project that brings to mind Hilbert's larger research program. For some time, philosophers made a farce of Hilbert's mathematical views by reducing them to the procrustean bed of his proof theory. Surely anyone who has skimmed Leo Corry's study (Corry 2004) of Hilbert's long-standing interests in classical as well as modern physics would realize that his was no narrow philosophy. Neugebauer clearly knew very well that Hilbert's driving motivation called for the mathematization of everything that fell within the scope of number, where the concept of number itself was to be understood in the most abstract possible sense. Neugebauer apparently took no great interest in the formalist ideology that informed Hilbert's research program, but he nevertheless shared its ambitious global viewpoint.

A central point for Neugebauer was that rigorous axiomatic reasoning in the style of Euclid arose rather late. He liked to call on Archytas, a contemporary of Plato, as someone bearing witness to the primacy of algebraic content over the geometrical form in which the Greeks dressed their mathematics. That tradition, so he argued, began in Mesopotamia over a thousand years before Euclid was born. This revisionist view also aimed to debunk the notion of a "Greek miracle" that sprang up during the sixth century from the shores of Ionia. Neugebauer was convinced that most of the sources that reported on the legendary feats of ancient heroes—Thales, Pythagoras, and their intellectual progeny—were just that: legends that had grown with the passing of time. So his constant watchword remained skepticism with regard to the accomplishments of the early Greeks, whereas Toeplitz, Becker, and others began to analyze extant sources with a critical eye toward their standards of exactness.[58]

[58] See (Christianidis 2004) for a recent account of older as well as the newer historiography on Greek mathematics.

In 1949 Neugebauer was invited to Cornell University to deliver six lectures on ancient sciences as one of the distinguished speakers in its Messenger series. He was the first historian of mathematics to be so honored. Afterward, he went over his notes to produce the carefully sculpted six chapters of his *The Exact Sciences in Antiquity*, published in 1951 by Munksgaard in Copenhagen with high-quality plates. For the second edition, Neugebauer updated the material and added two technical appendices, but he still hoped to have "avoided … converting my lectures into a textbook" (Neugebauer 1957, ix). Evidently, he still very much valued the less formal form of exposition associated with oral exposition, perhaps with Klein's Göttingen lectures still in mind. Significantly, Neugebauer dedicated this now classic book to "Richard Courant, in Friendship and Gratitude." Elaborating on that dedication in the preface, he wrote that it was Courant who enabled him to pursue graduate studies in ancient mathematics, and he went on to remark: "more than that I owe [to him] the experience of being introduced to modern mathematics and physics as a part of intellectual endeavour, never isolated from each other nor from any other field of our civilization" (Neugebauer 1957, vii).

Neugebauer tended to choose his words carefully, so we may be sure that this public acknowledgement of his debt to Courant was far more than just a friendly gesture. His allusion to physics brings to mind the famous Courant-Hilbert volume from 1924 which gave physicists the tools they needed to handle Schrödinger's equation and related problems in quantum mechanics. Yet, clearly, what Neugebauer had in mind here went far beyond the usual appeal to the unity of mathematical and physical ideas, for he wrote that Courant's vision saw these fields of intellectual endeavor as "never isolated from each other nor from any other field of our civilization." This brief remark comes very close to capturing the essence of Neugebauer's own understanding of what it meant to study the history of mathematics. Regarding Courant's own vision, he said this on the occasion of his 75th birthday: "… the real core of his work [consisted] in the conscious continuation and ever widening development of the ideas of Riemann, Klein, and Hilbert, and in his insistence on demonstrating the fundamental unity of all mathematical disciplines. One must always remain aware of these basic motives if one wants to do justice to Courant's work and to realize its inner consistency" (Neugebauer 1963, 1).

Shortly after the first edition of *The Exact Sciences in Antiquity* was published in Copenhagen in 1951, George Sarton wrote a lengthy review for *Isis* in which he noted that no one but Neugebauer could have written such a volume. Sarton also paid tribute to Cornell University for its role in helping the author produce this idiosyncratic *synthesis* based on his six Messenger Lectures from 1949. This opportunity, Sarton felt sure, gave Neugebauer just the incentive he needed to address a broader set of historical issues, something he was otherwise loathe to do. In his review, Sarton put the matter this way: "as he does not like synthetic work and even affects to despise it, he would probably not have written this book without that flattering invitation, and we, his readers, would have been the losers" (Sarton 1952, 69).

One can easily read between the lines here, since Sarton, the doyen of American historians of science, certainly saw himself as a leading representative of the genre of scholarship to which he here alluded. Nor was his review particularly positive. He voiced skepticism when it came to Neugebauer's claims regarding the historical impact of Babylonian mathematics and astronomy. Noting that neither Hipparchus nor Ptolemy made mention of earlier Babylonian theoretical contributions, he wondered how historians could ever know that they drew on such sources? As for Babylonian algebra, why should we assume that this knowledge survived long after the period of Hammurabi when there is no extant evidence for a continuous tradition of high mathematical culture in Mesopotamia? And if such mathematical knowledge persisted, how was it transmitted? After all, the complexity of the Babylonian algebraic and astronomical techniques required an expertise similar to Neugebauer's own. Sarton also took sharp issue with Neugebauer over the "centrality of Hellenistic science," especially his claim that this melting pot of ancient science later spread to India before entering Western Europe, where it held sway until the time of Newton. In Sarton's view, the Hellenistic period marked the final phase of Babylonian science, though he admitted some faint influences on both the Indian and Islamic cultural spheres. For the most part, however, he contrasted the larger long-term impact of Greek science with the relatively meager legacy of the Babylonian tradition. For him, this was the gravest shortcoming of all; how could Neugebauer write a book called *The Exact Sciences in Antiquity* and virtually ignore the achievements of the Greeks? Doing that was comparable to writing a play entitled *Hamlet* while leaving out the figure of Hamlet himself. With that, he ended by chiding Neugebauer's Danish editors—identified as Zeuthen's countrymen—for allowing their distinguished friend to make such a blunder.

Sarton's criticisms reflect the views of a generalist who clearly found Neugebauer's overall framework far from convincing. He had the highest respect for the author's specialized contributions to research on the ancient exact sciences—work that required not only formidable mathematical abilities but also immense discipline—but this review makes plain that he saw Neugebauer's book as the product of a remarkable specialist. His overall verdict—seen from his personal vantage point as someone who hoped to open inroads for the history of science within the curriculum of American higher education—echoed Neugebauer's own forthright opinion that he "did not like synthetic work." *Exact Sciences*, he opined, was of limited value for introductory courses and it should not be taken as a model for teaching the history of ancient science. Though full of nicely chosen anecdotes and a good deal of general information, it simply could not pass muster as a global account of the history of the exact sciences in ancient cultures. Noel Swerdlow later expressed a very different opinion: "Neugebauer here allowed himself the freedom to comment on subjects from antiquity to the Renaissance. The expert can learn something from it, and from its notes, every time it is read, and for the general reader it is, in my opinion, the finest book ever written on any aspect of ancient science" (Swerdlow 1993, 156).

George Sarton saw himself as a champion of what he called a synthetic approach to the history of mathematics (Sarton 1936b, 11). What Neugebauer thought about this can well be surmised by what he wrote in the preface to the first edition of *Exact Sciences in Antiquity*: "I am exceedingly skeptical of any attempt to reach a "synthesis"—whatever this term may mean—and I am convinced that specialization is the only basis of sound knowledge" (Neugebauer 1957, vii–viii). Paging through Sarton's booklet, *The Study of the History of Mathematics*, one can easily understand Neugebauer's dismissive attitude. There one reads that:

> The main reason for studying the history of mathematics, or the history of any science, is purely humanistic. Being men, we are interested in other men, and especially in such men as have helped us to fulfill our highest destiny. As soon as we realize the great part played by individual men in mathematical discoveries—for, however these may be determined, they cannot be brought about except by means of human brains—, we are anxious to know all their circumstances. (Sarton 1936, 12)

Sarton's humanistic approach to the history of mathematics thus derives from simple human curiosity, which he admits is the same instinct that feeds public fascination with murderers. Whereas newspapers skillfully exploit this "insatiable desire to know every detail of a murder case, those who are more thoughtful wish to investigate every detail of scientific discoveries or other creative achievements" (*ibid.*). This loftier interest apparently has much to do with Sarton's sympathy for hero worship: "One soon realizes that mathematicians are much like other men, except in the single respect of their special genius, and that genius itself has many shapes and aspects" (*ibid.*).

Not surprisingly, Neugebauer drew a sharp line between his work and that of generalists like Sarton, though he never launched a frontal attack on the latter's own works. He did, however, occasionally publish critical responses to Sarton's opinions in *Isis*, one of which sheds much light on the intellectual fault lines that divided them. In a review of B. L. van der Waerden's *Science Awakening*, Sarton expressed dismay over the author's "shocking ingratitude" towards Moritz Cantor, whom he called "one of the greatest scholars of [the] last century, a man to whom every historian of mathematics owes deep gratitude." After citing this passage, Neugebauer went on to write a "Notice of Ingratitude" (Neugebauer 1956):

> Since I must conclude that this statement in its generality would also apply to myself, I should like to point out that I never felt a trace of indebtedness to Cantor's voluminous production. I do not deny, of course, the fact that it had a great influence, though in a direction quite opposite to what Professor Sarton's statement implies. I always felt that its total lack of mathematical competence as well as its moralizing and anecdotal attitude seriously discredited the history of mathematics in the eyes of mathematicians, for whom, after all, the history of mathematics has to be written. In methodological respects, Cantor's work might be of some value for historians of science since it contains so many drastic examples of how one should not approach a problem... If Cantor had not philosophized about a goose counting her young or about oriental mathematics, which was equally inaccessible to him, but instead had studied the texts themselves, he would have avoided countless misinterpretations and inaccuracies which have become commonplace. It was with good reasons that the *Bibliotheca Mathematica* for years ran a special column devoted to corrections of errors in Cantor's *Geschichte der Mathematik*. But no amount of corrections can ever remedy consistent mediocrity. (Neugebauer 1956, 58)

Given that Neugebauer's academic career was decisively shaped by his training and background as a mathematician, one can easily understand his aversion to the writings of Cantor and Sarton. He was most definitely not a "synthetic" historian in the sense of Sarton, but we can say just as assuredly that his work was guided by a larger view of the history of mathematics. His was an approach to history deeply grounded in the mathematical culture he grew to know, and as we have seen, his intellectual journey from Graz to Göttingen—while marked by a few rough spots or unexpected transitions—had a clear and consistent trajectory. Looking at the two seminar lectures he delivered in Graz on the geometrization of physics—where he described developments that spanned from pre-historic times right up until Weyl's latest speculations on a new physical theory that united gravity and electromagnetism—one cannot escape the impression that Neugebauer's sensibilities as a historian were from the very beginning guided by a grandiose vision. He worked on details, but always with a larger landscape in mind, and his approach to history of science was certainly far more ambitious and revolutionary than anything Sarton ever conceived. A passage from the early essay (Neugebauer 1927b) captures his personal *Weltanschauung* very nicely: "… every historical investigation can only be counted as a useful preliminary contribution to a further synthesis if it is guided by two viewpoints: to see the history of mathematics in the framework of general history and to understand mathematics itself not as a collection of formulas to be continually increased, but as a living unity." (Neugebauer 1927b, 60).

Noel Swerdlow gave a most apt description of the "zwei Seelen" that dwelled within Otto Neugebauer and that colored all his work:

> At once a mathematician and cultural historian, Neugebauer was from the beginning aware of both interpretations and of the contradiction between them. Indeed, a notable tension between the analysis of culturally specific documents, whether the contents of a single clay tablet or scrap of papyrus or an entire Greek treatise, and the continuity and evolution of mathematical methods regardless of ages and cultures, is characteristic of all his work. And it was precisely out of this tension that was born the detailed and technical cross-cultural approach, in no way adequately described as the study of "transmission," that he applied more or less consistently to the history of the exact sciences from the ancient Near East to the European Renaissance.
>
> But if the truth be told, on a deeper level Neugebauer was always a mathematician first and foremost, who selected the subjects of his study and passed judgment on them, sometimes quite strongly, according to their mathematical interest. (Swerdlow 1993, 141–142)

Neugebauer generally avoided historical or methodological controversies, despite the fact that his name was often invoked by others. He found such disputes accomplished nothing; worse still, they distracted scholars from doing serious work. This, too, reflected his background and special place within the discipline. He had many friends and allies within the world of mathematics, most of whom deeply admired his achievements as an historian. Already in 1936, he was invited to deliver a plenary lecture at the International Congress of Mathematicians held in Oslo.

Neugebauer's own attitude toward his work contained elements of irony and playfulness. When he came to the end of his Messenger lectures on the exact sciences in antiquity, he offered this telling simile to describe his understanding of the historian's craft:

> In the Cloisters of the Metropolitan Museum in New York there hangs a magnificent tapestry which tells the tale of the Unicorn. At the end we see the miraculous animal captured, gracefully resigned to his fate, standing in an enclosure surrounded by a neat little fence. This picture may serve as a simile for what we have attempted here. We have artfully

erected from small bits of evidence the fence inside which we hope to have enclosed what may appear as a possible, living creature. Reality, however, may be vastly different from the product of our imagination; perhaps it is vain to hope for anything more than a picture which is pleasing to the constructive mind when we try to restore the past. (Neugebauer 1957, 177)

Acknowledgements This paper was made possible by support from the Johannes Gutenberg University, Mainz, and the Institute for Studies of the Ancient World. My work on the Otto Neugebauer Papers, part of The Shelby White and Leon Levy Archives Center at the Institute for Advanced Study in Princeton, was very ably assisted by the archivists there, Jillian Matos Wolf and Erica Mosner. I also benefitted from input on and criticisms of earlier drafts from Alexander Jones, Norbert Schappacher, Erhard Scholz, Reinhard Siegmund-Schultze, and Noel Swerdlow.

References

(Unpublished) Otto Neugebauer Papers, The Shelby White and Leon Levy Archives Center, Historical Studies-Social Science Library, Institute for Advanced Study, Princeton, NJ, USA. http://library.ias.edu/finding-aids/neugebauer.

Becker, H., H.-J. Dahms, and C. Wegeler (eds.). 1987. *Die Universität Göttingen unter dem Nationalsozialismus. Das verdrängte Kapitel ihrer 250jährigen Geschichte*. München: Saur.

Bergmann, B., M. Epple, and R. Ungar (eds.). 2012. *Transcending tradition: Jewish mathematicians in German-speaking academic culture*. Berlin: Springer.

Bieberbach, L. 1934. Die Kunst des Zitierens. Ein offener Brief an Herrn Harald Bohr in København. *Jahresbericht der Deutschen Mathematiker-Vereinigung* 44, 2. Abteilung 1–3.

Biermann, K.-R. 1973. *Die Mathematik und ihre Dozenten an der Berliner Universität 1810–1933 – Stationen auf dem Wege eines mathematischen Zentrums von Weltgeltung*. Berlin: Akademie-Verlag.

Bohr, H., and O. Neugebauer. 1926. Über lineare Differentialgleichungen mit konstanten Koeffizienten und fastperiodischer rechter Seite. *Nachrichten der Gesellschaft der Wissenschaften zu Göttingen, Mathematisch-Physikalische Klasse* 1926: 8–22.

Born, M. (ed.). 1969. *Albert Einstein und Max Born, Briefwechsel*. München: Nymphenburger.

Born, M. 1978. *My life: Recollections of a Nobel Laureate*. New York: Scribner.

Brouwer, L.E.J. 1928. Intuitionistische Betrachtungen über den Formalismus. *Sitzungsberichte der Preussischen Akademie der Wissenschaften, Physikalisch-Mathematische Klasse*: 48–52.

Christianidis, J. (ed.). 2004. *Classics in the history of Greek mathematics*. Dordrecht: Kluwer.

Corry, L. 2004. *David Hilbert and the Axiomatization of Physics (1898–1918): From Grundlagen der Geometrie to Grundlagen der Physik*, Archimedes, vol. 10. Dordrecht: Kluwer.

Courant, R., and D. Hilbert. 1924. *Methoden der mathematischen Physik*. Berlin: Julius Springer.

Dahms, H.-J. 1999. Die Universität Göttingen 1918 bis 1989. Vom 'Goldenen Zeitalter' der Zwanziger Jahre bis zur 'Verwaltung des Mangels' in der Gegenwart. In *Göttingen, Geschichte einer Universitätsstadt*, ed. R. von Thadden, 395–465. Göttingen: Vandenhoeck und Ruprecht.

Deimel, A. 1924. *Sumerische Grammatik archaistischen Texte*. Rome: Pontificium Institutum Biblicum.

Dick, A. 1981. *Emmy Noether, 1882–1935*. Trans. H.I. Blocher. Boston: Birkhäuser.

Eckert, M. 2013. *Arnold Sommerfeld, Atomphysiker und Kulturbote, 1868–1951. Eine Biografie*. Göttingen: Wallstein.

Engel, F., and P. Stäckel. 1895. *Die Theorie der Parallellinien von Euklid bis auf Gauss*. Leipzig: Teubner.

Frei, G., and U. Stammbach. 1992. *Hermann Weyl und die Mathematik an der ETH Zürich, 1913–1930*. Basel: Birkhäuser.

Gillings, R.J. 1982. *Mathematics in the time of the Pharaohs*. New York: Dover.
Greenspan, N. 2005. *The end of the certain world: The life and science of Max Born*. New York: Basic Books.
Hankel, H. 1869. *Die Entwicklung der Mathematik in den letzten Jahrhunderten*. Tübingen: L. Fr. Fues'sche Sortimentsbuchhandlung.
Heath, T.L. (ed.). 1956. *The thirteen books of Euclid's elements*, vol. 1. New York: Dover.
Hieronimus, E. 1964. *Theodor Lessing, Otto Meyerhof, Leonard Nelson: Bedeutende Juden in Niedersachsen*. Hannover: Verlag für Literatur und Zeitgeschehen.
Hilbert, D. 1917. Axiomatisches Denken. *Mathematische Annalen* 78: 405–415.
Hilbert, D. 1922. Neubegründung der Mathematik. Erste Mitteilung. *Abhandlungen aus dem Mathematischen Seminar der Hamburgischen Universität* 1: 157–177.
Hilbert, D. 1925. Über das Unendliche. *Mathematische Annalen* 95: 161–190.
Hilbert, D. 1926. Sur l'Infini. Trans. A. Weil. *Acta Mathematica* 48: 91–122.
Hilbert, D. 2009. *David Hilbert's lectures of the foundations of physics, 1915–1927*, ed. T. Sauer and U. Majer. Heidelberg: Springer-Verlag.
Hilbert, D. 2013. *David Hilbert's lectures of the foundations of arithmetic and logic, 1917–1933*, ed. W. Ewald and W. Sieg. Heidelberg: Springer-Verlag.
Hurwitz, A., and R. Courant. 1922. *Vorlesungen über allgemeine Funktionentheorie und elliptische Funktionen*. Berlin: Julius Springer.
Jungnickel, C., and R. McCormmach. 1990. *Intellectual mastery of nature. Theoretical physics from Ohm to Einstein, Volume 2: The now mighty theoretical physics, 1870 to 1925*. Chicago: University of Chicago Press.
Kellogg, O. 1923. The Hurwitz-Courant Funktionentheorie. *Bulletin of the American Mathematical Society* 29: 415–417.
Klein, F. 1926. *Vorlesungen über die Entwicklung der Mathematik im 19. Jahrhundert. Teil I. Für den Druck bearbeitet von R. Courant und O. Neugebauer*. Berlin: Springer.
Klein, J. 1936. Die griechische Logistik und die Entstehung der Algebra. *Quellen und Studien zur Geschichte der Mathematik, Astronomie und Physik, B: Studien* 3: 18–105 and 122–235.
Klein, J. 1968. *Greek Mathematical Thought and the Origin of Algebra*. Trans. Eva Brann. Cambridge, MA: MIT Press.
Kneser, H. 1921. Untersuchungen zur Quantentheorie. *Mathematische Annalen* 84: 277–302.
Knorr, W.R. 1982. Techniques of fractions in ancient Egypt and Greece. *Historia Mathematica* 9: 133–171.
Mehra, J., and H. Rechenberg. 1982. *The historical development of quantum theory, Vol. 1: The quantum theory of Planck, Einstein, Bohr and Sommerfeld. Its foundation and the rise of its difficulties (1900–1925)*. New York: Springer.
Mehrtens, H. 1987. Ludwig Bieberbach and "Deutsche Mathematik". In *Studies in the history of mathematics*, MAA studies in mathematics, vol. 26, ed. E.R. Phillips, 195–241. Washington, DC: Mathematical Association of America.
Neugebauer, O. 1925. Litteraturanmeldelse. Review of *The Rhind Mathematical Papyrus*, by T. E. Peet. *Matematisk Tidsskrift* A: 66–70.
Neugebauer, O. 1926. *Die Grundlagen der ägyptischen Bruchrechnung*. Berlin: Julius Springer.
Neugebauer, O. 1927a. Zur Entstehung des Sexagesimalsystems. *Abhandlungen der Akademie der Wissenshaften zu Göttingen* 13: 1–55.
Neugebauer, O. 1927b. Problemkreise der Mathematik in historischer Entwicklung. *Verhandlungen der Physikalisch-Medizinischen Gesellschaft zu Würzburg* 52(1): 56–64.
Neugebauer, O. 1927c. Über Geschichte der Mathematik. *Universitätsbund Göttingen* 9(1): 38–45.
Neugebauer, O. 1930. Das Mathematische Institut der Universität Göttingen. *Die Naturwissenschaften* 18: 1–4.
Neugebauer, O. 1931. Zur vorgriechischen Mathematik. *Erkenntnis* 2: 122–134.
Neugebauer, O. 1932. Apollonius-Studien. *Quellen und Studien zur Geschichte der Mathematik, Astronomie und Physik, B: Studien* 2: 215–254.
Neugebauer, O. 1934. *Vorlesungen über Geschichte der antiken mathematischen Wissenschaften, Erster Band: Vorgriechische Mathematik*. Berlin: Springer.

Neugebauer, O. 1936. Zur geometrischen Algebra (Studien zur Geschichte der Algebra III). *Quellen und Studien zur Geschichte der Mathematik, Astronomie und Physik, B: Studien* 3: 245–259.

Neugebauer, O. 1951. *The exact sciences in antiquity*. Copenhagen: Munksgaard.

Neugebauer, O. 1956. A notice of ingratitude. *Isis* 47: 58.

Neugebauer, O. 1957. *The exact sciences in antiquity*, 2nd ed. Providence: Brown University Press.

Neugebauer, O. 1963. *Reminiscences on the Göttingen Mathematical Institute on the Occasion of R. Courant's 75th Birthday*. Otto Neugebauer Papers, Box 14, publications vol. 11.

Neugebauer, O. 1969. *The exact sciences in antiquity,* 2nd rev. ed. New York: Dover.

Noether, E. 1919. Die arithmetische Theorie der algebraischen Funktionen einer Veränderlichen, in ihrer Beziehung zu den übrigen Theorien und zu der Zahlkörpertheorie. *Jahresbericht der Deutschen Mathematiker-Vereinigung* 28: 182–203.

Pyenson, L. 1995. Inventory as a route to understanding: Sarton, Neugebauer, and sources. *History of Science* 33: 253–282.

Reid, C. 1970. *Hilbert*. New York: Springer Verlag.

Reid, C. 1976. *Courant in Göttingen and New York: The story of an improbable mathematician*. New York: Springer Verlag.

Remmert, V., and U. Schneider. 2010. *Eine Disziplin und ihre Verleger. Disziplinenkultur und Publikationswesen der Mathematik in Deutschland, 1871–1949*. Bielefeld: Transcript.

Rowe, D.E. 1986. 'Jewish Mathematics' at Gottingen in the Era of Felix Klein. *Isis* 77: 422–449.

Rowe, D.E. 1989. Interview with Dirk Jan Struik. *The Mathematical Intelligencer* 11(1): 14–26.

Rowe, D.E. 2012. Otto Neugebauer and Richard Courant: On exporting the Göttingen approach to the history of mathematics. *The Mathematical Intelligencer* 34(2): 29–37.

Rowe, D.E. 2013. Otto Neugebauer's vision for rewriting the history of ancient mathematics. *Anabases – Traditions et réceptions de l'Antiquité* 18: 175–196.

Rowe, D.E. 2015. Transforming tradition: Richard Courant in Göttingen. *The Mathematical Intelligencer*, published online 16 January, 2015. doi:10.1007/s00283-014-9522-9.

Rowe, D.E., and R. Schulmann. 2007. *Einstein on politics: His private thoughts and public stands on nationalism, zionism, war, peace, and the bomb*. Princeton: Princeton University Press.

Sarkowski, H. 1996. *Springer-Verlag: History of a scientific publishing house*. Berlin: Springer.

Sarton, G. 1936. *The study of the history of mathematics*. Cambridge, MA: Harvard University Press.

Sarton, G. 1952. Review of O. Neugebauer. *The Exact Sciences in Antiquity, Isis* 43: 69–72.

Schappacher, N. 1987. Das Mathematische Institut der Universität Göttingen 1929–1950. In *Die Universität Göttingen unter dem Nationalsozialismus Das verdrängte Kapitel ihrer 250jährigen Geschichte*, ed. H. Becker, H.-J. Dahms, and C. Wegeler, 345–373. München: Saur.

Schappacher, N. 1991. Edmund Landau's Göttingen: From the life and death of a great mathematical center. *The Mathematical Intelligencer* 13(4): 12–18.

Schappacher, N., and M. Kneser. 1990. Fachverband-Institut-Staat. In *Ein Jahrhundert Mathematik, 1890–1990. Festschrift zum Jubiläum der DMV*, ed. G. Fischer et al., 1–82. Braunschweig: Vieweg.

Scholz, E. 2001. Weyls Infinitesimalgeometrie, 1917–1925. In *Hermann Weyl's Raum-Zeit-Materie and a general introduction to his scientific work*, DMV Seminar, vol. 30, ed. E. Scholz, 48–104. Basel/Boston: Birkhäuser Verlag.

Siegmund-Schultze, R. 1993. *Mathematische Berichterstattung in Hitlerdeutschland*. Göttingen: Vandenhoeck & Ruprecht.

Siegmund-Schultze, R. 2001. *Rockefeller and the internationalization of mathematics between the two world wars: Documents and studies for the social history of mathematics in the 20th century*, Science Networks, vol. 25. Basel/Boston/Berlin: Birkhäuser.

Siegmund-Schultze, R. 2008. Antisemitismus in der Weimarer Republik und die Lage jüdischer Mathematiker: Thesen und Dokumente zu einem wenig erforschten Thema. *Sudhoffs Archiv* 92(1): 20–34.

Siegmund-Schultze, R. 2009. *Mathematicians fleeing from Nazi Germany: Individual fates and global impact*. Princeton: Princeton University Press.

Struik, D.J. 1948. *A concise history of mathematics*. New York: Dover.

Swerdlow, N.M. 1993. Otto E. Neugebauer (26 May 1899–19 February 1990). *Proceedings of the American Philosophical Society* 137(1): 138–165.

Tollmien, C. 1990. 'Sind wir doch der Meinung, daß ein weiblicher Kopf nur ganz ausnahmsweise in der Mathematik schöpferisch tätig sein kann...' – eine Biographie der Mathematikerin Emmy Noether (1882–1935) und zugleich ein Beitrag zur Geschichte der Habilitation von Frauen an der Universität Göttingen. *Göttinger Jahrbuch* 38: 153–219.

Van der Corput, J.G. 1920. Über Gitterpunkte in der Ebene. *Mathematische Annalen* 80: 1–20.

Weyl, H. 1921. Über die neue Grundlagenkrise der Mathematik. *Mathematische Zeitschrift* 10: 39–79.

Weyl, H. 1932. Zu Hilberts sechzigsten Geburtstag. *Die Naturwissenschaften* 20: 57–58.

Zeuthen, H.G. 1896. *Geschichte der Mathematik im Altertum und Mittelalter*. Kopenhagen: Verlag A. F. Hoest.

“Not in Possession of Any Weltanschauung”: Otto Neugebauer’s Flight from Nazi Germany and His Search for Objectivity in Mathematics, in Reviewing, and in History

Reinhard Siegmund-Schultze

Introduction and Aims of the Article

In the spring of 1938, shortly after the occupation of Austria by Nazi Germany, the political situation in Europe was tense, also for mathematics. On 16 May 1938, Warren Weaver, head of the division for Natural Sciences of the U.S. American Rockefeller Foundation, had an interview at the Institute for Advanced Study in Princeton with one of the leading American mathematicians, Oswald Veblen. Weaver made the following note in his diaries:

> The Zentralblatt was originally set up as a Göttingen enterprise under the editorship of Neugebauer and as a competitor of the Berlin-sponsored Jahrbuch. [...] N. is an Austrian, served in the Army with distinction, and is not Jewish, but left immediately after the Hitler regime was installed because of personal principles. He receives a small salary from Springer and the remainder from Carlsbad Foundation.[1] V. considers him ‘the past master in the world of history of science’. For some time N. has feared that the continued publication of the Zentralblatt by Springer was in danger. This spring Blaschke complained to N. concerning the decrease in German reviews and the increase in others. V. refused editorship when the journal was founded because he ‘avoids stuffed-shirt jobs’ but accepted editorship last year[2] at N.’s warm urging because V.s relations with all German mathematicians are very good.
>
> Should difficulties develop, they would hope to shift publication to the [a] Scandinavian firm without attempting to move Neugebauer. Ultimately V. thinks that publication should

[1] What is meant here is the Carlsberg Foundation, which was financed from breweries. The sources of funding for Neugebauer’s position in Copenhagen were manifold, and supported his activities in historical research and in editing. Jessen (1993, 128) notes that Neugebauer was financed between 1934 and 1936 by the Rockefeller and Rask-Ørsted Foundations, and from 1937 to 1939 by the latter and by Carlsberg. See also Ramskov 1995, 345.

[2] Veblen appears on the title page of *Zentralblatt* for the first time in volume 14 (1936).

R. Siegmund-Schultze (✉)
Faculty of Engineering and Science, University of Agder,
Gimlemoen 25a, Postboks 422, 4604 Kristiansand S, Norway
e-mail: reinhard.siegmund-schultze@uia.no

A. Jones et al. (eds.), *A Mathematician’s Journeys*, Archimedes 45,
DOI 10.1007/978-3-319-25865-2_2

and must occur in this country. V. leaves with WW two letters from N. for our information. All of these facts are intended merely to serve as a background if and when an emergency arises.[3]

This note sums up some of the major events pertinent to the discussion in the following article, in particular Otto Neugebauer's involvement in mathematical reviewing (abstracting). Already during World War I Oswald Veblen (1880–1960) had pleaded for an American abstracting journal in mathematics, criticizing, together with other Americans, German 'control' of the abstracting system (Siegmund-Schultze 1994). However, the prohibitive costs of a publication in the U.S., alongside improvements in the German abstracting system beginning in 1931 led the Americans to postpone founding a journal of their own. That year the new "Zentralblatt für Mathematik und ihre Grenzgebiete" (Central journal for mathematics and its bordering disciplines) was founded in Göttingen by Richard Courant and Otto Neugebauer. Supported by the Springer publishing house, they pursued an approach that was sufficiently "modern" both with regard to their editorial principles as well as the international languages employed. Since then, as Weaver described in his diaries, the editorial offices of the *Zentralblatt* had moved to Copenhagen due to Nazi pressure, although the journal continued to be published by a German firm, Julius Springer, in Berlin. In spite of the mass emigration of mathematicians from Germany, during the early Nazi years this mathematical reviewing (abstracting) could be carried on in a rather undisturbed way from Copenhagen, although moving to America also remained an option. But "ultimately", as Weaver reports in his diaries, there was no doubt in Veblen's mind that, for both political and for mathematical reasons, reviewing would and should come to America. He, Veblen, would be proven right very soon.

This article is primarily a biographical essay devoted to Otto Neugebauer (1899–1990), probably the most influential historian of mathematics and astronomy of the twentieth century, certainly for exact sciences in antiquity. It complements other earlier accounts, above all Swerdlow (1993) and Pyenson (1995), by drawing on unpublished documents, described in detail in the next section. These help shed some light on the working conditions and aims of a scholar who did much to conceal his life and certainly his motivations and emotions from the public. It was in keeping with his character that toward the end of his career he chose to destroy all his personal and scientific correspondence. By pointing to Neugebauer's various and versatile activities during a long and productive life, both in historical research and as an editor of journals and book series, this article explores his work in the context of tendencies to "modernize" and "internationalize" mathematics while taking note of his shifting views on the relative importance of these two activities. In order to understand Neugebauer's political and philosophical views after leaving Nazi Germany (the period of primary focus in this paper), one must first take into account his work during the Weimar Republic when he helped support the organization of mathematics in Göttingen as well as launching the *Zentralblatt*. These activities will

[3] Rockefeller Archive Center (RAC), Weaver diary 1938, vol. 5, 106/107.

be discussed in some detail in sections "Otto Neugebauer in Göttingen: Aiming at Modernizing Mathematics, Neugebauer at the End of the Republic of Weimar: Introducing Modern Forms of Mathematical Reviewing via Zentralblatt in 1931, A Preliminary Appraisal of Neugebauer's Weltanschaung in Weimar".

The main thesis of this article is that Neugebauer, acting in an age of extremist ideologies and dictatorships, tried to maintain and expand the rationalistic and internationalist ideals long associated with Göttingen science and mathematics, and that this commitment was central both in his organizational work and in his approach to history. A useful interpretative scheme in this context is provided by Paul Forman's influential article (1971) on Weimar culture,[4] although the arguments there have to be specified for the case of Neugebauer, who stood between the scientific and humanistic cultures. By underlining certain apparent and real contradictions and divergent tendencies in Neugebauer's activities, while taking note of his disillusionments as he grew older, this article will try to draw tentative conclusions and offer suggestive explanations to account for these. By so doing, it also aims to explain some apparent inconsistencies in Neugebauer's statements about his own "world view". The "main thesis" of this article cannot be fully established here, partly for lack of sources about Neugebauer's personal life, for instance his experiences as a soldier during World War I. The analysis could be deepened, though, by systematically examining Neugebauer's often quite sharp-tongued reviews of historical work in *Zentralblatt* and other places, and – not least – by looking at his historical research itself. For example, Neugebauer published several very interesting, more detailed reviews in journals such as *Die Naturwissenschaften*. These definitely put to rest any conjecture that Neugebauer did not reflect on the role of "interpretation" or "synthesis" in the sense of a general cultural history.[5] These topics, however, remain largely outside the scope of this essay.

Unpublished Sources Used[6]

The present article draws on a variety of unpublished various sources, including some personal documents from Neugebauer's hand. Relatively little correspondence has survived, however, particularly from the pre-war period.[7] Nevertheless,

[4] I am here citing Forman's general argument about the role of cultural pessimism, not the "strong Forman thesis" about its impact on those who pursued quantum physics.

[5] Neugebauer's reviews of historical books and articles, in particular in the three relevant reviewing journals, *Jahrbuch* (1927/28), *Zentralblatt* (1931–1938), and *Mathematical Reviews* (1940–1953) comprise in total about 300 publications. I only found reviews by Neugebauer in MR during the years up until1953.

[6] I am grateful to the archives listed below which granted permission for me to quote from their sources. Translations from German are mine if not otherwise indicated. For lack of space original German text will only be quoted in exceptional instances.

[7] According to Pyenson 1995, 274, Neugebauer left instructions for his personal correspondence to be destroyed after his death. David Rowe informs me that Neugebauer told him back in 1982 that

an important source is the Nachlass of Neugebauer's friend, Erich Bessel-Hagen (1898–1946), which is located at the University Library in Bonn. This contains 160 pieces of correspondence between Bessel-Hagen and Neugebauer.[8] This correspondence will be used for the first time in this paper although it cannot be exhausted by this analysis.[9] Another major source for this article is the correspondence between Neugebauer and his former mentor and friend in Göttingen, Richard Courant (1888–1972), as far as this has been preserved.[10] The Rockefeller Archives in Sleepy Hollow, outside New York City, contain documents reflecting on Neugebauer's historical and organizational work; those cited here are, for the most part, hitherto unpublished, at least in English.[11] Sources used to a lesser extent include the Otto Neugebauer Papers at The Shelby White and Leon Levy Archives Center of the Institute for Advanced Study at Princeton,[12] the Manuscript division of the Göttingen University Library (Hasse Papers), the historical archives of the Springer publishing house, [13] the Oswald Veblen Papers at the Library of Congress, the R.G.D. Richardson Papers at Brown University, the Niels Bohr Archives in Copenhagen, the George David Birkhoff Papers at the Harvard University Archives, and the papers of Hermann Weyl, Heinz Hopf, and Bartel L. van der Waerden at the ETH Zürich. The

he intended to burn all his correspondence from ZB. Swerdlow wrote to Rowe that he personally witnessed him throwing out letters at the IAS. Clearly, he himself ordered his own Nachlass there, and aside from the Kennedy letters there are only manuscripts and notes in it.

[8] To be cited in the following as NLBH Bonn. Almost all the letters are from Neugebauer to Bessel-Hagen.

[9] Bessel-Hagen served as a kind of assistant to O. Toeplitz in their work for "Quellen and Studien zur Geschichte der Mathematik", co-edited from 1929 by Toeplitz, Neugebauer and J. Stenzel (see below). The correspondence with Bessel-Hagen contains interesting judgments made by Neugebauer about the importance of his historical discoveries. In one discussion from 1931 about C.L. Siegel's planned edition of Riemann's manuscripts in "Quellen and Studien", which was viewed very favorably by Neugebauer, he insists on the need for close collaboration between historians and mathematicians (NLBH, nos 60 and 61).

[10] Courant's correspondence is located in the Archives of New York University, Elmar Bobst Library. However, to this date it is not yet registered with call numbers. It will be cited as Courant Papers New York City. One of the few biographical documents from Neugebauer's hand is his address (Neugebauer 1963) to Courant's 75th birthday. (available online at https://sites.google.com/site/neugebauerconference2010/web-exhibition-neugebauer-at-goettingen).

[11] These Archives will be cited in the following as RAC. Some documents have been published earlier in a German book (Siegmund-Schultze 1993) that focuses on mathematical abstracting journals. The monograph on Rockefeller's contribution to the internationalization of mathematics (Siegmund-Schultze 2001) stressed the role of pure mathematical research, which was at the time the primary concern of the Rockefeller philanthropies.

[12] In the following referred to as SWLLA Princeton. The correspondence in the Otto Neugebauer Papers at SWLLA is mainly restricted to Neugebauer's exchange of letters with E.S. Kennedy between 1950 and 1990. However these Papers also contain Neugebauer's handwritten war diaries (Tagebuch 1917–1919) yet to be analysed. See several remarks in David Rowe's article in the present collection.

[13] These archives are now deposited at the Zentral- und Landesbibliothek Berlin (ZLB) and will be referred to as SVA (Springer-Verlagsarchiv). They do not seem to contain much correspondence on the journals and series edited by Neugebauer.

personal file kept on Neugebauer at the University Archives in Göttingen for the period 1925–1966, which comprises about 150 pages, could be possibly used more systematically for a full biography of Otto Neugebauer.[14]

Otto Neugebauer in Göttingen: Aiming at Modernizing Mathematics

Neugebauer's historical research, which was of course his primary personal interest, took place while he worked as an editor of abstracting journals (especially the *Zentralblatt*) and books (among them the '*Ergebnisse*' series). The latter work was partly done to make a living, but still these two activities were less opposed to each other than it might appear at first glance. After all, and as a first observation, reviewing and historical research are both based on critical analyses of texts, be they modern mathematical texts or cuneiform tablets. Both activities are thus in a sense "conservative" and so they differ strongly from doing original mathematical research.

As a matter of fact, Neugebauer's career as a historian of ancient mathematics began in 1924 during his stay in Copenhagen with the mathematician Harald Bohr (1887–1951), brother of Niels Bohr. It was there that he wrote a review of T. Eric Peet's edition (1923) of the Papyrus Rhind (Neugebauer 1925). The proposal to write this review (in German) for the Danish *Matematisk Tidsskrift* came from Bohr, who had once been a student of the famous Danish mathematician and historian of mathematics, Hieronymus Georg Zeuthen (1839–1920).[15] Harald Bohr, who happened to be a good friend of Richard Courant in Göttingen, was also Neugebauer's co-author in the latter's only mathematical publication (1926). Upon returning to Göttingen the following year Courant set the next task that led to Neugebauer's involvement in the historiography of mathematics, to co-edit with him the famous "Vorlesungen über die Entwicklung der Mathematik im 19. Jahrhundert" (1926), which had been delivered during the war by Courant's former mentor Felix Klein (1849–1925), now deceased.

As is well known, Courant was at the time heavily involved in several activities aimed at "modernizing" the infrastructure of mathematics in Göttingen and throughout Germany, both in cooperation with Ferdinand Springer (focusing on publications)

[14]The file from the Göttingen University Archives will be used in section "Neugebauer's Attitudes Toward Germany Before and After the War" at the end of this article for a discussion of Neugebauer's compensation claims after the war.

[15]The episode is reported in the unpublished Danish biography of Harald Bohr, written by Kurt Ramskov for his Ph.D. (Ramskov 1995, 256). Compared to later reviews written by Neugebauer, the one on Peet's edition is more in the nature of a report. However, it served Neugebauer as a starting point for his dissertation, which he defended 2 years later (Rashed and Pyenson 2012, 4).

and together with the Rockefeller philanthropies[16]; his latter activities culminated in 1929 with the erection of a new mathematics building in Göttingen (Siegmund-Schultze 2001). Indeed one can look at these activities as a realization of Klein's dreams. It is well known that the two leading figures in Göttingen, Felix Klein and David Hilbert (1862–1943), complemented each other quite nicely and deliberately in a kind of double strategy for modernizing mathematics (Rowe 1989). On the one hand, they were pursuing pure mathematical research (including modern axiomatics and foundational research), a direction promoted by Hilbert and later by Edmund Landau, Emmy Noether, and others. On the other hand the 'double strategy" secured and strengthened the connections to applications and teaching, tasks which fell more to Klein and Courant. Work on the mathematical foundations of modern physics, in particular relativity theory, was one of several connecting links that joined these two tendencies. At the same time the leading Göttingen mathematicians shared a commitment to the ideal of internationalization in mathematics.

"Modernizing" mathematics in this broad sense, however, did not imply the modernizers were unconcerned about losing touch with older traditions of German mathematics or that they overlooked the dangers of exaggerated specialization within mathematics. Indeed, these were concerns which Neugebauer shared with other mathematicians and historians of mathematics. Like Courant, he took a critical stance toward exaggerated practical and presentist views. Such concerns sometimes resonated with warnings against a superficial "mechanization" of modern life which were repeatedly uttered in the Republic of Weimar (Forman 1971). People like Courant who – in contrast to Oswald Spengler with his warnings against a 'decline of the West' – continued to believe in mathematical and scientific progress, recommended increased attention to teaching and to applied mathematics, but also to historical research. In this context Courant and Neugebauer proposed that more attention also be paid to historical perspective in mathematical research. In the preface to their edition of Klein's historical lectures, they wrote in August 1926:

> At a time when also within science the practitioners are too much impressed by the present and tend to look at the particular as in unnatural enlargement and as being of exaggerated importance, Klein's work enables to reopen the eyes of many for the connections and lines of development of our science as a whole. (Klein 1926, v)

In addition to Klein's lectures, Courant engaged Neugebauer, who from 1927 was also a "private docent", in other organizational work. Neugebauer thus contributed much to the development of the library and the collection of mathematical models. Courant once described his assistant as "having all the virtues of pedantry and none of the vices" (Reid 1976, 94). So he entrusted Neugebauer with the details for the planning of the new Rockefeller-sponsored mathematical institute in Göttingen, which opened in November/December 1929, as described by Constance Reid:

[16] Support for mathematics during the 1920 came mostly from the "International Education Board," which was one of several Foundations owned by the Rockefeller family.

> For this work Courant turned to Neugebauer, whose artistic gifts and knowledge of engineering, draughtsmanship, and the needs of mathematicians made him the perfect choice. Courant and Neugebauer were eleven years apart in age and completely different characters. "Courant's word was maybe," Hans Lewy said to me. "Neugebauer's was yes, yes, no, no." But the two were fast becoming an administrative team, the purposeful disorganization of Courant balanced by the efficiency of Neugebauer. (Reid 1976, 108)

One can find ample and clear evidence, particularly in the correspondence between Neugebauer and Bessel-Hagen, who was also an assistant to Courant at the time, that Neugebauer personally identified with the competitive spirit and internationalist outlook in Göttingen. Not only did he forge strong friendships with both Bessel-Hagen and Courant, but also with B.L. van der Waerden (who attended his historical lectures), and particularly with the topologists Heinz Hopf and Pawel Alexandroff, both of whom visited Göttingen regularly.[17] Neugebauer, who had already during his early studies in Graz shown a keen interest in the general theory of relativity,[18] certainly enjoyed such company. He also shared some of the hubris of these young men, an attitude particularly common in Göttingen, where it was fashionable to belittle those considered dilettantes, be they philosophers, politicians, philologists without scientific education, or teachers.[19] At the same time, Neugebauer strongly felt the inadequacy of his own mathematical talent, particularly at the beginning of his career, when he was still looking for a topic to work on.[20] As late as 1963, in his address to Courant's 75th birthday, one senses the relief Neugebauer must have experienced when he found a field of study far removed from the work of the leading and often awe-inspiring Göttingen mathematicians, but still tolerated and acknowledged by them: his research in the history of ancient mathematics.[21]

Neugebauer not only identified with modernist Göttingen; he was also certainly aware of a certain degree of envy toward it on the part of German mathematicians outside, sometimes coupled with accusations – using partly historical arguments –

---

[17] In the summer 1926 Neugebauer was on vacation in France with the two latter, writing regularly joint postcards to Bessel-Hagen. See the report by Alexandroff (Aleksandrov) (1976). Neugebauer and his friends also shared the habit of addressing each other with special nicknames. See the quote from the letter to Heinz Hopf, given below.

[18] As a 22 year old, Neugebauer gave an extensive talk (34 pages), entitled "Über die Erweiterung der allgemeinen Relativitätstheorie durch Hermann Weyl" during two sessions of the physical seminar at Graz. A scan of the script is available from the website of SWLLA Princeton (http://cdm.itg.ias.edu/cdm/). David E. Rowe discusses Neugebauer's training as a physicist in his contribution to the present volume.

[19] In a letter to Bessel-Hagen dated 16 April 1924 he opposed recent tendencies towards a "Studienrats-Wurstmaschine", i.e. giving priority to the training of teachers. (NLBH Bonn, corr. Neugebauer, no.2) One finds further evidence in Bessel-Hagen's correspondence with Neugebauer of the latter's contempt for teachers who published on the history of mathematics without sufficient competence in the subject.

[20] On 21 April 1924 Neugebauer wrote to Bessel-Hagen: "The constant and constantly growing doubt about my mathematical talent tends to paralyze my energy." (NLBH Bonn, corr. Neugebauer, no.3)

[21] One reads there: "Even Courant could not exercise any direct influence whatever on my work." (Neugebauer 1963, 9).

to the effect that Göttingen replaced contemplation and research by "organization" and money. In a 1928 publication Neugebauer alluded to this indirectly, reacting to a speech delivered earlier that year by the topologist Max Dehn (1878–1952). Dehn was formerly one of Hilbert's most prominent students and now a professor at Frankfurt, where he ran a well-known seminar on the history of mathematics (Siegel 1965). In January 1928, Dehn spoke to a broader audience at Frankfurt University about the "Peculiar Nature of the Mathematician's Mind".[22] In his address he referred to a "somewhat skeptical attitude of many a contemporary mathematician [...] reinforced by what is going on in the neighboring field of physics. [...] Out of this skepticism there develops a certain resignation, a kind of mistrust in the power of the human mind in general."[23] Dehn warned that this resignation could *not* be overcome by a mere 'organized enterprise':

> Not by scientific mass production [Massenarbeit], not by ever more papers which contain investigations of meaningless special cases or generalizations – there are appearing now several thousand mathematical papers a year – is our science able to really progress but only by individual creative accomplishments. Such accomplishments will hardly occur in an organized enterprise [organisierter Betrieb].[24]

Neugebauer concludes his article of late 1928 with the following words, referring clearly to Dehn's talk:

> In my paper I had to use repeatedly the word 'organization'. This has nearly become a specter so that recently [...] there was talk about the 'desperation' which overcomes the mathematician 'in view of the havoc wreaked by him' for which allegedly 'resignation' was the only remedy. Certainly, such an opinion also has much to it; however, it seems to me that the 'havoc' is not so much caused by the scientist himself. It is rather the expression of a deeper historical process, which leaves but little space to the ideal (and sometimes also very convenient) life of a scholar who only lives for his own thoughts, thereby imposing on the scholar duties for the broader public. The deeper meaning of all organization therefore seems to me not mechanization of science but the positive, responsible effort, to reveal on a broader basis that scientific thought has a right to exist. To successfully serve this purpose, is the aim of the institutions at Göttingen.[25]

While Forman emphasizes Neugebauer's admission that Dehn's "opinion also has much to it", it seems to me that his remark has an ironic undertone; it is rather, at best, a kind of polite concession vis-à-vis Dehn. In fact, Neugebauer puts much more emphasis on the 'responsibility' of the modern scholar to engage in developing the connections of science and mathematics to society at large. And, indeed, in

[22] Dehn 1928. Dehn's speech is partly quoted and analyzed by Forman 1971, 54–55. I owe to Forman also the reference to Neugebauer's response of 1928 below, which Forman, however, does not quote in detail.

[23] Dehn 1928, 15, using a translation by Forman 1971, 54. Dehn's former teacher, David Hilbert of Göttingen, had opposed such sentiments, in particular the alleged "ignorabimus" in mathematics, since his famous speech on mathematical problems in Paris 1900. He reiterated his argument with side-swipe at Spengler in his 1930 talk in Königsberg "Logic and the Knowledge of Nature" (Hilbert 1930).

[24] Dehn 1928, 18. My translation.

[25] Neugebauer 1928, 111. My translation from German.

his work that led to the new mathematical building financed by Rockefeller, Neugebauer tried to combine "old models" (which, like Klein's library, still had the potential to be emulated elsewhere) with the new demands for communication and research. As he put it in his paper for *Die Naturwissenschaften*, published on the occasion of the official opening of the mathematical institute on 9 December 1929:

> In some respect we deviated from old models, not least with respect to the intent to make the new building not just functional but also comfortable – a tendency which was basically foreign to an earlier era ... We hope and believe that the new institute does not contribute to the often prophesied 'mechanization' of science but offers instead a working place *to be liked* for teaching and learning, and, above all a place for pure science.[26]

Not surprisingly, it was in the context of Neugebauer's work for the Göttingen institute building that the Rockefeller officials became aware of him for the first time. The Princeton physicist and Rockefeller man, Augustus Trowbridge, wrote in his diaries about his visit to Göttingen in 1928:

> Mathematical Institute of the University of Göttingen.
>
> March 21st, A.M. Courant and Dr Neugebauer (an attractive young mathematician, who seems to have the details of the new building in charge and who is specialist in History of Mathematics, having had training in Chaldean Archeology[27] before he took up Mathematics) came to hotel while A.T. was still at breakfast. (RAC, Trowbridge Log 905, vol. 4, fol. 119)

Neugebauer at the End of the Republic of Weimar: Introducing Modern Forms of Mathematical Reviewing via *Zentralblatt* in 1931

Neugebauer's really substantial disagreement, however, did not arise with Dehn, who after all had close connections to Göttingen and, as a Jew, would soon be expelled from Germany. Real conflicts, however, arose with more conservative mathematicians, particularly those associated with Ludwig Bieberbach in Berlin. This is revealed in Neugebauer's activities following the formation of the new *Zentralblatt* in 1931. This journal was a response to the problems resulting from the strong increase in mathematical publication worldwide, as mentioned by Dehn. At the latest by the First World War, the traditional German mathematical reviewing journal, *Jahrbuch über die Fortschritte der Mathematik* (Annual on mathematical progress), which had always been centred in Berlin, had problems coping with the flood of publications. The *Jahrbuch* was traditionally bound to rigid editorial

---

[26] Neugebauer 1930, 4. My translation from German.

[27] A systematic study of ancient cultures by Neugebauer before he took up mathematics appears unlikely based on existing biographical accounts. He studied electrical engineering and physics in Graz (Austria) and Munich between 1919 and 1922. Then he continued with mathematics in Göttingen, while his interest in Babylonian mathematics apparently occurred only after his Ph.D. on Egyptian mathematics of 1926 (See Rowe's essay in this volume, Swerdlow 1993 and Rashed and Pyenson 2012).

principles. It published systematic annual volumes ("Jahrbuch") that reported on production in mathematics world-wide for the respective calendar year. This implied that the editors could only start publishing the first issue of the annual when the calendar year had ended. To be sure, the careful, year-by-year reviewing of the *Jahrbuch* reflected its value as a mirror for the history of mathematics during the late nineteenth and early twentieth centuries. But scientific abstracting had also become increasingly important as a tool for research over the course of this period, whereas the *Jahrbuch* continued to fall further behind in its publication schedule. Thus timely publication, a requirement of "modernity," had become highly problematic, but the *Jahrbuch* was also an old-fashioned enterprise since it published its reviews exclusively in German.

When in 1930 rumors of the impending foundation of a new abstracting journal reached Berlin, the *Jahrbuch*'s publisher, Walther de Gruyter, realized that this represented a serious threat to his business interests. Springer had become the leading publisher in mathematics after the War, and now wanted to expand further into reviewing. Some Berlin mathematicians considered the founding of the *Zentralblatt* to be a continuation of the old institutional competition between the two German mathematical centers, a conflict which had for instance come up half a century before between the traditional *Crelle Journal* in Berlin and the younger *Mathematische Annalen*. The latter had originally been published by Teubner in Leipzig, but had strong ties with Göttingen after 1886, when Klein joined the faculty there. These institutional frictions were further aggravated by ideological problems. Many German mathematicians of the older generation considered the *Jahrbuch* part of a venerable German mathematical tradition that had to be saved at all costs. Those mathematicians saw the rapidity and internationality of the *Zentralblatt* not so much as signs of progress but rather as marking the decline of German mathematical culture (Siegmund-Schultze 1994, 319).

The managing editor of the *Jahrbuch*, Georg Feigl, was employed at the Prussian Academy of Sciences in Berlin. He was perplexed by the aggressiveness of Neugebauer and Springer in trying to entice German reviewers away from the *Jahrbuch* to the new *Zentralblatt*. In an internal report to the Academy's commission for the edition of the *Jahrbuch*, headed by Bieberbach, Feigl wrote on 15 December 1930:

> The leaders of Springer's Zentralblatt, Prof. Courant and Dr Neugebauer, Göttingen, kept utmost secrecy about the planned enterprise, although they are personally well acquainted with the mathematicians at the Berlin Academy and with the managing editor of the Jahrbuch. Thus negotiations with the new enterprise have been made impossible. It was only around 1 November, after Neugebauer had approached collaborators of the Jahrbuch, both orally and through letters, in order to win them over as reviewers, that the managing editor received official information from Göttingen. The promise to come to Berlin as soon as possible for discussions was only fulfilled on 9 December. There was a longer discussion on that day between E. Schmidt, Bieberbach and the managing director on the one hand, and Neugebauer on the other.[28]

[28] Siegmund-Schultze 1993, 55–56, my translation from German.

There was no doubt some embarrassment on both sides in the discussion between the Göttingen and Berlin mathematicians in late 1930. It even appears plausible that the mathematicians responsible for the *Jahrbuch* felt deceived by Courant and Neugebauer.[29]

It seems the Göttingen mathematicians deemed collaboration or agreement with the *Jahrbuch* as unrealistic from the outset, probably because the latter abstracting journal showed no signs or promise of changing its obsolete editorial principles. (Indeed the *Jahrbuch* would continue to publish until the end of the war in 1945 with clear conservative overtones, given the political conditions in Nazi Germany). Neugebauer apparently did not have much success in his efforts to win over former collaborators (reviewers) from the *Jahrbuch* for the *Zentralblatt*.[30] The role of foreign collaborators, particular for the English language, therefore became even more important for the *Zentralblatt*, whereas Neugebauer at one point harboured doubts that the *Jahrbuch* would continue to have any foreign reviewers at all.[31]

At the time the *Zentralblatt* was founded in 1931 some American mathematicians saw this confusing situation in Germany as another incentive to go forward with older plans for an American abstracting journal. Indeed, Neugebauer, in a letter to Courant from 24 March 1931, alluded to "new plans for the foundation of an American abstracting journal." But, in the same letter, Neugebauer expressed confidence in his ability to convince the Americans of the *Zentralblatt*'s qualities: "Veblen is going to come to Germany and will try to reach an agreement with us. Personally I am very much in favor of the idea of an American branch of the Zentralblatt".[32]

From the start the *Zentralblatt* was an international undertaking, as reflected by its editorial board: P. Alexandroff (Moscow), H. Hahn (Vienna), G.H. Hardy (Oxford), G. Julia (Versailles), O. Kellogg (Cambridge, Mass), T. Levi-Civita (Rome), R. Nevanlinna (Helsinki) and H. Thirring (Vienna). These foreign co-editors appeared on the title page, together with four scholars from Germany, alongside Neugebauer as managing editor. Originally, in addition to abstracts, *Zentralblatt* printed occasional "Vorläufige Mitteilungen", short pre-prints in the manner of the French *Comptes Rendus*. But this practice was terminated with volume 3 (1932) due to "lack of space" [*Zentralblatt* 3 (1932), 1]. This and similar innovations were discussed in some detail for instance with American mathematicians.[33]

---

[29] There was a rather apologetic letter written by Courant to Feigl, 31 October 1931, quoted in Siegmund-Schultze 1993, 56.

[30] On 10 December 1938 Neugebauer wrote to Veblen that "due to the original boycott [of the Zentralblatt] due to its fight against the 'Fortschritte' [i.e. the Jahrbuch] German reviewers never played an important role at the Zentralblatt." (Siegmund-Schultze 1993, 107. My translation from German).

[31] This in a German letter by Neugebauer to T. Levi-Civita to this effect, dated 21 July 1932, and quoted in Siegmund-Schultze 1993, 107.

[32] Siegmund-Schultze 1994, 318. In my publication of 1994 the letter was misdated 3 March 1931. The letter is now also available at https://sites.google.com/site/neugebauerconference2010/web-exhibition-neugebauer-at-goettingen.

[33] See the discussion in Pyenson 1995, 266–267, based on the Tamarkin Papers at Brown University.

What materialized in the long run in addition to *Zentralblatt* was, however, the important "Ergebnisse"- series. This series presented accounts of recent research findings in specialized mathematical monographs, many published in English, and all with the following note appended to the title: "Published by the managing office of the *Zentralblatt für Mathematik*." By 1938, when Neugebauer gave up editing both the *Zentralblatt* and the *Ergebnisse*, the latter series could boast several very successful publications, though their real influence would only become visible in the decades to come. Probably first among these in terms of its world-wide influence was A. N. Kolmogorov's "Grundbegriffe der Wahrscheinlichkeitsrechnung" (Basic Notions of the Calculus of Probability) of 1933. By 1938 some 23 of the *Ergebnisse* had appeared in print, five of them in English. Neugebauer played an important part in winning authors for the *Ergebnisse*; clearly his daily work as managing editor for *Zentralblatt* helped him to make the necessary contacts.

It would be misleading to assume that Neugebauer's work for *Zentralblatt* was of no import for his historical work, at the very least for its distribution and reception. As a matter of fact, Neugebauer's commitment to objectivity did not rule out using *Zentralblatt* and later the *Mathematical Reviews* for propagandizing his own new historiography of ancient mathematics and astronomy, in particular his notion of the algebraic nature of Babylonian mathematics. Referees, like his friend Erich Bessel-Hagen, but also Dijksterhuis, Tropfke, and Becker helped to spread the results of his research, and occasionally Neugebauer published self-abstracts in *Zentralblatt* as well. In at least one instance, Neugebauer sent the exact wording of a review for a paper of his own to the referee for confirmation, including in his draft already the signature of the prospective referee, Bessel-Hagen.[34] On a later occasion Neugebauer returned a positive review of one of his articles to Bessel-Hagen with the words that "I would like to have stressed in the review the points of view I consider most important"; he included an alternative version of the review with his message. In this case Bessel-Hagen responded with the suggestion that this alternative version be published as Neugebauer's self-abstract. He added that he found Neugebauer's article "too apodictic," adding that it falsely claimed to have proven its results "like a mathematical theorem". In this case Neugebauer accepted Bessel-Hagen's concerns and printed the latter's original version, claiming that this was about matters of taste anyway.[35]

Apart from this beneficial side-effect of running an abstracting journal, it seems to be a real mystery, given the huge amount of organizational work, how Neugebauer

[34] This was the case for the review in *Zentralblatt* which appeared with the same wording in ZB 15 (1937), p. 53. In his letter to Bessel-Hagen, dated 1 November 1936, Neugebauer gave as excuse for his intervention that the paper was "boring", because it was meant just as preparatory work for another publication. Therefore Bessel-Hagen should not bother reading it. NLBH Bonn, correspondence Neugebauer, no. 120.

[35] This discussion is in NLBH Bonn, correspondence Neugebauer, nos. 137: Brief 26.3.38 (ON to BH), 138: 30.3.1938 (BH to ON), and 139: 2.4.38 (ON to BH). It was about Neugebauer's paper "Untersuchungen zur antiken Astronomie III." (1938)

was able to pursue his historical research and to have a family life as well.[36] The fact remains that he had to maintain his bread and butter job as an editor to finance his research. Another editorial job was, however, much closer to his own research interests. In 1929 he founded, with Otto Toeplitz (1881–1940) and Julius Stenzel (1883–1935) as co-editors, "Quellen und Studien zur Geschichte der Mathematik, Astronomie und Physik (QS)." This Springer series was devoted to the history of the mathematical sciences, divided into two parts, Abteilung A, for the publication of sources and B, for studies, distinguishing and uniting philological and technical research.

Neugebauer did not have the option, enjoyed by many research mathematicians at Göttingen, to pursue further study abroad by obtaining one of the prestigious though rather frugal Rockefeller grants, even though he had helped plan the institute building. Indeed, Neugebauer's research in the history of mathematics had at the time[37] not yet entered the realm of interests of the Rockefeller people. This becomes clear when reading the following note by W.E. Tisdale, Trowbridge's successor as head of the Rockefeller Office in Paris. Its contents are related to a meeting with Courant in Göttingen, 12 January 1930, shortly after the Rockefeller-sponsored mathematical institute had been opened.

> Courant also wanted to know if we would consider his first ass't, Dr. Neugebauer, in the field of History of Mathematics, who wants to go to Cairo to work through old Arabic documents. Courant says the man is the best of his 8 ass'ts,[38] and if proposed would want about six months. I told him to write to me if and when he made up his mind as to what he wanted. Neugebauer is a very able fellow, has had real charge of the details of the building of the new institute here, forceful, pleasing to meet, efficient, and probably worthwhile if his program is at all good.[39]

While Courant had managed to get grants for mathematicians with ties to Göttingen including Werner Fenchel (1905–1988), Hans Lewy (1904–1988), Isaac Schoenberg (1903–1990), B.L. van der Waerden (1903–1996), and even for applied mathematicians[40] such as Wilhelm Cauer (1900–1945) and Alwin Walther (1898–1967), almost all of them younger than Neugebauer, he was unable to muster Rockefeller support for the latter. Indeed, Tisdale's commentary sounded a skeptical note ("if his program is at all good"). As a matter of fact, Neugebauer would never obtain a traditional RF grant for study abroad, nor would he ever travel to Cairo either.

---

[36] Neugebauer married fellow student Grete Brück with whom he had two children, born in 1929 and 1932. There is evidence in Neugebauer's correspondence with Bessel-Hagen, that he received much secretarial help from his wife.

[37] This changed later, see below.

[38] He was actually the only "Oberassistent" and as such indispensable, as Reid 1976, 129 reports from her conversations with Neugebauer and with other assistants such as Hans Lewy.

[39] RAC, Tislog 4 (1930) Göttingen, 12 January 1930, pp. 6/7, partly quoted already in Siegmund-Schultze 2001, p. 212.

[40] RF was originally hesitant with respect to support for applied mathematics and it required Courant's power of persuasion to reach support for Cauer and Walther.

Nor would Neugebauer find additional support by the German University system which could have supplemented his modest remuneration as Oberassistent. One has to consider that around 1930 it was the time of the so-called "Notverordnungen" (emergency decrees) in Germany, when even the salaries of professors were reduced. Thus it is not surprising that Courant's letter in favor of Neugebauer, written to an official in the German ministry of education, was rather timid and defensive.[41] In a letter to Wolfgang Windelband, dated 27 May 1931, Courant stressed that Neugebauer had earlier ruled out accepting two offers of professorships at the Technical Universities in Brunswick and Darmstadt: "After a careful check he found that the teaching duties at a Technical University cannot be reconciled with his scientific program. He has therefore declined any involvement in the appointments in spite of his very insecure and restricted economic situation." Courant then continued: "We know that the only way worthy of Neugebauer's accomplishments is creating a personal professorship for his field. But we also know that we have to shelve this wish in view of the general situation."

Courant therefore asked instead for some modest support for Neugebauer to finance literature and "smaller trips". Otto Toeplitz, Neugebauer's co-editor of the *Quellen und Studien*, who wrote a letter to the same official on 8 December 1931, supported the application, stressing the role of the history of mathematics "in a time of a total decay of mathematics into isolated disciplines". But nothing except the formal title as extraordinary professor for Neugebauer (from 1932) appears to have come out of Courant's and Toeplitz's initiatives. Neugebauer had to continue to rely on private money, coming from publishers and, somewhat later, from foundations. The negligible status of the history of mathematics as an academic subject in Germany would even pose a problem when Neugebauer applied for compensation after the war, as we will see.

It was not just the economic situation that was precarious around 1930 for young scholars like Neugebauer. Also politically they saw the "writing on the wall" with the rise of extremism, particularly from the right. One finds a rare outburst of political anxiety on the part of Neugebauer at the end of one of his letters, written to Erich Bessel-Hagen on the eve of the 1930 elections for the German Reichstag. At this time Erich Bessel-Hagen held a modest position at the University of Bonn. The elections in September 1930 would lead to a sensational gain in votes for the Nazi party NSDAP. Neugebauer, in the highly emotional and condensed passage of his letter dated 20 August 1930, was apparently still optimistic about the outcome of these elections. He alludes to the already existing Nazi influence in the government of the German state of Thuringia and to the failed campaign of the extreme right in favor of having a referendum against the Young plan for paying reparations for the lost war. A picture from Neugebauer's hand (Fig. 1) alludes to recent efforts by the German communists to jump on the nationalist bandwagon against Versailles and

[41] A copy of this letter as well as of the following one by Toeplitz is in the Courant Papers at the New York University Archives (without call number).

Fig. 1 Political drawing from Neugebauer's letter of 20 August 1930 to Bessel-Hagen. The text reads "Hand in Hand fürs Vatterland!," thereby mocking the sanctified notion of "Vaterland"(fatherland) by misspelling. Courtesy University Library Bonn, Nachlass E. Bessel-Hagen

the reparations, efforts which did not pay off in the end for the communists in the following elections.[42]

Neugebauer's letter gives a certain impression of the kinds of desperate and quite cynical feelings many ambitious young scholars in Germany – and not just those who were Jewish – must have harbored around 1930 when the economic and political situation threatened to thwart their plans for the future. Indeed, the situation for Courant, Neugebauer, and other German mathematicians with internationalist inclinations changed considerably for the worse in 1933.

A Preliminary Appraisal of Neugebauer's Weltanschauung in Weimar

Before entering Neugebauer's time of emigration I try to sum up his efforts for "modernizing" mathematics in the late 1920s and early 1930s and to assess to what extent this was an expression of his world view. I will have to complement and modify the picture in the conclusions (section "Conclusions: Another Attempt at

[42] I omit here giving an English translation of the German text, which anyway is only understandable for readers with thorough familiarity with the politics of the time. The original German letter from 20 August 1930, kept in NLBH Bonn (correspondence Neugebauer, no. 79), reads in its final political part as follows:

"Falls in Thüringen wirklich die Sonne scheinen sollte, so besteht ja die Hoffnung, dass die dortige Sumpfflora verdorrt und so den Bauern die Sorge um eine zu gute Ernte erspart bleibt, die ja die Preise drücken könnte trotz aller Zollmassnahmen die uns vor den fremdländischen (nur mit Abscheu schreibe ich dieses welsche Wort) Produkten (Verzeihung: 'Vielfachen' wo wir doch Deutsche sind) abschliessen. Nieder mit dem Schmachfrieden von Fersaa-illes! Der liebe Gott scheint übrigens auch seine Markguthaben endgültig abgestossen zu haben, denn selbst die Stahlhelmpfarrer haben den Fuiksentscheid [sic; R.S.] nicht herbeiführen können. Oder steht er noch auf der Basis Seines [sic! added by ON; R.S.] ersten Bandes [apparently Old Testament; R.S.] und liebt nicht die Antisemiten. Weiss Gott? [sic for 'Weiss Gott!"; R.S.] Amen Dein O.N."

[In the margins: "Bitte als Frakturtypen zu lesen, ich habe nur eine welsche Schreibmaschine."]

Tracing Neugebauer's "Weltanschauung""), after taking into account Neugebauer's later views as they developed in the wake of the "Zentralblatt Affair" in 1938 and the aftermath of the war.

At first glance there appears to be a contradiction between Neugebauer's attitude as a historian and his effort for a modern abstracting journal which did not pay much respect to the traditional annual and German-centered reporting in the manner of the *Jahrbuch*. However, having Courant on his side and aiming at the unity of mathematics on an international level,[43] Neugebauer's effort appears to be a "realistic" one, not only in the obvious sense that he was personally in need of means for financing his historical research by earning money through editing, but also in the broader sense of not irrationally clinging to the past. In fact there existed various and conflicting motives among German mathematicians of the time to look into the mathematical past, as indicated in the case of Dehn. However, unlike philosophers, only few German mathematicians were impressed by the theses of Oswald Spengler of the alleged "Decline of the West" ("Untergang des Abendlandes"), more adequately to be translated as "The Downfall of the Occident", and these theses were clearly opposed by Göttingen mathematicians such as David Hilbert.

When Forman in 1991applauded "contemporary [i.e. of the 1990s; R.S.] humanistic scholars that [...] have responded to dysfunctional and inequitable social relations" of modern science, he noted:

> A less creditable case is that of German scholars following World War I. Following World War I, joining a general antirational retreat from the unhappy reality of defeat, German historians of science immersed themselves in Paracelsus, Arabic alchemy, Egyptian mathematics, and Babylonian astronomy. (Forman 1991, 81)

It seems to me that Forman's illuminating analysis applies only to those German scholars of the 1920s who supported anti-modernist positions. Neugebauer, however, was very critical towards the irrationalist or "lebensphilosophische" attitudes that dominated much of the philosophical and political discourse of the Weimar Republic. For instance, he concluded in a critical review of Sir Thomas Heath's book *Greek Astronomy* (London 1932), which, according to Neugebauer "ignored all proper astronomical and mathematical ideas," that even beyond Germany "exact thinking is usually not counted as thinking at all in a higher sense." [ZB 6 (1933), p. 3]. In any case, Otto Neugebauer, whose political views were always very liberal,[44] cannot, in my opinion, be subsumed under Forman's verdict, although he later, after the Second World War, surely would have had little patience with modern humanists and sociologists who proposed "deconstructionism" or similar sociological theories. If anything, Neugebauer's interest in the history of ancient cultures should be seen against the background of a general opposition to "Lebensphilosophie." Clearly, he had no sympathy for mathematically often incompetent philosophers who gave free rein to speculations about "rising" and "falling" cultures. In this

[43] Of course "internationalism" is not necessarily impartial by itself. There were other efforts in the 1930s to establish a German (Nazi-)-dominated 'internationalism': see Siegmund-Schultze 2002.

[44] This liberal outlook according to Neugebauer's biographer Swerdlow 1993, 145. One has to add that Neugebauer's 'liberalism' certainly did not tolerate political extremism of any persuasion, as revealed, for instance, in his letter to Bessel-Hagen of August 1930.

respect Neugebauer was well within the traditional internationalist outlook of Göttingen, especially with the views of Hilbert, Courant and the leftist neo-Kantian philosopher Leonard Nelson.[45]

Neugebauer's Flight from Nazi Germany to Copenhagen in the 1930s

The catastrophic losses for mathematics and physics in Göttingen and elsewhere in Germany, soon after the Nazis took power in early 1933, are well known.[46] Due to his work for the Göttingen mathematical institute and the *Zentralblatt*, Neugebauer had apparently acquired a reputation as an able organizer. Not being Jewish, he was initially considered by some a viable candidate to assume the directorship of the Göttingen institute after Courant's departure, all the more so since he had meanwhile acquired a professorial title, if not a proper chair.

However, he held this position just for 1 day, voluntarily resigning in late April 1933 with the declaration that he did not enjoy the confidence of the (Nazified) students.[47] That same month, after having suffered a nervous breakdown, Neugebauer, together with K. Friedrichs (both non-Jewish), drafted a letter to the minister of education, asking for a reconsideration of Courant's temporary dismissal.[48] The letter was signed by 28 colleagues. Although it opposed – after mentioning Courant's great merits for Göttingen's mathematics and physics – also "rumors about his political positions," the letter had no chance preventing Courant's "voluntary retirement" a few months later. Courant went first to England and finally settled in New York City in August 1934.

On 12 September 1933, after visiting his friend Harald Bohr in Copenhagen, Courant sent a letter to Neugebauer, who was then on vacation in Italy.[49] Writing

---

[45] For the discussion between philosophers and mathematicians at Göttingen during the 1920s, see Peckhaus 1990. One could possibly go a step further and examine to what extent Neugebauer's Austrian background came into play. Note that there were other Austrians with a critical look at German "Lebensphilosophie," for instance the applied mathematician Richard von Mises in Berlin, who was inspired by the Austrian Mach and had connections to the Vienna Circle. I do not believe, however, that Neugebauer and von Mises agreed on many points, considering the differences in their personalities and interests.

[46] For the overall situation in German mathematics at the time see Segal 2003, and Siegmund-Schultze 2009.

[47] Schappacher 1987, 364. In a letter to the dean dated 29 April 1933, Neugebauer refused to talk about his "Gesinnung" (political attitude) because he did not want to give the impression of defending himself. University Archives Göttingen, Personalia 1929–1946, N-Z. The claim in Reid 1976, p. 146, according to which Neugebauer refused to sign the oath of loyalty to the new government, cannot be substantiated, as already remarked in Schappacher 1993, l.c.. Such a claim would be hardly compatible with the fact that Neugebauer was finally granted unpaid leave to Copenhagen with the formal option to return to Göttingen (see below).

[48] Schappacher 1987, 350. The text of the letter is published in *Exodus Professorum*, pp. 22–24.

[49] According to Neugebauer's correspondence with Bessel-Hagen, NLBH Bonn.

from Göttingen in a "secret mathematical language" – apparently due to the surveillance of mail in Nazi Germany – Courant encouraged him to emigrate: "Generally Harald [Bohr] tends to emphasize the transformation theory instead of fixed point theorems. That method is also more rewarding for your work."[50]

Neugebauer finally went to Copenhagen in January 1934 after having negotiated with the Nazi officials to obtain an unpaid leave.[51] In spite of this formal agreement can be no doubt that Neugebauer never had a realistic chance to return to Göttingen under Nazi rule. Not only had the Mathematics Institute been cleansed of all his friends, but its new leadership would not have been inclined to offer him opportunities to pursue his historical research either.[52]

Soon after Neugebauer had stepped down as institute leader in Göttingen deliberations began in the U.S. to consider whether he could be brought to that country in order to help establish a mathematical abstracting journal there. Thus on 4 August 1933, Princeton's Oswald Veblen, now a member of the recently established Institute for Advanced Study (IAS), wrote to the secretary of the AMS, R.G.D. Richardson at Brown University: "It would seem not impossible that we may have, before many years, to undertake the continuation of the Zentralblatt. The simplest way would be to import Neugebauer. Have you ever talked about this contingency with Weaver or Mason?"[53] The reference to W. Weaver and M. Mason here pertains to possible support by the Rockefeller Foundation. Veblen received a skeptical reply 5 days later, on 9 August 1933: "I know that these things take a great deal of time and money so I am not enthusiastic about transferring the Zentralblatt to America. It would seem to me [...] that the disadvantages would outweigh the advantages."[54] In another letter (dated August 23), Richardson stressed the priority of mathematical research over reviewing (abstracting), noting that "the money involved would support two or three good mathematicians."[55]

Neugebauer's competence as a historian of mathematics was not even mentioned in the discussion between Veblen and Richardson. However, on 6 September 1933, the historian of science George Sarton (1884–1956), who edited the journal *Isis* from his modest position at Harvard University, wrote a letter to the director of the IAS, Abraham Flexner.

> I have just received a letter from Prof. O. Neugebauer of Göttingen, wherein he speaks of the great project which is now engaging his attention, to wit, the edition of a <u>corpus</u> of all

---

[50] Siegmund-Schultze 2009, 162. In the book I incorrectly assumed that Neugebauer was already in Copenhagen at the time.

[51] This follows from Neugebauer's letters to Helmut Hasse in the Manuscript Division of the Göttingen University Library. In his letter to Hasse dated 29 September 1934 Neugebauer states that he has been granted leave until the end of 1936. Cod Ms. Hasse, 1:1179, fol. 4.

[52] This is important to keep in mind in order to recognize the legitimacy of Neugebauer's compensation claims after the war, to be discussed at the end of this paper.

[53] Brown University Archives, Richardson Papers, Box Correspondence 1933 (German-Jewish Situation), file Oswald Veblen.

[54] Siegmund-Schultze 1994, 321.

[55] Ibid.

> the original documents of Babylonian mathematics. [...] As compared with Neugebauer I am only a dilettante. He works in the fronttrenches while I amuse myself wayback in the rear – praising the ones, blaming the others; saying this ought to be done, etc. – and doing very little myself. What Neugebauer does is fundamental, what I do, secondary. [...] Dr. Neugebauer did not ask me to write to you and made no appeal whatever to me.[56]

Flexner was apparently impressed and wrote to the leading American mathematician at his Institute, Oswald Veblen, on 8 September 1933:

> My disposition would be to invite Neugebauer, because he would bring to this country something absolutely new, namely the historical and humanistic side of mathematics. The success of the Institute of the History of Medicine at the Johns Hopkins with its liberalizing influence over the faculty as well as the students encourages me to try this novelty. Mathematics is something more than an affair of today and yesterday. It is a part of the cultural history of the race.[57]

Veblen, in his reply dated 11 September 1933, also expressed sympathy for the history of mathematics as a field of scholarly endeavor. However, he had to remind Flexner, who was an authority on higher education in general, of Neugebauer's primary importance for the community of mathematicians:

> I feel no doubt that Neugebauer is better in the Hist. of Math. than anything [sic] we have in this country. [...] From our point of view the chief difficulty in Neugebauer's case would be that he is Editor-in-Chief of the Zentralblatt für Mathematik, published by Springer, which has continued without interruption. I had thought that if the Z. should be stopped we in this country ought to take it and Neugebauer over. But it would be very expensive, and I am not ready to recommend using the Institute money for the purpose. On the other hand it would be a pity to take N. away from this job. Perhaps the best solution would be [...] to move N. to Copenhagen. [...]
>
> However, I have long had it in my mind that we ought to do something better in the History of Science [...] If you feel that this is the time to take a step towards a permanent start, using Neugebauer, I think the first step would be to get all possible information from Harald Bohr. [...][58]

Flexner apparently soon got cold feet and thus quickly agreed with Veblen, writing him on 14 September, 1933: "I think you are correct in saying that the Institute cannot take over the Zentralblatt; if that should be involved in offering the year's engagement to Neugebauer, Copenhagen would be far better for him." So nothing came of these early American plans to engage Neugebauer, whether as an editor of *Zentralblatt* or as a historian.

Meanwhile the *Zentralblatt* came under political attack. In a letter to Springer from 24 October 1933 the German Mathematical Society (DMV) demanded the closing of *Zentralblatt* in favour of the *Jahrbuch*, using a slogan frequently uttered by the Nazis, namely "common benefit precedes individual benefit."[59] Neugebauer, writing letters to Tamarkin, Bohr, Hardy and other foreign mathematicians in early November 1933, tried, somewhat naively, to use international protests as a propa-

---

[56] Basically quoted already in Pyenson 1995, 268 from SWLLA Princeton. Emphasis by Sarton.

[57] Siegmund-Schultze 2009, 307, quoted from SWLLA Princeton.

[58] SWLLA Princeton, Veblen, Box 32, folder 1933.

[59] Siegmund-Schultze 1994, 320–321. The German original is "Gemeinnutz geht vor Eigennutz".

gandistic tool to protect the *Zentralblatt* in Germany.[60] In the end, Springer was not forced by the Nazi government to terminate *Zentralblatt*, though the publisher deemed it wiser to have the journal edited from abroad. Neugebauer thus followed Courant's advice with respect to the "transformation theory": he stayed in Copenhagen after January 1934 and for the next 5 years directed the *Zentralblatt* from there.

This is not to say that there were no ramifications for *Zentralblatt* from the political situation in Nazi Germany during these years. In July 1936 Neugebauer left Copenhagen to attend the International Congress of Mathematicians in Oslo. Of course he was no longer regarded as a German mathematician in the eyes of the Germans who attended. The German delegation was led by the Göttingen didactician Walter Lietzmann, who had to stay in close contact with the Nazi authorities both before and during the Congress. Participation in international conferences was at that time very difficult for German mathematicians, not least due to problems obtaining foreign currency (Siegmund-Schultze 2002). Jewish mathematicians from Germany had no chance of being included in the delegation or of receiving any material support from the German government. Neugebauer offered financial support to his co-editor of the *Quellen und Studien*, Otto Toeplitz, for a possible stay in Oslo. But the latter found it too difficult even to pay for a train ticket, let alone the problem of having to cope with his political fears.[61]

Neugebauer's presence in Oslo primarily reflected his prominence as an editor, but he also received there broad recognition for his historical work as well, and so he was invited to give one of the 20 plenary talks. He spoke in German on "Greek Mathematics and its relationship to pre-Greek mathematics" (Neugebauer 1937). In this talk Neugebauer emphasized that any attempt at "explaining the contrast between the Babylonian numerical methods and the Greek geometrical models by the typical intuitive talents of the Greeks" would amount to "renouncing the entire basis of our scientific methodology" (Neugebauer 1937, 160). One has probably to read this as an indirect, but clear allusion to the alleged superiority of the "intuitive" and geometrical spirit of "German mathematics" in contemporary political discussions in Germany. Indeed, already in July 1933, half a year after the Nazis had come to power, Neugebauer had commented in a book review[62] about the positive effects of an "intermingling of cultures and races (in the cuneiform cultures) as opposed to the narrowly restricted Egypt development." Now in Oslo, however, Neugebauer

[60] See Library of Congress Washington, Oswald Veblen Papers, cont. 13, folder Springer, for instance copy of the letter by Neugebauer to J.D. Tamarkin, 8 November 1933 and copy of Veblen's letter to Springer, 24 November 1933 in support of *Zentralblatt*.

[61] See Otto Toeplitz in a letter to Courant, dated March 11, 1936 (Courant Papers New York). Toeplitz was concerned about being watched by Nazis in Oslo in his contacts with emigrants. Together with Christopher Hollings and Henrik Kragh Sørensen I am preparing a larger publication on the Oslo Congress of 1936 which will include Lietzmann's politically colored report to the Nazi authorities.

[62] This was a review in "Die Naturwissenschaften" of Johannes Tropfke's "History of Elementary Mathematics". See Neugebauer 1933, 563.

stopped short of using the word "race," perhaps because it had been misused so often by Nazi propagandists in Germany.

While Neugebauer was generally not given to political talk in public, he was a bit more outspoken in an interview published in the Norwegian Newspaper *Arbeiderbladet* on 14 July 1936. Neugebauer spoke in Danish; what he had to say in this interview is reproduced here in English translation from the Norwegian[63]:

Neugebauer-Interview in Oslo

Neugebauer is a small fair German who – as so many others in Germany – has become homeless; but he has been received with open arms in Copenhagen, where he works together with such illustrious scientists as the two Bohr brothers. Now he speaks Danish like a native.

Neugebauer was the first to begin investigating the history of pre-Greek mathematics and he has shown that Greek mathematical science builds on Babylonian foundations. The Greeks introduced the modern perspectives into mathematics, but the foundation belongs to the Babylonians. This is a discovery that Neugebauer has made through studies of cuneiform texts on clay tablets taken from excavations in Mesopotamia and now stored in museums in Berlin, London, Paris etc. The science of mathematics thus has a close connection to the science of language.

- [Question] One wonders why the Babylonians were so disposed to mathematics?
- Presumably through the specific blend of different types of people down there. The Babylonians built on the Sumeric culture which the English have studied very extensively through excavations at Ur that you have probably heard about."

Neugebauer's explanation for the fertility of Babylonian mathematics clearly resonates with the special spirit of the times (Zeitgeist) after Hitler had come to power in Germany. While avoiding any reference to this climate of opinion, he was stressing a viewpoint that flew in the face of Ludwig Bieberbach's racist "Deutsche Mathematik," namely that collaboration of different cultures ("races" as Neugebauer had said in 1933, now replaced by "different types of people" ["forskjellige folketyper"] in 1936) is vital for mathematical progress (Fig. 2).

The serious disturbance of international communication that followed the mass emigration from Germany was certainly very visible in Oslo. Still this did not exclusively concern restrictions faced by the Germans, as there was an absence of the Italians and the Russians as well.[64] The Russian mathematicians were barred from participation by Communist political authorities, whose politics also affected their publications abroad. Neugebauer pointed out this problem in a letter he wrote to Courant, who was then in New York, a few months later, on 14 March 1937 (translation from German):

Dear Courant,
You will certainly be interested to learn that Kolmogoroff and Khintchine had big scandals in Russia due to their Ergebnisse-reports, published in Germany. As a matter of fact, in

[63] See Anon. 1936. I thank Henrik Kragh Sørensen (Århus) for pointing me to this newspaper article.

[64] The Italians were excluded due to sanctions in connection with the Italian occupation of Ethiopia, the Russians did not take part without giving reasons. However these reasons can be documented from political discussions and decisions at the time in Russia. See my forthcoming study on the Oslo Congress.

Tirsdag 14. juli 1936. Arbeiderbladet

700 berømte matematikere

Lyn-intervjuer om universet som utvider sig, om hvad kileskriften fra Ur kan fortelle og mange andre ting

Amerika det førende land på matematikens område -- takket være nazismen.

Greta Keller på Saga.

Citater

Kongressens åpning i Aulaen imorges.

Fig. 2 Article on the Oslo International Congress of Mathematicians with "Lightning Interviews" with participants by an Oslo daily *Arbeiderbladet* on 14 July 1936, p. 5. The passages and the photo related to Neugebauer are put in frames. Courtesy Henrik Kragh Sørensen (Århus)

a
ZENTRALBLATT FÜR
MATHEMATIK
UND IHRE GRENZGEBIETE
REINE UND ANGEWANDTE MATHEMATIK
MATHEMATISCHE PHYSIK · ASTROPHYSIK
HERAUSGEGEBEN VON
P. ALEXANDROFF MOSKAU W. BLASCHKE HAMBURG H. BOHR KOPENHAGEN R. COURANT NEW YORK G. H. HARDY CAMBRIDGE
F. HUND LEIPZIG G. JULIA VERSAILLES H. KIENLE GÖTTINGEN T. LEVI-CIVITA ROM R. NEVANLINNA HELSINKI
J. D. TAMARKIN PROVIDENCE R. I. H. THIRRING WIEN O. VEBLEN PRINCETON B. L. VAN DER WAERDEN LEIPZIG
REDAKTION:
O. NEUGEBAUER
KOPENHAGEN
17. BAND
BERLIN
VERLAG VON JULIUS SPRINGER
1938

b
ZENTRALBLATT FÜR
MATHEMATIK
UND IHRE GRENZGEBIETE
REINE UND ANGEWANDTE MATHEMATIK
MATHEMATISCHE PHYSIK · ASTROPHYSIK
HERAUSGEGEBEN VON
W. BLASCHKE HAMBURG H. BOHR KOPENHAGEN R. COURANT NEW YORK G. H. HARDY CAMBRIDGE F. HUND LEIPZIG
G. JULIA VERSAILLES H. KIENLE GÖTTINGEN R. NEVANLINNA HELSINKI J. D. TAMARKIN PROVIDENCE R. I. H. THIRRING WIEN
O. VEBLEN PRINCETON B. L. VAN DER WAERDEN LEIPZIG
REDAKTION:
O. NEUGEBAUER
KOPENHAGEN
18. BAND
BERLIN
VERLAG VON JULIUS SPRINGER
1938

c
ZENTRALBLATT FÜR
MATHEMATIK
UND IHRE GRENZGEBIETE
REINE UND ANGEWANDTE MATHEMATIK
MATHEMATISCHE PHYSIK · ASTROPHYSIK
HERAUSGEGEBEN VON
K. BECHERT GIESSEN W. BLASCHKE HAMBURG E. BOMPIANI ROM H. GEPPERT GIESSEN H. HASSE GÖTTINGEN
F. HUND LEIPZIG G. JULIA VERSAILLES H. KIENLE GÖTTINGEN R. NEVANLINNA HELSINKI F. SEVERI ROM
W. SIERPIŃSKI WARSCHAU E. C. G. STUECKELBERG VON BREIDENBACH GENF
H. THIRRING WIEN B. L. VAN DER WAERDEN LEIPZIG
REDAKTION:
E. ULLRICH
GIESSEN
19. BAND
BERLIN
VERLAG VON JULIUS SPRINGER
1939

Fig. 3 (**a–c**) "The three title pages of Zentralblatt from 1938 and 1939 show the political changes which culminated in the Zentralblatt affair of October/November 1938. They reflect the withdrawal of P.S. Alexandroff in May 1938, the dismissal of T. Levi-Civita in October 1938, and the subsequent resignation of Neugebauer and all American, English and Danish members of the board."

> Russia there is flourishing now the same idiotic nationalism as in the Third Reich. Of course you should not write about these things to Russia, but because of Yellow Books you should be interested to know. For instance I do not believe that either of the two would now be able to write a Yellow Book without danger. [...]
>
> Many cordial greetings in haste
>
> Your O.N.[65]

Beginning with volume 18 (1938), the Moscow topologist Pawel Alexandroff resigned from the editorial board of *Zentralblatt* (See Fig. 3). Veblen commented on this in a letter from 10 May 1938 to Warren Weaver in the following words: "Whether this is because they [Alexandroff and other Russian reviewers] regard it as being too German or too international I do not know."[66]

Neugebauer, who was a good friend of Alexandroff from their days together in Göttingen, explained the situation to their mutual friend Heinz Hopf in Zürich, to whom he wrote on 2 June 1938 as follows:

> Dear K.T. [= Kleines Tier = Little Animal; R.S.],
>
> M.A.'s [= Meerarsch = sea-ass; R.S.] letter did not contain more than prattle about Weltanschauung and such – apparently to be officially approved by his superiors – which

[65] Courant Papers NYC. Neugebauer was of course alluding to the two influential booklets on probability theory of Kolmogoroff's and Khintchine's which appeared 1933 and 1934 in German in Springer's Ergebnisse ("Results") series. Although written before 1933 the two short monographs officially appeared in the Nazi years. Courant's old plans to win one of the two Russians for a more detailed text on probability theory for the "Grundlehren"-series (the "Yellow Series" as in the letter) were not realistic after the Stalinist interference into mathematics made communication difficult.

[66] Siegmund-Schultze 1994, 318/19.

allegedly caused him to leave the board of a non-Marxist journal. This should not, by the way, affect our personal relationship.[67]

Within Germany, there was always the potential for political interference by influential Nazis on *Zentralblatt* at this time. Veblen remained concerned about this, and he apparently still considered the possibility of moving *Zentralblatt* to the U.S. During the Congress in Oslo, he had asked Neugebauer to keep him informed about the situation, particularly about the costs for printing and distributing the abstracting journal in the event that "Zentralblatt would cut its connections to Springer."

This is revealed in a letter which Neugebauer sent to Veblen on 2 April 1937 with the requested details,[68] as mentioned in the introduction. However, Neugebauer also wrote in that letter:

> As a matter of fact so far I have not experienced a single interference into my journal editing. However, one never knows whether something happens out of the blue. Of course I should ask you to treat the whole matter confidentially. As long as Springer is able to carry on things in the current manner, I wish of course to help him maintaining the enterprise.[69]

While the political situation in early 1937 seemed to remain relatively calm, at least as far as *Zentralblatt* was concerned, it changed considerably for the worse 1 year later. On 12 March 1938 Hitler's troops marched into Austria to implement the so-called "Anschluss," an event that heightened nationalistic sentiments among some of the mathematicians in Germany.

The Austrian-born geometer Wilhelm Blaschke (1885–1962), who taught in Hamburg, and was a member of the editorial board of *Zentralblatt*, publicly welcomed the annexation of Austria, as the fulfillment of a "dream from my younger years."[70] Two days after the "Anschluss," on 14 March 1938, Blaschke wrote to Neugebauer in Copenhagen saying that "[i]t seems that the number of German collaborators, and even the role of the German language in Zentralblatt is constantly diminishing. If this continues, the publisher is going to face difficulties sooner or later."[71] Neugebauer replied angrily on 19 March 1938, saying that "[i]f in fact the role of the English language may have increased in time, then this is easy enough to explain. You know that in America especially mathematical production has grown considerably in the recent past." He also reminded Blaschke of the circumstances that led to the founding of the *Zentralblatt* in 1931 as an international journal in competition with the *Jahrbuch* (see above), while further pointing out that he had been unable to win additional German reviewers at the International Congress in Oslo. That same day he wrote to Veblen, complaining about Blaschke's "subversion" [Wühlarbeit] and inquiring whether the Americans could contribute in the range of 10,000 dollars should it be necessary to move the *Zentralblatt* to a Scandinavian

67 ETH Zürich, Heinz Hopf Papers. Hs 621: 1029.

68 The letter is quoted with some of the financial details also in Pyenson 1995, 270.

69 Siegmund-Schultze 1993, 160, there quoted from the Oswald Veblen Papers, Library of Congress, and here translated from German.

70 Siegmund-Schultze 2009, 87.

71 Siegmund-Schultze 1994, 322.

publisher. In another letter to Veblen, dated 10 December 1938, Neugebauer wrote that out of 300 *Zentralblatt* reviewers only 60 came from Germany.[72]

Neugebauer observed closely the political behavior of German scientists after the "Anschluss," and particularly of former Austrians (like himself) such as Blaschke and Erwin Schrödinger, as correspondence with his friend Courant in New York reveals.[73] On 16 May 1938, Warren Weaver of the Rockefeller Foundation was briefed by Veblen in Princeton about the situation at the *Zentralblatt*, as mentioned in the introduction.

On 21 May 1938, Neugebauer, in another letter to Courant, revealed his concern and efforts for the refugees Fritz Noether (1884–1941) in the Soviet Union and Willy Feller (1906–1970) in Sweden. As to the *Zentralblatt* Neugebauer finished the letter in a gloomy mood:

> Except for England and America there is basically no region [for recruiting reviewers, R.S.] left. The only thing which remains is working on our scientific projects and waiting for a bomb falling on our heads. (Courant Papers NYC)

The "Zentralblatt Affair," the Foundation of *Mathematical Reviews*, and Neugebauer's Role in Both

The situation became unbearable for Neugebauer in October 1938, after racist legislation had been promulgated in Italy with consequences for the Italian co-editor of the Zentralblatt, Tullio Levi-Civita (1873–1941). After P.S. Alexandroff had been forced by the Russians to withdraw from the editorial board in May 1938 (see above), Levi-Civita was excluded by Springer in October that same year even without informing the managing editor.[74] Neugebauer sent a printed postcard to several reviewers of *Zentralblatt*:

---

[72] Siegmund-Schultze 1994, 318.

[73] On May 11, 1938 Neugebauer sent Courant an excerpt from the Austrian and now German newspaper "Grazer Tagespost," 30 March 1938, with a "Declaration of university professor Dr. E. Schroedinger". After leaving Nazi Germany in 1933, the famous physicist Schrödinger, Austrian and non-Jewish like Neugebauer, had accepted in 1936 a professorship in Austrian Graz, at Neugebauer's alma mater between 1919 and 1921. Now in 1938 Schrödinger was apparently unwilling to move a second time. He "confessed shame to have missed the right way [apparently when opposing the occupation of Austria; R.S.] and promised obedience to the will of the Führer". Neugebauer wrote dryly (translated from German): "Dear Courant, Attached Schrödinger's latest publication. I would be grateful indeed if you sent it back after notice. I would not like to lose this document." (Courant Papers NYC). One has to assume – and the tone of the letter is in accord with this – that Neugebauer would never forgive his compatriot Schrödinger, although the latter would try to interpret his declaration later as purely tactical and although he left Graz for Ireland soon anyway. The evaluation of Schrödinger's publication in the "Grazer Tagespost" is still controversial in the biographical literature on Schrödinger. See Moore 1989.

[74] The last issue 10 of volume 17 (1938) of *Zentralblatt* and thus the title page is dated 7 May 1938. The last issue 10 of volume 18 (1938) of *Zentralblatt* is dated 8 October 1938 (433ff.). The first

> Since one of the editors of the Zentralblatt für Mathematik [Tullio Levi-Civita] has been eliminated [gestrichen] without communicating with him, with me or with the other editors, since further it has been demanded of me to consider other than purely objective points of view in the distribution of reviews, I have resigned the editorship of the Zentralblatt. I have to thank all my contributors most warmly for their many years of distinguished service and, above all, for the understanding with which they have accommodated themselves to the demands, not always convenient, that had to be placed upon them. O. Neugebauer.[75]

In his reference to "other than purely objective points of view in the distribution of reviews" Neugebauer was alluding also to demands made by German mathematicians. For instance F. K. Schmidt in a letter to Springer, dated 18 May 1938, found it "totally impossible to let books by German authors be reviewed by émigrés".[76]

It is not clear whether Neugebauer sent copies of this postcard to all reviewers within Germany. He certainly sent one to Hasse[77]; this could hardly have caused political troubles for Hasse as he was close to the regime (see below). However, in Bessel-Hagen's papers in Bonn which include extensive correspondence with Neugebauer about *Zentralblatt*, the postcard could not be traced. Instead, one finds there the following postcard from Neugebauer to Bessel-Hagen, dated 22 November 1938: "If you now at the same time teach analytical number theory and continue intensively historical research, I should have full understanding if you do not find time anymore for writing abstracts. I am myself no longer capable of doing it and neither are many of my friends."[78]

This so-called "Zentralblatt affair" finally led to the foundation of *Mathematical Reviews* in the U.S. in 1940 – its import was considerable, as emphasized by Nathan

---

Italian "Royal Decree" including "Measures for the defence of race in fascist school" had been promulgated on 5 September 1938.

[75] The translation follows Swerdlow 1993, 149–150. Since this important document has apparently not yet been published in German, I give here the original wording. The postcard can be found, for instance, in the Heinz Hopf Papers, ETH Zürich, Hs 621: 1031, and in the Helmut Hasse papers in Göttingen:

"København Ø

Blegdamsvej 15

November 1938

Da einer der Herausgeber des Zentralblattes für Mathematik gestrichen worden ist, ohne ihm, mir oder den anderen Herausgebern Mitteilung zu machen, da ferner von mir verlangt wurde, bei der Verteilung der Referate andere als rein sachliche Gesichtspunkte zu berücksichtigen, habe ich die Redaktion des Zentralblattes niedergelegt. Allen meinen Mitarbeitern habe ich für die jahrelange ausgezeichnete Tätigkeit wärmstens zu danken und vor allem für das Verständnis, mit dem sie sich in die nicht immer bequemen Anforderungen gefügt haben, die an sie gestellt werden mussten.

O. Neugebauer"

[76] SVA, Abteilung C, no. 778 (Neugebauer).

[77] Now in Cod Ms. Hasse, 1:1179, fol. 12, in the Manuscript Division of the Göttingen University Library

[78] NLBH Bonn, correspondence Neugebauer, no. 157, my translation from German. Bessel-Hagen was not able to completely follow Neugebauer's suggestion, although he tried repeatedly to excuse himself vis-à-vis the new German editors E. Ullrich and H. Geppert, claiming to be over-worked. However, he continued to publish in *Zentralblatt* for instance reviews of Neugebauer's works.

Reingold: "of all the reactions to Nazism of the American mathematicians, [this was] by far the most significant" (Reingold 1981, 327). The affair and its effects have been described in some detail by Reingold (1981), Swerdlow (1993) and Pyenson (1995). It may here suffice to recall that the well-known German number theorist Helmut Hasse in a letter to the American Marshall Harvey Stone, dated 15 March 1939, said the following, quoted in part by Reingold: "Looking at the situation from a practical point of view, one must admit that there is a state of war between the Germans and the Jews. Given this it seems to me absolutely reasonable and highly sensible that an attempt was made to separate within the domain of Zentralblatt the members of the two opposite sides in this war."[79]

Hasse failed to understand why the Americans withdrew their collaboration from the *Zentralblatt* in favor of what he referred to as "Neugebauer's pro-Jewish policy" (Reingold 1981, 331). Hasse's letter left a very unfavorable impression on the community of American mathematicians and this was bound to lend support to those who worked to found the *Mathematical Reviews*.

As far as Neugebauer was concerned he severed his connections with Springer totally. In a letter from 9 November 1938 to Neugebauer, Springer had expressed lack of understanding for Neugebauer's withdrawal both from *Zentralblatt* and *Ergebnisse*.[80] But the publisher then still hoped for a continuation of the *Quellen und Studien*. However, on 3 December 1938 Neugebauer received a letter from Oskar Becker (1889–1964), the philosopher and historian of mathematics in Bonn. In it Becker demanded the resignation of the co-editor Otto Toeplitz from *Quellen und Studien* because of his Jewish descent. He proposed Bessel-Hagen in Bonn as a replacement. Neugebauer, who appreciated Becker as a scholar but despised his politics, responded with the announcement of his own resignation and proposed Becker and Bessel-Hagen as the new editors. He then continued:

> I have inferred from your letter that also scientists in Germany consider the fact of extensive pogroms a sufficient reason for making it for a scholar of outstanding merit impossible even to work scientifically.... You are writing that you as a National Socialist apparently have an opinion differing from mine, in spite of your personal respect for Mr. Toeplitz. I can only reply that I am not in the happy possession of any 'Weltanschauung' and I am therefore in need to consider in each case individually, what to do, without being able to retire to a previously given dogma. This disadvantage in practical life is perhaps made good by sparing me to separate from people for whom I have respect simply because they are unhappy enough to be tortured by other people.[81]

The result was the termination of *Quellen und Studien* that same year 1938. It was that incident which inspired me to the title of this article. It is typical of Neugebauer's caustic humor and growing cynicism how he would later, after the war, comment on that affair with Becker.[82]

---

[79] Siegmund-Schultze 1993, 164. The letter is written in English in the original.

[80] SVA, Abteilung C, no. 778 (Neugebauer).

[81] Siegmund-Schultze 2009, 163–164.

[82] In a review of O. Becker and J.E. Hofmann: *Geschichte der Mathematik*, Bonn 1951, Neugebauer criticized philosophical prejudices and the failure to mention O. Toeplitz's historical book "Entwicklung der Infinitesimalrechnung" (1949). He then follows this with a grotesque remark:

While it became clear very soon that Springer would continue *Zentralblatt* even without Neugebauer and without the American, English and Danish reviewers, the foundation of a competing American journal, which would later become the *Mathematical Reviews*, was no foregone conclusion.[83] The leading American mathematician George David Birkhoff (1884–1944) was among those who as late as January 1939 were skeptical about taking on this burden.[84] He and another Harvard mathematician, Wilhelm Graustein, were also concerned that the founding of the *Reviews* might endanger plans for the next International Congress of Mathematicians. This was to be held in 1940 at Cambridge, Massachusetts, but later had to be postponed until 1950 because of the outbreak of the war.[85]

Birkhoff had an influential voice, but the majority of the American mathematicians felt otherwise. A Committee of the AMS for the Mathematical Reviews was founded and headed by mathematician C.R. Adams from Brown University in Providence, R.I. On 18 February 1939, the Adams committee interviewed Neugebauer, who had arrived in the U.S. for a temporary stay of 10 weeks on 13 February 1939 (Swerdlow 1993, 150), in detail about his experiences and for advice.

Also the officials of the Rockefeller Foundation, in particular President Raymond B. Fosdick, were gradually won over to provide financial support for mathematical abstracting. As late as 23 November 1938 Warren Weaver had strong doubts as is revealed in a letter to Veblen.[86] However a discussion between President Fosdick and Weaver on 23 February 1939, as reproduced in Weaver's diary, seems to have brought the change:

> Discussion of Neugebauer-"Zentralblatt" situation. The possible transfer of the 'Zentralblatt' or its equivalent to this country is one instance of a general situation of considerable importance and interest, – namely, the transference to this country of responsibility for the maintenance and protection of certain cultural values which historically have been chiefly located in Europe. This journal, moreover, is more accurately viewed as an international coordinating and synthesizing influence in mathematics than as a mere mechanical bibliographical aid, its reviews being critical and prepared by the leading specialists of the world. WW is not as yet prepared to make any recommendation, since it is not clear whether a new American journal will actually be founded, nor what the financial necessities will be; but RBF suggests that the division not be too much concerned over the fact that this proposal is outside of program.[87]

"Only as a minor detail of bibliographical accuracy I wish to state that I was not the sole editor of the Quellen und Studien (p. 323): I share the honor of having founded this series with J. Stenzel and O. Toeplitz and the pleasure of having exploded it with O. Becker." (Neugebauer 1953, 366–367).

[83] On the Foundation of *Mathematical Reviews* see Reingold 1981 and Pyenson 1995, 270–273. The latter goes into more detail in the case of Neugebauer, quoting the Neugebauer-Richardson correspondence at the Brown University archives.

[84] Siegmund-Schultze 1994, 313.

[85] Reingold 1981, 332.

[86] The letter is quoted in Pyenson 1995, 271.

[87] RAC, Weaver diary 1939, vol. 6, 37/38. Siegmund-Schultze 1994, 323 and Siegmund-Schultze 2001, 211. The Rockefeller Foundation voted in favor of a subsidy of $ 12,000 for the Reviews.

The same 23 February, Weaver talked to T.C. Fry, Neugebauer und Courant. Neugebauer made it clear to the Americans that hopes for an apolitical and objective running of journals with publishers in German dominated countries had become illusionary:

> Thursday, February 23, 1939
> Dr. T.C. Fry, Dr. O. Neugebauer, and Professor R. Courant
> Luncheon and general discussion of "Zentralblatt" situation. WW reports to the group the episode [...] of the visit of a consultant of Springer and a representative of the German Ministry in Berlin to Ruzicka's laboratory[88] and their report that an editor of a journal published by Springer is free to make use of Jewish or émigré assistance in countries which are not closely politically associated with Germany [here obviously Switzerland; R.S.]. N. points out first of all that this is both intensely disagreeable and thoroughly impracticable for an editor to make the delicate decision as to whether or not a given country is closely associated politically with Germany. What would one say, for example, about Hungary at the present time, or for that matter Denmark? Furthermore, this situation is in a constant state of flux, and the editor of a scientific journal can hardly be expected to turn himself into a prophetic student of current political affairs. Moreover, N. says, the statement was in effect wholly inaccurate. The editors of the new edition of the 'Encyklopädie der Mathematischen Wissenschaften' wished to have certain articles revised by émigrés now located in England. This was absolutely forbidden, although it could hardly be argued that England is a country closely politically associated with Germany. Moreover, N. points out that one must always take into account unofficial regulations as well as official regulations. A Government official will say to Springer, 'I wish to emphasize that this is completely unofficial, but I would suggest the possible desirability or advisability that you do so and so.' If Springer were so unrealistic as to overlook this advice, he would find that his publishing business would become impossible as by, for example, a mysterious reduction in the amount of paper which he was able to buy.[89]

Growing Attention in America and Support for Neugebauer's Historical Research during the 1930s and 1940s, in Particular by the Rockefeller Foundation

As discussed before, the possibility of "importing Neugebauer", either by offering him a position at the IAS in Princeton or at Brown University was already discussed, but quickly rejected in 1933, despite his importance as a historian. The reasoning then was that it was preferable to let *Zentralblatt* continue to function with the resources available in Copenhagen. While back in 1930 Courant had not been successful in obtaining a Rockefeller Fellowship for Neugebauer (see above), the Foundation had since become increasingly aware of the importance of Neugebauer's historical research. Warren Weaver, in an interview with Harald Bohr in Berlin 24 May 1933, was informed that, along with other Göttingen mathematicians, also "Neugebauer has been told [by the Nazi authorities] not to lecture."

---

[88] This is about the Rockefeller sponsored chemical laboratory of Leopold Ružička (1887–1976) in Zürich, who would receive the Nobel prize that same year 1939.

[89] RAC, Weaver diary 1939, vol. 6, p. 39.

Weaver added in his diaries: "Neugebauer is very able and has been making important research in the pre-Grecian history of mathematics. Bohr thinks he will be the greatest mathematical historian who has ever lived." (RAC, Weaver diary 1 (1933), 88/89)

Support by Rockefeller for Neugebauer's stay in Copenhagen followed. The Foundation was constantly reminded of Neugebauer's increasing fame as a historian, for instance through a letter written by R.C. Archibald (1875–1955) to Warren Weaver on 3 December 1936.[90] In his letter, Archibald, a historian of mathematics at Brown University in Providence, called Neugebauer "the young Austrian genius" who deserved more financial support, in particular "for photographs and an assistant".[91]

In 1937, i.e. still in relatively peaceful times, Weaver found Neugebauer's historical project even more interesting than Courant's local plans for applied mathematics in New York. After a conversation with Niels Bohr, Harald's brother, that took place near New York City on 14 February 1937, Weaver wrote in his diary:

> B. talks with great enthusiasm and considerable feeling concerning the fundamental contribution which Neugebauer is making to our knowledge of ancient civilizations ... There seems to be no question that this is the most significant contribution that has ever been made to the history of mathematics and thus directly to our knowledge of the historical development of science in general.... WW thinks that this is the kind of exception to program which is desirable. B[ohr] also talks to WW about Courant's ambition to obtain financial assistance in his plan of developing education and research in applied mathematics in the New York area. WW points out that this also would have to be treated as an exception to program, and that it does not seem to me that such a proposal deserves exceptional treatment in at all the same sense as N.s does. B. entirely agrees with this decision and accepts it. (RAC, Weaver diary 1937, vol. 4, 34/35)

History of mathematics was apparently seen at that point, unlike Courant's project, as being in the tradition of pure research, which was the traditional concern of the philanthropists.[92] About a year later, on 17 January, 1938, Weaver met Neugebauer in person in Copenhagen:

> Brief discussion with O. Neugebauer of his work. His researches on ancient astronomy are exceedingly laborious inasmuch as they sometimes involve numerical calculations covering periods of one hundred years or more in order to fill in gaps. N. greatly appreciates the assistance already given, but feels that he has very great need of an assistant,[93] particularly inasmuch as he now wishes to investigate ancient Hindu methods. This will necessitate learning Sanskrit, one of the few ancient languages which N. does not know.

Weaver continued discussions with the Bohr brothers, in which he raised the:

[90] Archibald sent a copy to Birkhoff, which is now in the Birkhoff Papers at Harvard University Archives 4213.2.2 box 1, file 1936.

[91] Neugebauer on his part appreciated for instance Archibald's "great bibliography of oriental mathematics," a spare copy of which he sent with these words to Bessel-Hagen in 1930 (NLBH Bonn, no.44, 13 June 1930).

[92] The interests of the Foundation would, of course, change once again when war-preparedness set in in 1940.

[93] Neugebauer wrote to Courant 21 May 1938 that Archibald had told him confidentially about imminent support from the RF, in particular for his assistant Olaf Schmidt.

> [...] possibility of an assistant for Neugebauer, although WW points out that he can be far less definitive and far less encouraging in this case, since any assistance of this work constitutes a definite exception to our stated program. WW pointed out that exceptions to our regular program are made only in instances of very great interest and importance, and that therefore our support to Neugebauer, although modest in amount, could properly be interpreted as indicating a very special conviction of N's ability and the importance of his work. Niels Bohr seemed very gratified at this statement [...] (RAC, Weaver diary 1938, vol. 5, 3/4)

Not just because of his fear of the burden of reviewing for the Americans, but also with respect to Neugebauer as a historian, Harvard's Birkhoff was not very supportive of the latter's move to the U.S. When in late 1938 Archibald wrote to Birkhoff in favor of creating a professorship for Neugebauer in the history of mathematics at Harvard University, he received a mixed response with nationalistic overtones.[94] Birkhoff replied on 2 December 1938 that Harvard President Conant was interested in the history of science and mathematics, but with a focus on the modern periods. As ever skeptical towards immigration of foreign scientists, Birkhoff remarked about Neugebauer's situation at Copenhagen: "I cannot believe that the pressure for him to leave there is very great", adding that "these temporary subsidies either of visiting foreigners or of enterprises like the *Zentralblatt* are really in the end a concealed trap." With respect to the history of mathematics in particular, Birkhoff remarked (somewhat flatteringly) further: "As an American I would, of course, prefer to see such a Chair as that of D.E. Smith at Columbia kept in the American tradition and held by you or someone else in this country."

However Birkhoff left it to the discretion of Brown University "to take on Neugebauer in order to develop the historical center at Providence still further. In that case I see no reason why he should receive an extravagant salary and I feel he would be peculiarly fortunate to be invited."

Two months later, Neugebauer's fortunes changed quite dramatically. In January 1939, while still in Copenhagen, he reported to Bessel-Hagen about the very favorable conditions for his impending research professorship at Brown.[95] During his first stay in the U.S., Neugebauer met Birkhoff on 29 March 1939 to discuss with him the question of the "usefulness of a general catalogue of cuneiform texts in this country." [96] Neugebauer found Birkhoff's proposal calling for such a general catalogue unrealistic; he insisted instead, in a follow-up letter, on "my own narrower plans, ... namely to make a complete list of practically all mathematical and astronomical texts." [97]

---

[94]The letter quoted is in copy in the Harvard University Archives, G.D. Birkhoff Papers 4213.4.5. Box 1, file personal 1938/39. I have not found a copy of Archibald's letter.

[95]NLBH Bonn, Neugebauer to Bessel-Hagen, 16 January 1939, Neugebauer Correspondence, no, 160. Neugebauer wrote about the promising library conditions and to be allowed to bring his "excellent student Olaf Schmidt."

[96]Letter Neugebauer to Birkhoff, 30 March, 1939, Harvard University Archives, G.D. Birkhoff Papers 4213.2, box 13, file J-N. I thank June Barrow-Green (London) for providing a copy of Neugebauer's letter, which is already written with a letter head showing his affiliation with Brown University.

[97]See previous note.

Although Neugebauer thanked Birkhoff in the same letter for a "very interesting and inspiring" meeting, this was very likely the same meeting to which Neugebauer alluded in an undated postcard to Courant: "I have just been at Birkhoff's. Now I am very depressed because he uses formulations which I would have expected from an extremely German-friendly circle, but never here in this place."[98]

However, both in his attitudes toward immigrants as well as in his rather unconcealed anti-Semitism, Birkhoff was not representative of the majority of opinion among American mathematicians (Siegmund-Schultze 2009).

After the war ended, Neugebauer's historical research was supported not only by Brown University, but above all by the Rockefeller Foundation and the Institute for Advanced Study in Princeton, where Neugebauer's Nachlass is now preserved. Indeed the collaboration of these three institutions during the war apparently had long-lasting effects for the historiography of ancient mathematics and astronomy.

The Rockefeller Foundation paid a 10-year grant from 1943 for Neugebauer's research associate, the young Assyriologist from Chicago, Abraham Sachs. The latter had already received a fellowship from the Foundation in 1941, as Weaver noted in his diaries following a meeting with Neugebauer on 27 April 1943:

> Dr. Sachs, of the Oriental Institute, is just completing the second year of an RF fellowship with Neugebauer. S. was well trained at the O.I., and had had just enough mathematics so that he was trying to apply some of N's techniques of chronology. The fellowship has been altogether too much of a success, since N., S., Brown, and the O.I. now apparently all agree that a combination has been formed which must not be broken up.
>
> N. sees before him large possibilities, but possibilities which he cannot exploit alone. It is, moreover, a long slow program which must have reasonable assurance for years or it is not worth starting. N. is giving us a new understanding of the earliest history of mathematics, or rather of the earliest history of quantitative and analytical thinking. He studies mathematical astronomy at a time when that was all the science there was, and at a time when astronomy, cosmology, philosophy, and religion were all interwoven.
>
> We estimate that about $40,000 would stabilize their program for about ten years; and N. thinks B. [sic, probably for Brown University; R.S.] would carry on from there. (RAC, Weaver diary 1943, vol. 7, p. 110)

Neugebauer at *Mathematical Reviews* and His Skepticism regarding "Modernization" of Mathematics at Brown University

The protocol of Neugebauer's 18 February 1939 interview with the Adams commission shows that he was anxious that the responsibilities for the new journal should be distributed on broader shoulders than had been the case with *Zentralblatt* in Copenhagen. There Neugebauer had basically acted alone (together with his wife, of course) and supported by Werner Fenchel, as he described in his interview. In the protocol one reads: "N. agrees with Adams that the chief editor had better be a

[98] Courant Papers NYC.

relatively young man." This accords with a remark made by Adams in his covering letter to Birkhoff: "Effectively, he said eight years on the Zentralblatt were enough for a man who preferred not to have a life sentence to abstract journal editing."[99]

Nevertheless, Neugebauer kept busy with the *Mathematical Reviews* (MR) in the first 2 or 3 years of its existence. After the Adams Committee had decided on 30 May 1939 to proceed with the establishment of MR (Price 1990), Neugebauer was appointed editor, together with J. Tamarkin. However, as Pitcher (1988) reports, the immigrant from Germany and Sweden, Willy Feller, 7 years Neugebauer's minor, was appointed as "technical assistant" from 1939 and signed himself managing editor in 1943. At that time, Neugebauer and Tamarkin were no longer spending much time at MR. However, Neugebauer remained influential even later and prevented occasional attempts at influencing the reviewing process by personal and partial points of view.[100]

Brown University and its graduate school faculty under long-standing AMS-secretary R.G.D. Richardson aimed to establish a large-scale, multi-functional mathematical center in Providence. This was to include applications (e.g. the well-known summer school at Brown during the war), publishing, and history of mathematics. Richardson's institute, parallel to Courant's institute at New York University, thus became a nucleus of modernization of American mathematics in the early 1940s, partly repeating or catching up with developments which had taken place in Göttingen and Berlin[101] two decades or so before. The faculty at Brown in applied mathematics was overwhelmingly composed of European refugees (Reingold 1981, 335).

Among the by now large group of émigré mathematicians, Neugebauer was a late arrival to the U.S.. During his first stay in Providence in the spring of 1939, he apparently received a request from his friend Bessel-Hagen to enquire about a possible position at Brown University for Felix Hausdorff (1868–1942), who was in Bonn. Hausdorff, the famous topologist and author of "Grundzüge der Mengenlehre" (Foundations of Set Theory, 1914), was Jewish and so he had been dismissed from his professorship there. Although he was already over 70 years old, he still hoped to

---

[99] C.R. Adams to G.D. Birkhoff, February 18, 1939, Birkhoff Papers, Harvard University Archives, Box 2, 4213.4.5. (correspondence and talks 1937–1943). The protocol of the interview with Neugebauer is there attached in copy.

[100] In late 1944, after the death of George D. Birkhoff, Neugebauer did not comply with a wish of Birkhoff's son, the mathematician Garrett Birkhoff, to give G.D. Birkhoff's work on the general theory of relativity in flat space to Harry Bateman for review. Instead, a review by Hermann Weyl appeared in Mathematical Reviews (MR0008365), that was quite critical. (Weyl was very strongly opposed to GDB's theory, about which they corresponded. One should assume that ON was well aware that Weyl felt this approach threw away Einstein's key insight, his equivalence principle linking gravitational and inertial effects. The political overtones here do indeed seem highly significant. I owe this remark to David Rowe). The incident is documented in the papers of Oswald Veblen, who originally supported Garrett Birkhoff's request (Library of Congress, Veblen Papers, General Correspondence, Container 2, folder Birkhoff, George D. 1929–47.)

[101] With respect to the foundation of the "Quarterly of Applied mathematics", edited at Brown University from 1943, deliberate comparisons were made with Richard von Mises' "Zeitschrift für Angewandte Mathematik und Mechanik" (from 1921) in Berlin. See Siegmund-Schultze 2009.

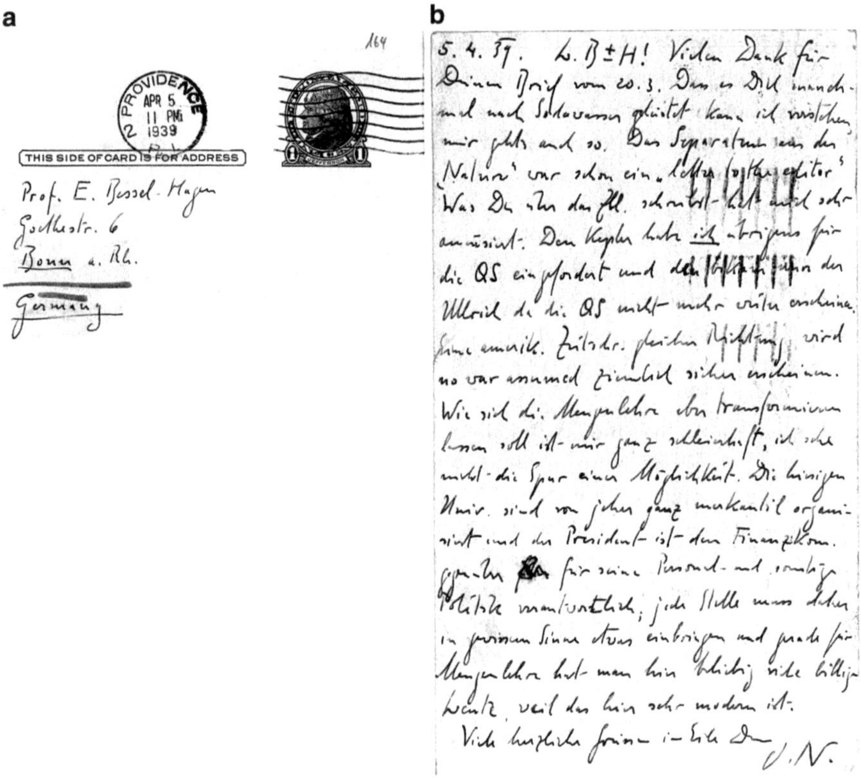

Fig. 4 (**a**, **b**) Postcard from Neugebauer to Bessel-Hagen, 5 April 1939. Courtesy University Library Bonn, Nachlass E. Bessel-Hagen (Front and back)

be rescued from Nazi Germany by an appointment abroad.[102] This hope turned out to be in vain; he and his family committed suicide 3 years later. In 1939 Neugebauer saw no chance for Hausdorff precisely because his research field had become so modern and fashionable in the U.S. that the Americans did not need older and more expensive foreign professors to cultivate it. Neugebauer wrote the following postcard to Bessel-Hagen on 5 April 1939 (Fig. 4), which beyond doubt alludes to Hausdorff[103]:

[102] These efforts are also documented in letters written by Courant in February 1939 and by G. Pólya in May 1939. Cf. Siegmund-Schultze 2009, 96–97. Neugebauer's postcard from April 1939, to be mentioned below, fits exactly into this time frame.

[103] NLBH Bonn, correspondence Neugebauer, no. 164. My translation is of the second part. There can be no doubt that the "transformation of the theory of sets" in the postcard refers to Hausdorff's emigration. Bessel-Hagen's efforts for his much older friend Hausdorff are documented elsewhere in his Nachlass. The word "transformation" as a pseudo-mathematical and secret code-word for "emigration" was well known to Neugebauer not least from Courant's letter to him in September 1933, which was quoted above.

The QS [Quellen und Studien; R.S.] do not appear anymore. An American journal of the same direction [Eudemus; R.S.] will most likely appear no war assumed [last three words here in English in the original; R.S.]. How the theory of sets could be transformed is totally beyond me, I do not see the slightest chance. The universities here are traditionally commercial, and the president is responsible to the financial committee for his politics, particularly with respect to personnel. Therefore each position has, in a way, to earn its own income, and as far as set theory is concerned they have here plenty of affordable people, because the topic is very modern here.

In its continued modernization in the early 1940s, Brown University received financial support from, among other institutions, both Rockefeller and Carnegie. The Carnegie Corporation gave even more money than Rockefeller, namely the colossal sum of 60,000 dollars, to support the *Mathematical Reviews*. This money was allotted in order to launch a huge microfilming (experiment) project,[104] an experiment that finally failed. The journal's initial offer to deliver photocopies of publications to readers – a kind of precursor to online publication of today – did not survive. There are signs that Neugebauer found these developments exaggerated and that he even considered them a kind of "over-modernization". In a letter to Courant from 26 December 1939, he complained that "we were forced to microfilm all papers under review, even the ones that were easily available."[105] In the same letter he called this microfilming project part of "Bush's idiotic plan"[106] to create a general catalogue of all mathematical works where one can find within seconds films of all relevant papers."[107] In this same connection, Neugebauer also criticized Thornton Fry as someone for whom "advertisement is everything" and "his only interest is 'glory'."

It is telling that Neugebauer had even become skeptical about seeking financial support for his own research and publications in the history of mathematics from political organizations such as UNESCO. Upon his arrival in the U.S. Neugebauer, together with Archibald, began editing a new journal entitled "Eudemus: an international journal devoted to the history of mathematics and astronomy",[108] which was produced by the Danish publisher E. Munksgaard. As a result of the war, however, only one volume of the journal appeared in 1941. When after the war the UNESCO official and historian of science J. Pelseneer offered help in reviving *Eudemus*, Neugebauer replied 3 December 1947 with the following letter:

Dear Dr. Pelseneer,
[...] It is very difficult for me to answer your question concerning plans for a future congress and concerning the support of Eudemus by UNESCO. I confess that I belong to the

---

[104] The project is described by Pyenson 1995, 272–273. See also Price 1941.

[105] Courant Papers, NYC. Translation from German.

[106] Vannevar Bush, the engineer and inventor of an analogue computer, the 'differential analyser', was then president of the Carnegie Corporation and a very influential American science organizer in the years to come. He became head of the war organization for research OSRD. According to Sarton, Bush showed open contempt for the history of science. Pyenson 1995, 281.

[107] Neugebauer's letter to Courant, 26 December 1939. Courant Papers New York, Translation from German.

[108] See also Neugebauer's postcard to Bessel-Hagen, dated 5 April 1939 and quoted above.

> dying-out tribe of scholars who deeply dislike organized work. I myself have never profited from a congress though I have spent time, money and headaches to participate in several of them. I am completely unable to work on problems that somebody else thinks up for me and I am also unable to propose problems for somebody else. I know that more of my remaining lifetime is necessary to complete the studies which interest me deeply. Thus as far as I am concerned I shall spend all my efforts in the continuation of my own research program and the only hope I have is that I am as little as possible disturbed by outside influences. I fully realize the practical importance of organized research but I myself am unfitted for such programs and I have decided to keep out of them. It is for this very reason that I would ask only as a last resort the support of UNESCO for Eudemus. I hope you will not consider this attitude as unfriendliness in any sense. It is merely the result of a long experience with all kinds of organizations that I hope I can spend the rest of my life without being attached to any officially recognized group of scholars.[109]

Eudemus, which had been planned as an American continuation of the long-since terminated *Quellen und Studien*, did not reappear, although the same source reveals that Neugebauer somewhat later supported and drafted an initiative in favor of *Eudemus*, which was sent to the Carnegie Corporation by Brown rector H.M. Wriston.[110]

Neugebauer's letter to Pelseneer can be read as a kind of personal assessment of his legacy and a statement of the final wishes of a man who had an excellent record promoting the "organization" of mathematics but who had become increasingly disenchanted with the relentless pace of modernization. Having achieved much for a broader culture of mathematics, including reviewing, Neugebauer was now eager to return to his core historical research; in all likelihood, he was disillusioned by the various instances of political interference with his work. Fortunately, the "rest of my life" for the then 48 years old Neugebauer would amount to over four decades with many fundamental publications to come, among them arguably the most important was the three-volume "Astronomical Cuneiform Texts" (ACT).

Neugebauer's Attitudes Toward Germany Before and After the War

To understand Neugebauer's political and philosophical views after the war one must certainly take into account his general attitude toward Germany and especially the insufficient or entirely uncritical manner in which the Germans went about "coping with the past". During his entire emigration Neugebauer had followed political events in Nazi Germany closely, particularly the political behavior of the scholars who remained behind. We have already noted his critical and sometimes cynical remarks about Becker, Blaschke and Schrödinger. Neugebauer showed little understanding for mathematicians who were perhaps politically naive, such as Heinrich Grell (1903–1974), who was once a student of Emmy Noether. Grell had

---

[109] Brown University Archives, H.M. Wriston, President, folder 'History of Math.'

[110] This initiative failed, since, according to Swerdlow 1993, 151, the Corporation refused to give money. I assume this happened under the influence of V. Bush (see above).

at first sympathized with the Nazis, but then he ran into political[111] problems with them. After having been personally approached by Grell for help, Neugebauer wrote to his friend Heinz Hopf in Zürich on 2 January 1936: "I find it outrageous of him now to seek support abroad after having had no luck with the Third Reich."[112]

On the other hand, Neugebauer apparently had little sympathy for those who took any extremist political positions, as revealed by his letter to Bessel-Hagen (in) from 1930. He shared this attitude with the former Göttingen mathematician and emigrant Herbert Busemann (1905–1994). Indeed, both refused to help the former communist (and mathematician from Göttingen), Rudolf Lüneburg (1903–1949), during his emigration. In 1935 Busemann wrote to Courant, who had been seeking to help Lüneburg, that he and Neugebauer felt it was unwise to let "people reach influence ... who would use the latter to curtail my resp. our freedom."[113]

Although not Jewish himself, Neugebauer, like many other refugees from Germany, felt bitter about the behavior of many of his former compatriots. This is clear from occasional remarks which are documented. Even regarding the Dutch mathematician B. L. van der Waerden, who had once attended Neugebauer's lectures in Göttingen on the history of ancient mathematics but who later stayed in Germany (at Leipzig) in spite of opportunities abroad,[114] Neugebauer grew increasingly critical. In 1940 he wrote to Bessel-Hagen: "What you write about Bartel is all well but for reasons I cannot explain here I don't give a damn for him."[115] In August 1945 Neugebauer wrote to his friend Heinz Hopf in Zürich about van der Waerden:

> I do not mind his remaining a German professor until the end. I do mind his remaining German professor at the beginning. However, I feel very differently than the Lord and I do not intend to do anything positive or negative. I know that Bessel-Hagen behaved excellently and I would like very much to hear about him as soon as he is again localized.[116]

To van der Waerden himself Neugebauer had written a week before: "What has happened to the Zentralblatt? I hope it is completely ruined."[117]

When a former colleague from Göttingen complained that Neugebauer continued to write personal letters in English, not using his mother tongue even though the war had ended, Neugebauer offered the oft-cited rejoinder, which, however, probably expressed only half of the truth: "I must remark that the language I use in my letters does not depend on my mother but on my secretary."[118]

---

[111] "Political" in a broader sense which in Grell's case seems to have included discrimination against "deviant" sexual orientation.

[112] ETH Zürich, Heinz Hopf Papers, H 621: 1013. My translation from German.

[113] Quoted in Siegmund-Schultze 2009, 184.

[114] See Siegmund-Schultze 2011.

[115] NLBH Bonn, correspondence Neugebauer, no. 179; 25 October 1940. Translation from German.

[116] Neugebauer to H. Hopf 15 August 1945, ETH Zürich, Hopf Papers, Hs 621: 1041. Original English.

[117] 9 August 1945, ETH Zürich, B.L. van der Waerden Papers, Hs 652: 11496. This hope of Neugebauer's would not be fulfilled as the future revival of *Zentralblatt* would show.

[118] Swerdlow 1993, 155. Swerdlow does not reveal the name of the mathematician; one may conjecture this was van der Waerden or perhaps F. Rellich.

Neugebauer, and even the more conciliatory Courant, continued to feel skeptical about the Germans in the following years. This shared skepticism was quite unlike the attitude of others, as for instance the non-Jewish emigrant Carl Ludwig Siegel, who would later return to Göttingen in 1951. When the latter proposed raising money for books that would be sent to the Göttingen institute library, once so dear both to Courant and Neugebauer, Courant distanced himself from this plan in a letter to Hermann Weyl, written 17 January 1947:

> My own feeling is that I rather help individuals [...] than institutions. Almost everybody whom I asked, including Neugebauer and Hopf, are disinclined to participate in Siegel's action. [...] Strangely enough, my general feeling toward Germany and German institutions, instead of mellowing, is becoming more and more irritated. The percentage of reassuring communications is so very small. It seems that not only I, but also Franck [physicist James Franck; R.S.] and Neugebauer, are feeling the same way.[119]

In sum, Neugebauer's relationship with Germany was unsentimental, to say the least. Neugebauer remained definitely less active or engaging than Courant in reconnecting to Germany in the years to come, not least of course, because he focussed more on his own research than Courant did.

One reason for Neugebauer's indifference to Germany might have been related to the fact that the University of Göttingen failed to apologize for Neugebauer's expulsion in 1933. In the administrative files of the university one finds a note, dated 28 March 1946, which omits Neugebauer's name from an earlier list of former faculty members (dated November 1945) who as refugees were to be reinvited to Göttingen. Only after an intervention by the British Military Government in February 1946 was Neugebauer's name discussed. However, according to the same document, the university authorities defended their original position with the following words: "He went 1933 to Copenhagen, without political pressure having been exerted on him, which means voluntarily. There he accepted the chair for history of mathematics which already had been prepared for him." This document contains an additional penciled-in comment about Neugebauer: "Out of the question, went voluntarily."[120]

Ten years later, in July 1957, when Neugebauer's claim for financial compensation was discussed on a higher administrative level, the ministry in Hanover recognized that Neugebauer had tirelessly supported Jewish colleagues and could therefore not return from Copenhagen. However, Neugebauer's claim was acknowledged only as "partially justified" ("zum Teil begründet"). Because there had been no chair for history of mathematics in Germany in the 1930s and since such a position did not even exist in the 1950s, according to the decision of the minister, Neugebauer could only claim compensation for a position as assistant professor (Diätendozent) with lower income.[121] This ministerial decision of 1957, which

[119] Courant Papers NYC.

[120] University Archives Göttingen, Universitätskuratorium, Durchführung des Gesetzes zur Wiederherstellung des Berufsbeamtentums IX 83, II: Teil (1935–1950), fol. 234–238.

[121] University Archives Göttingen, Universitätskuratorium Göttingen K XVI.IV, Ad 194 (O. Neugebauer). "Wiedergutmachungsbescheid", Niedersächsischer Kultusminister, Hannover 4. Juli

expressly recognized that Neugebauer had once been offered an appointment at the Technical University Darmstadt in 1931, shows, once again the kinds of problems which Neugebauer had to face throughout his life simply because his research field stood outside the traditional disciplines.

As Neugebauer's letter to Hopf of August 1945, in particular his remark concerning his friend Bessel-Hagen has shown, he was able to distinguish between various attitudes among the mathematicians who remained in Germany during the Nazi years. However, one probably does not go amiss in assuming that he was deeply troubled and disappointed by the indifference to suffering shown by most of his German colleagues. Probably his own brief experience with life in Nazi Germany contributed to his general doubts about the integrity of humanity and the honesty of personal motives throughout history. His colleague at Brown University, the mathematician Philip Davis, who knew Neugebauer well, reported in his "Reminiscences and Appreciation" the following remark made by him, which he considered typical: "If you never heard the sound of Nazi boots below you in the street you cannot understand the history of the period." (Davis 1994, 130)

Conclusions: Another Attempt at Tracing Neugebauer's "Weltanschauung"

Neugebauer's remark about "the sound of Nazi boots" surely indicates how deeply he was affected by the events of those years. Given the striking soberness of his mind (attested by Swerdlow, Davis, Pyenson), one can easily understand his inclination to withdraw to a purer realm of intellectual life. He would henceforth focus on a field of research grounded in eternal ideas rather than one dependent on the ephemeral lives and activities of human beings. Given further his well-known aversion to biographies (Pyenson 1995), one may safely assume that Neugebauer would never have approved of an attempt, as undertaken in this article, to trace the roots of his "Weltanschauung". Nevertheless, such an attempt seems to me worthwhile, not least due to certain contradictions in pronouncements he made but also because of various connections between his personality and his approach to research.

In 1993 Neugebauer's former student and collaborator, Noel Swerdlow acknowledged that Neugebauer's historical work reflected conflicting, though "complementary" perspectives:

> There are two principal interpretations of the history of the mathematical sciences that have an important role in Neugebauer's work, both true, both of value, but not entirely compatible. To borrow a term from Niels Bohr; they are complementary.
>
> On the one hand mathematics and the mathematical aspects of other sciences have a continuity and universality that is independent of time, place or the character of any particular mathematician. [...]

1957. The same personal file shows that Neugebauer in July 1962 received 1093 Deutsche Mark compensation per month.

> The other interpretation looks upon the mathematical sciences as a characteristic and fundamental product of each individual culture, and thus the differences between the mathematics of Babylonians, Greeks or modern Europeans are of the greatest significance in coming to understand the character, certainly the creative character, of each civilization.[122]

One should not be surprised that when mathematicians approach the history of their subject they tend to project present-day mathematics into the past. Neugebauer was a mathematician by training and, in addition, he depended on mathematicians for support and approval, which helps to explain why Swerdlow wrote further: "But if the truth be told, on a deeper level Neugebauer was always a mathematician first and foremost, who selected the subjects of his study and passed judgment on them, sometimes quite strongly, according to their mathematical interest." (Swerdlow 1993, 141)

This notwithstanding, Neugebauer's linguistic abilities and historical sense prevented him from falling into the traps or naïve historical views sometimes entertained by research mathematicians. Still, the fact that so little is known about the individual mathematicians and astronomers during ancient times, the period in which Neugebauer was most interested, tended to make discussion of the "human factor" in history somewhat pointless. Thus, his own special research interests reinforced Neugebauer's general historiographic views, and he repeatedly found strong words to deride work on the history of science which emphasized the importance of purely human factors. A particularly telling episode is connected with Neugebauer's sharp criticism of a review written by the Belgian theoretical physicist Leon Rosenfeld, which discussed Neugebauer's revised edition of H.G. Zeuthen's "Forelæsninger over Mathematikens Historie: Oldtiden" from 1893. Rosenfeld praised this as a "great work", but then added these remarks:

> Even the very best human endeavor, however, has its weaknesses. [...] I could not imagine those old mathematicians as men of flesh and blood. [...] Zeuthen was obviously too much of a mathematician and too little of a genuine historian to pay proper attention to this human side of the question.[123]

In a letter from 17 October 1950, Neugebauer protested to the editor of *Centaurus*, Mogens Pihl. Although this ultimately went unpublished, it deserves to be quoted almost in entirety:

> I wish to thank you for sending me the first issue of Centaurus. I think it is quite a nice issue and I wish you best success. If you allow me to voice a private opinion in one point, I may say that the only flaw I found is the review of Rosenfeld, which I can only consider as monumental nonsense. He has obviously not yet understood that historical research is a science and not poetry. He wants to learn something about the motives of men of whom we know hardly more than their time of life. Not to say that I seriously doubt that we know the motives of any scholar, even of our contemporary colleagues. I hope Rosenfeld's lack of

[122] Swerdlow 1993, 141. A related and more detailed analysis of the conflict of the mathematician and the cultural historian in Neugebauer, and of his relation to George Sarton, is given by David Rowe in the present volume, also partly based on testimony by Swerdlow.

[123] Rosenfeld 1950.

insight into the methodology in our field will not exercise too much influence on the editorial policy. [....][124]

One can safely assume that the Marxist physicist Rosenfeld was not as naïve as his review might appear at first glance. What he probably wanted to express was a desire to learn something more about the societal conditions under which ancient mathematics was produced, in particular the philosophical connections as well as possible applications. With the World War still vivid and in view of the developing atomic race, Rosenfeld probably also wanted to make another point – a bit mistaken for the topic in question – about the responsibility of scientists for their work. One may even assume that Neugebauer, the liberal-minded refugee from Germany, felt similarly disillusioned in the post-war environment.[125]

But while Rosenfeld reacted with political activity and increased historical interest (Jacobsen 2012), the disillusioned historian, writing from McCarthy stricken U.S., only felt a sense of embarrassment. Consequently he argued in the extreme opposite direction, denying that one could make judgments about motives "even of our contemporary colleagues." And disillusioned Neugebauer was, as we have already seen from his letter to Pelseneer. To Pihl in Copenhagen he wrote in a follow-up letter from 19 December 1950: "The only reasonable thing to do is to shoot oneself as soon as possible and not to wait for the politicians, the militaries, and the physicists."

Neugebauer reacted with similar impatience and intolerance to George Sarton's "new humanism" in the history of science.[126] Sarton, as we have seen, had ranked himself well below Neugebauer, when writing to Veblen in 1933. Neugebauer repeatedly confirmed that ranking, belitteling Sarton as that "eminent compiler of many volumes [...] 'reminiscent of the mentality of Isidore of Seville.'" (Swerdlow 1993, 155). In a letter to van der Waerden, then in Zürich, Neugebauer even went so far as to call Sarton an "Obertrottel" (great fool).[127] He then published "A notice of ingratitude" in *Isis*, in which he defended van der Waerden against Sarton's reproach for having shown "shocking ingratitude towards Moritz Cantor", the leading historian of mathematics in the nineteenth century. Instead, Neugebauer criticized Cantor's work for "its total lack of mathematical competence as well as its moralizing and anecdotal attitude [which] seriously discredited the history of mathematics in

[124] Neugebauer to M. Pihl, 17 October 1950, Niels Bohr Archives Copenhagen, Mogens Pihl Papers,"Otto Neugebauer, corr re., 1950." Thanks go to Felicity Pors (Copenhagen) for providing a copy.

[125] Many liberal minded immigrants to the U.S., among them physicists and mathematicians like J. Franck, H. Lewy, H. Weyl, R. von Mises, felt uneasy about the ideological climate in the U.S. after WWII, with the maintenance of secrecy regulations, increasing anti-Communist hysteria, and the threat of the atomic bomb. Their new experiences often collided with their feelings of gratitude to the host country. Richard von Mises, for one, was warning against the "extreme" ideologies of the "two strongest political powers" who threatened each other with "physical annihilation". (Mises 1951, 14/15).

[126] An interesting comparison between Sarton and Neugebauer with respect to their professional interests, attitudes, and emotions is contained in Pyenson 1995.

[127] Pyenson 1995, 282.

the eyes of mathematicians, for whom, after all, the history of mathematics has to be written."[128]

It is interesting to see how Neugebauer here shows his solidarity – on a professional level – with van der Waerden, a competent mathematician who had turned historian in recent years. This although the latter was politically – at least from the experiences of the war – not at all close to Neugebauer, probably less so than Rosenfeld and Sarton. As Pyenson rightly and sensitively remarked about Neugebauer's lack of interest in biographies: "This lack of interest in lives seems to relate less to Neugebauer's person (his early emotional experiences may not have been so different from George Sarton's) than to his chosen field." (Pyenson 1995, 274)

One has also to take into account that, to a certain degree, Neugebauer's frequently sharp utterings said more about himself and his passionate, but narrowly focused research interests than about the competing projects of fellow colleagues. Indeed, other sources indicate that behind the scenes Neugebauer acted in support of Sarton, whom he publicly criticized.[129]

Moreover, one finds several striking modifications or changes in Neugebauer's Weltanschauung over the years, some no doubt resulting from frustrating experiences and others simply age-related. In the introduction to his semi-popular 1951 book "The Exact Sciences in Antiquity" Neugebauer was somewhat apologetic about the simplifications he had to resort to in such a concise work: "I am exceedingly skeptical of any attempt to reach a 'synthesis' – whatever this term may mean – and I am convinced that specialization is the only basis of sound knowledge."[130]

What a difference to the preface he and Courant wrote in 1926 for Felix Klein's historical lectures, where they warned against exaggerated specialization (see above)! Similarly, one sees a world of difference when reading Neugebauer's letter to Pelseneer, in which he denounced 'organization,' and then compares this with his article on the Göttingen institute in 1928 (Neugebauer 1928). Times had changed and Neugebauer had changed within them. Indeed, the multi-talented Neugebauer, who had long been living in two worlds, research and organization, was now, after the war, trying to flee into the first almost entirely.

Open questions still remain, and some bear directly on the deepest level of Neugebauer's personality. For one, Neugebauer was never a conciliatory personality like Courant, as described by Courant's biographer Constance Reid. One might wonder how Neugebauer's personality was affected by the fact that he lost his parents at an early age, or to what extent the lack of parental love might have

[128] Neugebauer 1956. The "Notice of ingratitude" is more completely quoted and analyzed in David Rowe's article in this volume.

[129] "WW questions N. concerning George Sarton and also concerning [I.B.] Cohen. N. has met Cohen and has a superficial impression that he is deeply interested in the history of science and curious about it, but he has no basis for judging C.'s real ability. N. thinks very highly indeed of S., and thinks he is worth any support." (RAC, Weaver diary 1941, vol. 7, p. 60)

[130] Neugebauer 1957, vii/viii, from the preface to the first edition of 1951.

influenced his attitude toward other people.[131] Maybe he needed unambiguous answers and clear-cut decisions more than others, certainly more than Courant.

One may also wonder, though probably not enough documentary evidence survives to decide, how Neugebauer experienced the First World War. In Neugebauer's handwritten diaries from the First World War one finds commentary such as the following: "In my diary I have constantly avoided to express my own feelings or thoughts (Stimmungen oder Gedanken)."[132] Neugebauer's biographer and friend N. Swerdlow says: "He liked this [i.e. his duties as an artillery officer; R.S.] despite its danger since it gave him a good deal of independence, and later remarked mordantly that these were among the happiest days of his life." (Swerdlow 1993, 139) I conclude that the war probably sharpened Neugebauer's wish for independence, so often expressed in his letters. We recognize at the same time – for instance by Swerdlow's reference to his "mordant" remarks – Neugebauer's cynical wit.

Summing up, two factors have to be considered to account for Neugebauer's "Weltanschauung", in particular his apparent or real rejection of philosophical or political judgments: the "complementarity of his two perspectives" [133] on the history of mathematics, but also the vicissitudes of Neugebauer's long and eventful life, which was highlighted by a series of highs and lows and marked by many contradictory, often frustrating and sobering political experiences.

Disillusioned by political ideologies since his days in Göttingen, Neugebauer continued to trace the eternal truth of "exact thinking" throughout human history while consistently downplaying the role of personality and purely human motives in the history of science. In Göttingen Neugebauer had still hoped to pursue his own personal quest in a specific social context. He did so by engaging in organizational work, and cooperating with non-political enterprises such as Springer and internationalist foundations such as Rockefeller. However, experiences over the years (the *Zentralblatt* affair and probably the political atmosphere during the McCarthy era) had further disillusioned him and progressing age had taken its toll as well.

We have seen that Neugebauer did, indeed, pass philosophical and political judgment on people, theories, and on events, even while declaring (as in his letter to the Nazi Oskar Becker) that he himself was not "in the happy possession of any Weltanschauung". Yet even this declaration was itself a political act that expressed a "Weltanschauung" all the same. Neugebauer knew, of course, that historiography is not merely concerned with facts but also with values and "Weltanschauung" as well. The final words in Neugebauer's lecture course on "Vorgriechische Mathematik", which he offered during the summer of 1934 in Copenhagen, immediately following his flight from Nazi Germany, thus provide an apt conclusion to this paper:

[131] However, Neugebauer in his correspondence with Bessel-Hagen refers repeatedly to "the Mutter" (the mother) in Graz, whose demise in 1938 he mourns. The "Mutter" may have been his aunt or step-mother. David Rowe reminded me that Neugebauer dedicated his book "Vorgriechische Mathematik" (1934) to her ("der Mutter gewidmet"). (Neugebauer 1934, p. v).

[132] See Neugebauer's war diaries at SWLLA Princeton as mentioned above, p. 3, 6 October 1919.

[133] As described by Swerdlow, which is, for the full picture, in need of further investigation.

> Meaning and method of historiographical representation: to find out the facts as accurately as possible, but then form them into living constructs, such that we experience the historical processes as organically possible.[134]

Acknowledgment I am very grateful to David Rowe, who not only copy-edited my paper for the English language but gave numerous advice for sharpening its focus as well. All archives used, which are mentioned in section "Unpublished Sources Used", are thanked for allowing to quote from their sources.

References

Alexandroff, P. 1976. Einige Erinnerungen an Heinz Hopf. *Jahresbericht Deutsche Mathematiker-Vereinigung* 78: 113–125.

Anon. 1936. "700 famous mathematicians. Lightning interviews on the expanding universe, what cuneiform writings from Ur can tell us and many other things. America the leading country in mathematics – thanks Nazism" (in Norwegian). *Arbeiderbladet* (Oslo), Tuesday 14 July 1936, 5.

Davis, P. 1994. Otto Neugebauer: Reminiscences and appreciation. *American Mathematical Monthly* 101: 129–131.

Dehn, M. 1928. *Über die geistige Eigenart des Mathematikers. Rede anläßlich der Gründungsfeier des Deutschen Reiches am 18. Januar 1928*. Frankfurt: Werner und Winter.

Exodus Professorum. 1989. *Exodus Professorum: Akademische Feier zur Enthüllung einer Ehrentafel für die zwischen 1933 und 1945 entlassenen und vertriebenen Professoren und Dozenten der Georgia Augusta am 18. April 1989*. Göttingen: Vandenhoeck & Ruprecht.

Forman, P. 1971. Weimar culture, causality, and quantum theory, 1918–1927: Adaptation by German physicists and mathematicians to a hostile intellectual environment. *Historical Studies in the Physical Sciences* 3: 1–115.

Forman, P. 1991. Independence, not transcendence, for the historian of science. *Isis* 82: 71–86.

Hilbert, D. 1930. Logic and the knowledge of nature. In *From Kant to Hilbert*, vol. 2, ed. W. Ewald, 1157–1165. Oxford: Clarendon Press.

Jacobsen, A.S. 2012. *Leon Rosenfeld : Physics, philosophy, and politics in the twentieth century*. Singapore: World Scientific 2012.

Jessen, B. 1993. Mathematiker unter den deutschsprachigen Emigranten. In *Exil in Dänemark. Deutschsprachige Wissenschaftler, Künstler und Schriftsteller im dänischen Exil nach 1933*, ed. W. Dähnhardt and B.S. Nielsen, 127–133. Heide: Westholsteinische Verlagsanstalt.

Klein, F. 1926. *Vorlesungen über die Entwicklung der Mathematik im 19. Jahrhundert. Teil I. Für den Druck bearbeitet von R. Courant und O. Neugebauer*. Berlin: Springer.

Mises, R.V. 1951. *Positivism. A study in human understanding*. Cambridge, MA: Harvard University Press.

Moore, W. 1989. *Schrödinger. Life and thought*. Cambridge: Cambridge University Press.

Neugebauer, O. 1925. Litteraturanmeldelse (T.E. Peet, The Rhind Mathematical Papyrus). *Matematisk Tidsskrift* A 66–70.

[134] "Sinn und Methode einer geschichtlichen Darstellung: die Tatsachen als solche so sauber als möglich herauszuarbeiten, dann aber sie so zu lebendigen Gebilden auszugestalten, dass wir die geschichtlichen Prozesse als organisch möglich empfinden." The 56 page handwritten script is available at the website of SWLLA Princeton (http://cdm.itg.ias.edu/cdm/). I do not find this quote in Neugebauer's book with the same title (Neugebauer 1934); he may have found it too personal for publication.

Neugebauer, O. 1928. Über die Einrichtung des Mathematischen Instituts der Universität Göttingen. *Minerva-Zeitschrift* 4: 107–111.
Neugebauer, O. 1930. Das Mathematische Institut der Universität Göttingen. *Die Naturwissenschaften* 18: 1–4.
Neugebauer, O. 1933. Review of J. Tropfke, *Geschichte der Elementarmathematik*, Zweiter Band, 3. Auflage, 1933, *Die Naturwissenschaften* 21: 563–564.
Neugebauer, O. 1934. *Vorgriechische Mathematik*. Berlin: Springer.
Neugebauer, O. 1937. Über griechische Mathematik und ihr Verhältnis zur vorgriechischen. *Comptes rendus de congrès international des mathématiciens, Oslo 1936* 1: 157–170.
Neugebauer, O. 1953. Review of O. Becker and J. E. Hofmann. *Geschichte der Mathematik, Centaurus* 2: 364–367.
Neugebauer, O. 1956. A notice of ingratitude. *Isis* 47: 58.
Neugebauer, O. 1957. *The exact sciences in antiquity*, 2nd ed. Providence: Brown University Press.
Neugebauer, O. 1963. *Reminiscences on the Göttingen mathematical institute on the occasion of R. Courant's 75th birthday*, 9. Available online at https://sites.google.com/site/neugebauerconference2010/web-exhibition-neugebauer-at-goettingen
Peckhaus, V. 1990. *Hilbertprogramm und Kritische Philosophie*. Göttingen: Vandenhoeck & Ruprecht.
Pitcher, E. 1988. Mathematical reviews. In *AMS Centennial Publications, vol. 1. A history of the second fifty years*, 69–89. Providence: AMS.
Price, G.B. 1941. Mathematical reviews offers a reading machine for microfilm. *Bulletin of the American Mathematical Society* 47: 1–2.
Price, G.B. 1990. *The founding of mathematical reviews*. http://www.ams.org/publications/mathreviews/GBaleyPrice.pdf
Pyenson, L. 1995. Inventory as a route to understanding: Sarton, Neugebauer and sources. *History of Science* 33: 253–282.
Ramskov, K. 1995. *Matematikeren Harald Bohr*. Aarhus Universitet, unpublished Ph.D. dissertation.
Rashed, R., and L. Pyenson. 2012. Otto Neugebauer, historian. *History of Science* 50: 402–431.
Reid, C. 1976. *Courant in Göttingen and New York. The story of an improbable mathematician*. New York: Springer.
Reingold, N. 1981. Refugee mathematicians in the United States of America, 1933–1941: Reception and reaction. *Annals of Science* 38: 313–338.
Rosenfeld, L. 1950. Review of H.G. Zeuthen, *Forelæasninger over Mathematikens Historie; Oldtiden* (Lectures on the history of mathematics; antiquity), revised edition by O. Neugebauer. *Centaurus* 1: 88–89.
Rowe, D. 1989. Klein, Hilbert, and the Göttingen mathematical tradition. *Osiris* 5(2): 186–213.
Schappacher, N. 1987. Das Mathematische Institut der Universität Göttingen 1929–1950. In *Die Universität Göttingen unter dem Nationalsozialismus. Das verdrängte Kapitel ihrer 250jährigen Geschichte*, ed. H. Becker, H.-J. Dahms, and C. Wegeler, 345–373. München: Saur.
Segal, S.L. 2003. *Mathematicians under the Nazis*. Princeton/Oxford: Princeton University Press.
Siegel, C.L. 1965. *Zur Geschichte des Frankfurter Mathematischen Seminars*. Frankfurt: Klostermann.
Siegmund-Schultze, R. 1993. *Mathematische Berichterstattung in Hitlerdeutschland. Der Niedergang des Jahrbuchs über die Fortschritte der Mathematik (1869–1945)*. Göttingen: Vandenhoeck & Ruprecht.
Siegmund-Schultze, R. 1994. 'Scientific control' in mathematical reviewing and German-U.S.-American relations between the two world wars. *Historia Mathematica* 21: 306–329.
Siegmund-Schultze, R. 2001. *Rockefeller and the internationalization of mathematics between the two World Wars. Documents and studies for the social history of mathematics in the 20th century*. Basel: Birkhäuser.

Siegmund-Schultze, R. 2002. The effects of Nazi rule on the international participation of German mathematicians: An overview and two case studies. In *Mathematics unbound: The evolution of an international mathematical research community, 1800–1945*, ed. K.H. Parshall and A. Rice, 335–357. Providence/London: AMS/LMS.

Siegmund-Schultze, R. 2009. *Mathematicians fleeing from Nazi Germany*. Princeton: Princeton University Press.

Siegmund-Schultze, R. 2011. Bartel Leendert van der Waerden (1903–1996) im Dritten Reich: Moderne Algebra im Dienste des Anti-Modernismus? In *"Fremde" Wissenschaftler im Dritten Reich. Die Debye-Affäre im Kontext*, ed. D. Hoffmann and M. Walker, 200–229. Göttingen: Wallstein.

Swerdlow, N. 1993. Otto Neugebauer (26 May 1899–19 February 1990). *Proceedings of the American Philosophical Society* 137: 139–165.

Otto Neugebauer's Visits to Copenhagen and His Connection to Denmark

Lis Brack-Bernsen

Otto Neugebauer visited Copenhagen twice, on Harald Bohr's invitation. The first visit, during the year 1924/1925, resulted in two papers, one of which is the only paper on pure mathematics that Neugebauer ever wrote. Later, as Neugebauer had to leave Germany due to severe problems with the Nazi regime, he obtained a sponsored professorship during the years 1934–1939. This paper covers Neugebauer's first sojourn in Copenhagen and describes how his situation in Göttingen became unbearable and forced him to leave Germany. It also reports on the Bohr brothers' assistance for scientists who had to flee from Germany and on Neugebauer's friendship with Harald Bohr. Finally it focuses on Neugebauer's activities in Copenhagen during the years 1934–1939: his research and collaboration with Danish Egyptologists, his teaching, and his relationship to his first doctoral student, Olaf Schmidt, who was my teacher.

Neugebauer and Harald Bohr

Harald Bohr (1887–1951) was a driving force behind the international connection between mathematicians during the times of the two world wars.[1] In his master's thesis in mathematics from 1909, Harald Bohr had solved a problem presented by Edmund Landau: how to multiply Dirichlet series. Because of this, he was invited by Landau to visit the famous mathematical institute in Göttingen. Here he made contact with Felix Klein, Landau, David Hilbert, Henri Poincaré, and the other visitors and students of mathematics at Göttingen, and he began to work on his doctoral thesis on the summation of Dirichlet series. He also communicated with Hardy and

[1] On the biography of Harald Bohr see Ramskov 1995.

L. Brack-Bernsen (✉)
History of Science, University of Regensburg, D-93040 Regensburg, Germany
e-mail: lis.brack-bernsen@psk.uni-regensburg.de

A. Jones et al. (eds.), *A Mathematician's Journeys*, Archimedes 45,
DOI 10.1007/978-3-319-25865-2_3

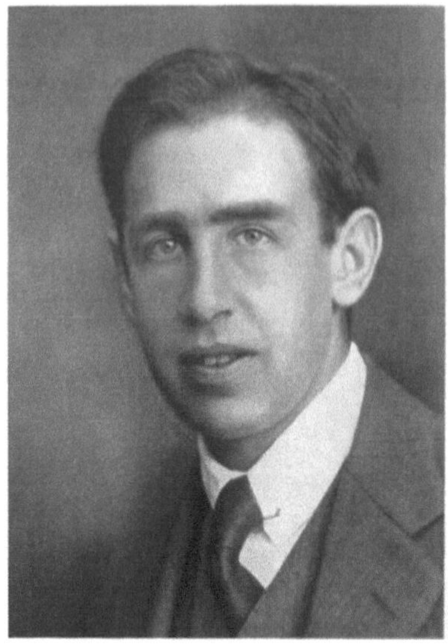

Fig. 1 Harald Bohr (22.4.1887–22.1.1951)

Littlewood in Cambridge. Later Bohr started a close collaboration with his mentor Landau on Riemann's "Zeta Function", ending up visiting Göttingen many times for shorter or longer periods. As a result, Bohr came into closer relations with Richard Courant, Hermann Weyl, Erich Hecke, and Constantin Carathéodory. His last and longest stay in Göttingen before the First World War took place during the winter semester 1913–1914, ending with a study visit to Paris where he got to know Borel, Lebesque, and Julius Franz Pál. Obviously, Harald Bohr (Fig. 1) had contact with the most important mathematicians of the time.

Direct collaborations were interrupted by the First World War, during which time Bohr acted as an intermediary for the correspondence between his friends on the two sides of the war. In addition, he was also part of an informal network of European and American mathematicians who gained influence in distributing grants by the International Education Board (1924–1928) of the Rockefeller Foundation, a foundation which would play a crucial role in the forced emigration after the Nazi takeover in 1933.[2]

Neugebauer's First Stay in Copenhagen 1924–1925

Neugebauer had changed his field of interest from physics to mathematics and moved to Göttingen in the fall of 1922, following the advice of Sommerfeld.[3] Having become an assistant at the mathematical institute in Göttingen in 1923, and

[2] For a comprehensive study of the German mathematicians during the Nazi regime, see Siegmund-Schultze 2009.

[3] For a detailed biography of Otto Neugebauer, see Swerdlow 1993.

Courant's special assistant in 1924, Neugebauer accepted the invitation of Harald Bohr to visit him in Copenhagen.[4] One purpose of this visit was to assist Bohr with the proofreading of his German-language papers on "almost" periodic functions. The visit started in March 1924 and lasted about a year, and it resulted in two published papers.

As a student in Göttingen, Neugebauer had become interested in the history of ancient Egypt and began studying Egyptian with Hermann Kees and Kurt Sethe, so he was well equipped to review the new edition of an old Egyptian mathematical papyrus. Therefore, Bohr asked him to write a review of T.E. Peet's edition of Papyrus Rhind for a Danish mathematical journal. This would become Neugebauer's first publication.[5] This publication was followed by a paper written together with Harald Bohr on differential equations with "almost" periodic functions.[6] This was to be Neugebauer's only work on modern mathematics.

The work with Papyrus Rhind may have incited Neugebauer's interest in ancient mathematics, resulting in a promotion in Göttingen 1926 with a dissertation on the foundation of calculating with Egyptian unit fractions. The mathematicians at the Institute in Göttingen had been quite alarmed by this change of interest towards ancient science by their very bright student, but Hilbert, Courant and others pleaded for letting him follow his interest – an intelligent student would always find his way. Much later, in a letter written to Jessen in February 1951, Neugebauer sees in Harald Bohr the one who encouraged him to investigate the sciences in Antiquity.

> [...] it was at my very first stay in Copenhagen (helping with the German of the Fastper. F. I) that he [Harald Bohr] gave me the Pap. Rhind to review for the [Matematisk] Tidsskrift — the start of all my later work. He had asked Poulsen to give me a key to the Ny Carlsbg. Glypto. and I have still my notebooks in which I sketched texts and reliefs from this collection. Without his encouragement I would have hardly dared to go into that field and he and Courant and Ehrenfest were the only ones who accepted such studies as 'legitimate' in an academic program.[7]

The Mathematical Institute in Göttingen

The mathematical Institute at the University of Göttingen had been the hub of the mathematical Universe from the end of the nineteenth century until it was disturbed by the Nazis in 1933.[8] During the 1920s Courant had been an excellent administrator of the Institute with Neugebauer assisting as his right hand man. The Rockefeller

[4] In 1915 Bohr had become a professor at the Technical University of Copenhagen (Den polytekniske Læreanstalt), a position he kept until, finally, a professorship was created for him at Copenhagen University in 1930.

[5] Neugebauer 1925.

[6] Bohr and Neugebauer 1926.

[7] Ramskov 2004, 254. Strangely, this letter is missing in files sent to me by Tage Gutmann Madsen whom I thank cordially for scanning material in Copenhagen and sending it to me.

[8] For the disastrous impact of the Nazis on the University of Göttingen, see Becker et al. 1998.

Foundation sponsored a new building for the mathematical Institute, which was inaugurated in 1929. Neugebauer had been essentially involved in the planning and construction of the new mathematical building.[9] With the well-equipped library and rooms for discussions, the new building was an ideal frame for the blooming institute.

But there seem to have been quarrel at the Institute due to conflicting political convictions. Neugebauer fought against the Nazis: at one point he had taken the student Oswald Teichmüller by the collar and thrown him out of the building, because he was distributing Nazi propaganda in the entrance hall of the Institute. Teichmüller was a brilliant young student but a fanatic Nazi. Perhaps, therefore, it was only very late that Courant proposed him for a stipend for excellent students.[10] As a result of pressure from Nazi students, six professors (state employees, mostly Jews) were exempted from all their duties at the University of Göttingen. Three of these were the "non-Aryan" mathematicians Felix Bernstein, Richard Courant, and Emmy Noether; a fourth, the physicist Max Born.

Later, nationalistic students organized a boycott of lectures held by Edmund Landau and Paul Bernays, and they ensured that Neugebauer, who supported his Jewish colleagues, was suspended as politically unreliable and intolerable. Neugebauer had refused to sign a loyalty oath to the new government, whereafter the president of the university forbade him to enter the Institute. The situation in Göttingen had become unbearable for Neugebauer. Instead of going to the USA, he accepted the invitation from Harald Bohr to come to Copenhagen University. Living in Europe, it would be possible for him to continue his work as founder and chief editor of the *Zentralblatt für Mathematik und ihre Grenzgebiete*.[11]

The Bohr Brothers and Their Assistance to Refugees

Harald Bohr, together with his brother Niels (their mother was Jewish), was very active in helping scientists and other refugees who had to flee from Germany. They were both members in the committee of *Den danske Komitée til Støtte for landflygtige Aandsarbejdere*, (The Danish committee for support for exiled intellectual workers). This was one of five organizations created to evaluate refugees in Denmark

[9] Neugebauer 1930.

[10] For more details see Schappacher 1998.

[11] See also Swerdlow 1993, 145 and Pyenson 1995. As noted above, Reinhard Siegmund-Schultze's stirring book, Siegmund-Schultze 2009, provides a detailed study of the situation of mathematicians living in or fleeing from Nazi Germany, in which many individual fates are treated, letters and other documents from the time are reproduced, and the global impact of the immigrating mathematicians is analyzed. Appendix 1 lists those 145 German-speaking mathematicians who emigrated during the Nazi period, and those 27 who were murdered or driven to suicide by the Nazis. It was not without risk to resist or criticize the new government, so colleagues from outside Germany tried to step in and help. Appendix 3.1 reproduces a report compiled by Harald Bohr "together with different German friends" on the conditions in German Universities.

and organize their support. Some of the friends and colleagues of Niels and Harald Bohr were able to stay in Denmark, while many others came to Denmark, merely using Copenhagen as an intermediate station before continuing to Sweden, England, or the USA. The Bohr brothers were popular and well known in Copenhagen, and they had good contact with the government and with sponsors. This enabled them to help many refugees. Due to their prestige and good connections with scientists around the world, they were also able to help many scientists obtain more permanent jobs in other countries. It is not possible to give a list of those who were helped by them, since they burned all their correspondence on the evening of April 9 1940, the day on which Denmark was invaded by the German army. However, several of the letters which Harald Bohr wrote to his colleagues, asking for help to mathematicians without job or money, were kept by the recipients.[12]

Denmark is a small country, and hence only a few refugees could be employed and stay in Denmark after their flight. Besides, the growing number of refugees resulted in restrictions for persons who wanted to come to Denmark. Olaf Schmidt told me the following anecdote about Harald Bohr's strategy for preventing the immigrations officers from sending refugees directly back to Germany when arriving in Copenhagen. When a mathematician was to arrive in Copenhagen without visa, Bohr knew the precise time of his arrival. He and Neugebauer were at the airport of Copenhagen at the moment of his arrival. Harald would welcome his "dear friend" cordially and talk incessantly and in a very friendly manner to the immigration officers. He would carefully fill one pipe after another with tobacco and, while smoking, involve the officers in small talk and argue very amicably and calmly why this fine person should be allowed to see Copenhagen and not be sent directly back to Germany. Suddenly it would be too late for a flight back. Harald Bohr pretended to be surprised by this fact, but he solved the problem of what to do with the stranded person by proposing that his wife could arrange their guestroom for him. The next day the mathematician was sent to Norway.[13]

---

[12] Excerpts from such letters to friends and colleagues in Sweden, Norway, England, and the USA, are quoted in Ramskov 1995.

[13] Much more on refugees in Denmark, and on those who helped them can be found in Steffensen 1986. Steffen Steffensen (1908–1984) was a professor in German literature at the University of Copenhagen. Through many years he collected a comprehensive material on the exile of German speaking intellectuals in Denmark, scholars as well as artists and politically engaged persons. One goal of his project was to illustrate what a richness these refugees brought to their host country Denmark. After his death in 1984, it was decided to conclude the project and publish those articles which were already finished or were planned by Steffensen. The book (only existing in Danish, though the German-language Dähnhardt and Nielsen 1993 is largely based on it) begins with a description of the situation in Denmark concerning the refugees fleeing from Nazi Germany followed by the presentation of different help organizations. Then the fates of persons, who could stay in Denmark for at least some time, are presented in essays. The essays are organized according to their field: Scientists, Humanities, Psychologists, Philosophers, Artists, Musicians, and Writers. Børge Jessen (1907–1993) wrote the contribution on the mathematicians who worked for some time in Denmark (Jessen 1986, also in German translation as Jessen 1993). These are Otto Neugebauer, Werner Fenchel (1905–1988) and his wife Käte (born Sperling, 1905–1983), Herbert Busemann (1905–1994), and Willy Feller (1905–1970). Jessen held a chair as a professor at the

In his obituary for Harald Bohr, Neugebauer writes[14]:

> On January 22, 1951 Harald Bohr died in Copenhagen. His name is familiar to mathematicians the world over and many of us will feel a debt of gratitude to him not to be forgotten to the end of our days.

After an account of Bohr's research and collaboration with numerous mathematicians, he continues:

> These investigations were carried out while ever darkening shadows of a new catastrophe were falling over Europe. When the drama began to unroll in 1933, Bohr exploited all his international connections to the advantage of refugees from dictatorship. There is scarcely a mathematician who had to leave his country who was not helped directly or indirectly by Bohr and his Danish friends. And perhaps equally great is the number of those unknown men and women who received his help but who did not have the good fortune of having academic connections in the outside world. He followed with distress the rapid development of barbarism in Germany.

It was Harald Bohr's good relationship to sponsors that enabled him to offer Neugebauer a professorship in Copenhagen when he had to leave Göttingen 1933. The professorship was granted half by the Rockefeller Foundation and half by the Danish Rask-Ørsted foundation. His payment during the years 1937–1939 were ensured by the Danish Carlsbergfonden and the Rask-Ørsted-Fonden.[15]

Neugebauer's Second Stay in Copenhagen 1934–1939

Neugebauer arrived in Copenhagen towards the end of 1933, and started working as a professor at the Mathematical Institute (Fig. 2) in January 1934. During his time in Denmark, Neugebauer learned Danish, lectured at the university,[16] directed the publishing offices of the *Zentralblatt für Mathematik*, and was extremely active in the research of ancient mathematics and astronomy. In December 1938 he was offered a professorship at Brown University, which he accepted in February 1939 during a 10 weeks' stay in Providence. He came back to Copenhagen in May, and returned to Providence with his family in midsummer. Soon he was followed by his and Bohr's pupil (and my teacher) Olaf Schmidt.

Neugebauer was a huge gain for Copenhagen, not only for those interested in ancient mathematics or astronomy, but also for the Mathematical Institute. His work as editor of the *Zentralblatt*, with the assistance of his wife and of Werner Fenchel, opened up steady contacts to mathematicians over the whole world. Neugebauer on

---

mathematical institute of the Copenhagen University; later Fenchel also became a professorship at the mathematical Institute, where both he and Jessen were my teachers. In 1929 Jessen had stayed in Göttingen for one semester, visiting Landau and Hilbert.

[14] Neugebauer 1952.

[15] The Rask-Ørsted-Fonden was establish after the first world war for supporting the scientific exchange and collaboration between Danish and foreign scientists.

[16] Jessen 1986, 88; Jessen 1993, 127–128.

Fig. 2 The building to the left housed the mathematical institute from 1933, the time when it was built, until 1963, when the institute was transferred to the newly constructed Ørsted Institute. It had the address Blegdamsvej 15 and was next to Blegdamsvej 17, the building to the right, which housed the university's Institute of Theoretical Physics, founded in 1921 by Niels Bohr, and since 1965 called the Niels Bohr Institute

his side was glad to live in Denmark, and would have liked to stay there. But as it became clear that the war would spread to Denmark, and that Neugebauer in that case would be drafted into the German army, he left Denmark in June 1939.

How much Neugebauer owed to Harald Bohr, and his gratitude towards him, is testified in the preface to the third volume of *Mathematische Keilschrift Texte* (MKT). Here one can read the words by which Neugebauer dedicated MKT to Harald Bohr:

> Die Ausarbeitung dieser nunmehr abgeschlossenen Bände der MKT hat den größten Teil der verfügbaren Zeit der drei Jahre in Anspruch genommen, die ich jetzt in Kopenhagen zugebracht habe. Dass sie mir vergönnt waren, habe ich vor allem der steten Hilfsbereitschaft von Harald Bohr zu verdanken. Ihm sei das ganze Werk in herzlicher Dankbarkeit zugeeignet.
>
> [In English translation: The compilation of these volumes of MKT, which have been completed now, has taken most of the available free time of my last three years which I have spent in Copenhagen. That this time was granted to me is due to the steady helpfulness of Harald Bohr. The whole opus is dedicated to him in cordial gratitude.] (O. Neugebauer, Kopenhagen, 22. Dezember 1936)

Neugebauer's Research and His Collaboration with Danish Egyptologists

Neugebauer's stay in Copenhagen was very fruitful: besides many reviews, he succeeded in publishing more than 30 papers or books on ancient astronomy and mathematics during his 5 years in Denmark. Very soon, on April 6 1934, Neugebauer became member of Det Kongelige Danske Videnskabernes Selskab (The Royal Danish Academy of Science). He is listed under his full name Otto Eduard Hermann Neugebauer. Many of his research papers were published in the proceedings of the Academy, in their *Historisk-philologiske* or *Mathematisk-fysiske Meddelelser*.[17]

I shall just mention a few of the works which Neugebauer published while living in Copenhagen.[18] First, we have the well known books on pre-Greek mathematics: the *Vorlesungen über Geschichte der antiken Mathematischen Wissenschaften, erster Band: Vorgriechische Mathematik*, and the three-volume *Mathematische Keilschrift Texte* (*MKT*), which was his most important work from his Danish period (Fig. 3).[19] It is also worth mentioning his paper arguing the worthlessness of the Sothic Period for establishing early Egyptian chronology, on the first page of which Neugebauer accentuates that the nucleus of this work aroused in discussions with Dr. W. Feller, Stockholm. After leaving Germany (Kiel University) 1933, Feller had stayed in Copenhagen 1 year before going to Stockholm.[20]

Neugebauer also started a fruitful collaboration with Danish Egyptologists. Dr. Aksel Volten (1896–1963) of the *Ægyptologisk Laboratorium* had drawn his attention to the Egyptian astronomical texts PBerol. 8279 in the Berlin papyrus collection, the Stobart tablets in Liverpool, and Papyrus Carlsberg 1 and 9. The latter two were parts of a large and very important collection of Egyptian papyri (mostly in Egyptian demotic and hieratic script) acquired by the Carlsbergfonden and handed over to the department of Egyptology.[21] Neugebauer, who was not aware of the existence of such texts, was excited and started working on the texts in collaboration with Volten and his colleague Hans Ostenfeld Lange (1863–1943).[22]

The first text Neugebauer worked on was Papyrus Carlsberg 9. With a lot of enthusiasm, and within an extremely short time, Neugebauer and Volten solved the riddles of the text which consisted of five separate sections. They were able to understand the individual sections and find the connections between the parts. Central for the understanding was the insight that section 3, a number scheme, and section 4, a calendar scheme, were based on the following period relation: 25 Egyptian years = 309 synodic months. Thereby they were able to determine the

[17] E.g. Neugebauer 1934b, 1938–1939.

[18] All papers and books which were publications published from Copenhagen can be found in Sachs and Toomer 1979.

[19] Neugebauer 1934a, 1935–1937.

[20] Neugebauer 1939.

[21] Information on the collection and its history is at http://pcarlsberg.ku.dk.

[22] See Schmidt 1950.

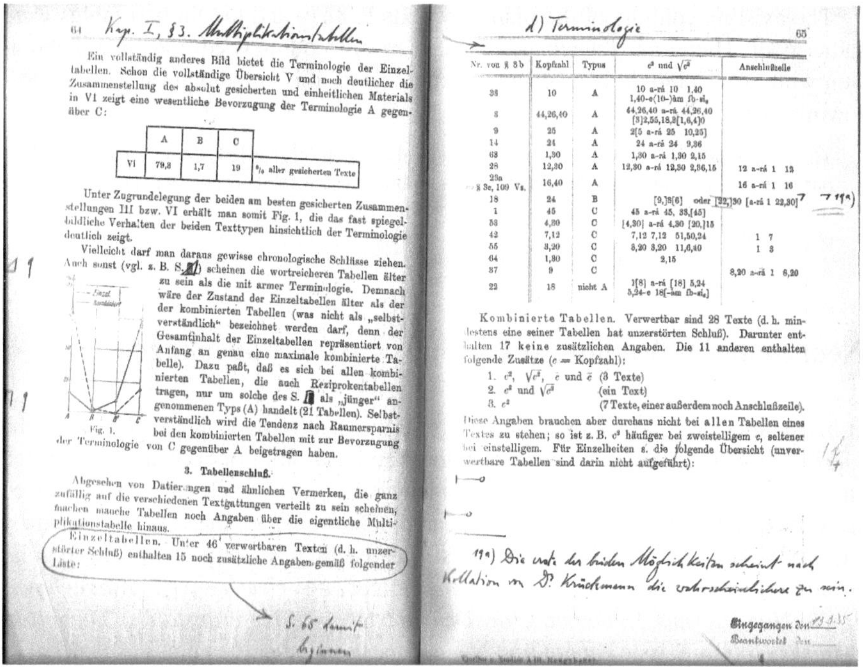
64

Ein vollständig anderes Bild bietet die Terminologie der Einzeltabellen. Schon die vollständige Übersicht V und noch deutlicher die Zusammenstellung des absolut gesicherten und einheitlichen Materials in VI zeigt eine wesentliche Bevorzugung der Terminologie A gegenüber C:

	A	B	C	
VI	79,8	1,7	19	% aller gesicherten Texte

Unter Zugrundelegung der beiden am besten gesicherten Zusammenstellungen III bzw. VI erhält man somit Fig. 1, die das fast spiegelbildliche Verhalten der beiden Texttypen hinsichtlich der Terminologie deutlich zeigt.

Vielleicht darf man daraus gewisse chronologische Schlüsse ziehen. Auch sonst (vgl. z. B. S.) scheinen die wortreicheren Tabellen älter zu sein als die mit armer Terminologie. Demnach wäre der Zustand der Einzeltabellen älter als der der kombinierten Tabellen (was nicht als „selbstverständlich" bezeichnet werden darf, denn der Gesamtinhalt der Einzeltabellen repräsentiert von Anfang an genau eine maximale kombinierte Tabelle). Dazu paßt, daß es sich bei allen kombinierten Tabellen, die auch Reziprokentabellen tragen, nur um solche des S. als „jünger" angenommenen Typs (A) handelt (21 Tabellen). Selbstverständlich wird die Tendenz nach Raumersparnis bei den kombinierten Tabellen mit zur Bevorzugung der Terminologie von C gegenüber A beigetragen haben.

Fig. 1.

3. Tabellenschluß.

Abgesehen von Datierungen und ähnlichen Vermerken, die ganz zufällig auf die verschiedenen Textgattungen verteilt zu sein scheinen, machen manche Tabellen noch Angaben über die eigentliche Multiplikationstabelle hinaus.

Einzeltabellen. Unter 46 verwertbaren Texten (d. h. unzerstörter Schluß) enthalten 15 noch zusätzliche Angaben gemäß folgender Liste:

65

Nr. von § 3b	Kopfzahl	Typus	c^2 und $\sqrt{c^2}$	Anschlußzeile
33	10	A	10 a-rá 10 1,40 1,40-e(10-)àm íb-si₈	
8	44,26,40	A	44,26,40 a-rá 44,26,40 [3]2,55,18,2[1,6,4]0	
9	25	A	2[5 a-rá 25 10,25]	
14	24	A	24 a-rá 24 9,36	
63	1,30	A	1,30 a-rá 1,30 2,15	
28	12,30	A	12,30 a-rá 12,30 2,36,15	12 a-rá 1 12
23a § 3e, 109 Vs.	16,40	A		16 a-rá 1 16
18	24	B	[9,]3[6] oder [22,]30	[a-rá 1 22,30]
1	45	C	45 a-rá 45, 33,[45]	
53	4,30	C	[4,30] a-rá 4,30 [20,]15	
42	7,12	C	7,12 7,12 51,50,24	1 7
55	3,20	C	3,20 3,20 11,6,40	1 3
64	1,30	C	2,15	
37	9	C		8,20 a-rá 1 8,20
22	18	nicht A	1[8] a-rá [18] 5,24 5,24-e 18[-àm íb-si₈]	

Kombinierte Tabellen. Verwertbar sind 28 Texte (d. h. mindestens eine seiner Tabellen hat unzerstörten Schluß). Darunter enthalten 17 keine zusätzlichen Angaben. Die 11 anderen enthalten folgende Zusätze (c = Kopfzahl):

1. c^2, $\sqrt{c^2}$, $\bar{c}$ und $\bar{\bar{c}}$ (3 Texte)
2. c^2 und $\sqrt{c^2}$ (ein Text)
3. c^2 (7 Texte, einer außerdem noch Anschlußzeile).

Diese Angaben brauchen aber durchaus nicht bei allen Tabellen eines Textes zu stehen; so ist z. B. c^2 häufiger bei zweistelligem c, seltener bei einstelligem. Für Einzelheiten s. die folgende Übersicht (unverwertbare Tabellen sind darin nicht aufgeführt):

Eingegangen den
Beantwortet den

Fig. 3 Two pages of the proofs of *MKT* from year 1935. Through my teacher Olaf Schmidt I know that Neugebauer's wife Grete (Bruck), who was herself a gifted mathematician, supported his work heavily, e.g., by proofreading and correcting manuscripts

meaning of the expressions "small" and "large" year: the normal "small" lunar year consisted of 12 synodic months, but regularly a 13th month was added, in which case the year would be called a "large year." The expression "large and small year" had previously been known from burials from the 12th dynasty, but nobody had figured out what was meant by such expressions.[23] The Lunar scheme of Papyrus Carlsberg 9 consisted of 9 large years of 13 lunar months and 16 small years of 12 lunar months. The joint edition was published in German in 1938.[24]

According to the preface to the edition of Papyrus Carlsberg 1,[25] this edition, too, is the result of a close and direct collaboration between the two authors. In July 1940 Lange added a second preface, writing that their collaboration regrettably had been interrupted by the developments in Europe, and that it was impossible to have correspondence with America. Therefore the authors decided to publish the edition before they had been able to solve all problems in the text.

---

[23] Different explanations are proposed in Ginzel 1909, 176–177.

[24] Neugebauer and Volten 1938.

[25] Lange and Neugebauer 1940.

The existing editions of the planetary texts P. 8279 and the Stobart tables were rather poor. Therefore Neugebauer decided to work out new editions in collaboration with Olaf Schmidt. Neugebauer acknowledges this collaboration with the following words:[26]

> The main burden of the very extensive numerical calculations carried out in order to compare the positions given by the ancient texts with the actual orbits of the planets rested on the shoulders of Mr. Olaf Schmidt who has therefore a very considerable share in this edition.

Neugebauer's Teaching at the University of Copenhagen

It has neither been possible to find out exactly when Neugebauer's lectures at the University of Copenhagen were given nor their exact titles or contents. All written material on lectures given at the Mathematical Institute by scholars not employed by the Danish government has disappeared. The books of these years issued by the University of Copenhagen only list the teaching announced by the regular professors and employees. In "*Forelæsninger og Øvelser ved Kjøbenhavns Universitet, Aarhus Universitet, og Den polytekniske Læreranstalt*", "*Danmarks tekniske Højskole*", one cannot find Neugebauer's name or the names of Fenchel or other scientists who were not state employees but paid by grants. "*Lektionskatalogen*" and other administrative sources on Neugebauer's lectures or on his students have disappeared from the Mathematical Institute. But according to Olaf Schmidt and Børge Jessen, Neugebauer's lectures on ancient mathematics and astronomy were followed by a very interested and enthusiastic circle of listeners.

Other sources give more information, allowing us to determine the subjects of Neugebauer's lectures over the 4 years 1934–1937 (Table 1). The preface to his *Vorlesungen über Geschichte der antiken Mathematischen Wissenschaften, erster Band: Vorgriechische Mathematik* is dated July 11, 1934. Here Neugebauer writes that it is lectures held in exactly this form in Copenhagen, which have been published in the book:

> Was hier veröffentlicht wird, sind wirklich Vorlesungen, die ich fast genau in dieser Gestalt in Kopenhagen gehalten habe.... Es ist also eine durchaus von persönlichen Ansichten getragene Auffassungsweise, die den Leitfaden der Darstellung abgibt.

Table 1 Lectures given by Otto Neugebauer at the University of Copenhagen

1934	Vorgriechische Mathematik
1935	[Babylonische Mathematik]
1936	Babylonische Astronomie
1937	Babylonische Astronomie

[26] Neugebauer 1942, 200.

This indicates that Neugebauer already gave his lectures on pre-Greek mathematics in the spring semester of 1934. And since *MKT* I and II came out in 1935, I assume that Neugebauer gave lectures on Babylonian mathematics during the two semesters of 1935. There still exist copies of Neugebauer's handwritten lectures on Babylonian astronomy (Fig. 4); therefore we know that they were held during the years 1936 and 1937. All lectures are written in German, so I assume that the lectures were given in German.

From these handwritten and duplicated lecture notes on Babylonian astronomy we know the contents of the lectures: The table of contents, Fig. 4, shows what was lectured on in 1936. Similarly, the list of contents in the notes from 1937 shows that the Babylonian theory of eclipses was the subject of Neugebauer's Copenhagen lectures during the following year.[27] The nucleus (structure and concise mathematical methods) of Neugebauer's treatment of the Babylonian astronomy is apparent in these notes. But it would take Neugebauer almost 20 more years to finish the enormous work on the astronomical cuneiform tablets, which was published in 1955.

In addition to lecturing, Neugebauer had also agreed to supervise students who wrote their master's theses on a subject from the history of mathematics.

From Olaf Schmidt and his daughter Inger, I know that Neugebauer had learned to speak Danish well. He always spoke Danish to Inger, both during his time in Denmark and many years later when Olaf together with his family visited Neugebauer in the USA. Some of Neugebauer's public lectures were undoubtedly held in Danish. A lecture "*Om babylonsk Matematik*," given in the fall session of 1937 for "*Matematiklaererforening*" (the association of the mathematics teachers), was subsequently published in Danish in the journal *Matematisk Tidsskrift*.[28] The same journal had previously invited Neugebauer to give an overview of the new results and insights into Babylonian mathematics, which his investigations of cuneiform texts had led to, and which had just appeared in German in *MKT* 1 and 2,[29] and this survey was followed by an article on the survival of Babylonian mathematical methods in Greek sources (Neugebauer 1937). He also published a review, of the Danish translation of Lancelot Hogben's popular *Mathematics for the Million*, in which he did not mince his words.[30]

It is also known that Neugebauer gave a very exciting and inspiring lecture on how he succeeded in deciphering Papyrus Carlsberg 9. It was held 1938, in Danish, in the "*Mathematisk Forening*" [the Mathematical Society]. The nucleus of the lecture was to show how Neugebauer was able to find the structures of the number scheme within 24 h through skilled analysis of numbers in section 3, resulting in a

---

[27] The lecture notes "*Vorlesungen über babylonische Astronomie Teil II, Die Theorie der Finsternisse*", Kopenhagen 1937, consist of pages numbered -1, 0, and 1-31 plus tables and corrigenda on pages numbered I-V.

[28] Neugebauer 1938a.

[29] Neugebauer 1936. The editors' introduction mentions the good attendance of the lectures which Neugebauer had held in the spring semester of 1934 which became his *Vorlesungen über Geschichte der antiken mathematischen Wissenschaften*.

[30] Neugebauer 1938b.

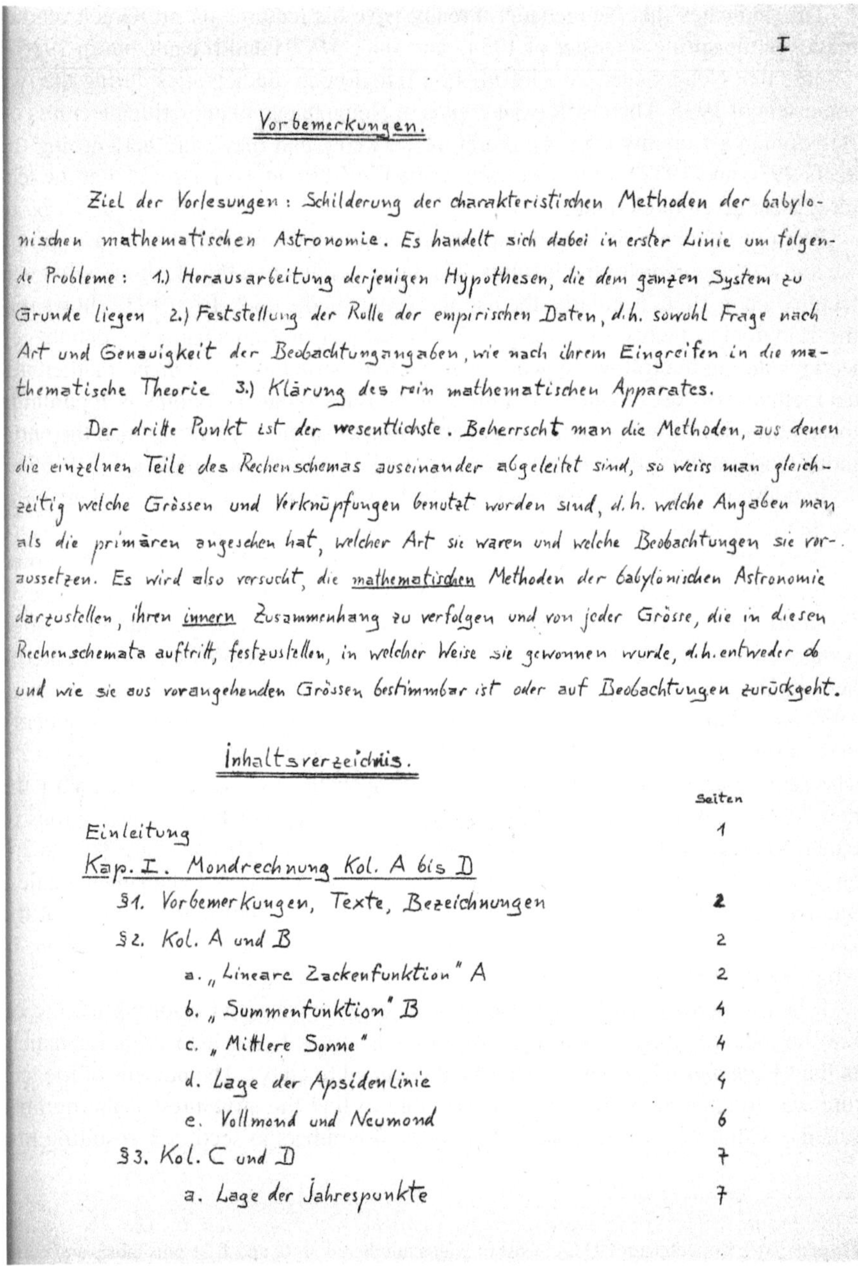
I

Vorbemerkungen.

Ziel der Vorlesungen: Schilderung der charakteristischen Methoden der babylonischen mathematischen Astronomie. Es handelt sich dabei in erster Linie um folgende Probleme: 1.) Herausarbeitung derjenigen Hypothesen, die dem ganzen System zu Grunde liegen 2.) Feststellung der Rolle der empirischen Daten, d.h. sowohl Frage nach Art und Genauigkeit der Beobachtungangaben, wie nach ihrem Eingreifen in die mathematische Theorie 3.) Klärung des rein mathematischen Apparates.

Der dritte Punkt ist der wesentlichste. Beherrscht man die Methoden, aus denen die einzelnen Teile des Rechenschemas auseinander abgeleitet sind, so weiss man gleichzeitig welche Grössen und Verknüpfungen benutzt worden sind, d.h. welche Angaben man als die primären angesehen hat, welcher Art sie waren und welche Beobachtungen sie voraussetzen. Es wird also versucht, die mathematischen Methoden der babylonischen Astronomie darzustellen, ihren innern Zusammenhang zu verfolgen und von jeder Grösse, die in diesen Rechenschemata auftritt, festzustellen, in welcher Weise sie gewonnen wurde, d.h. entweder ob und wie sie aus vorangehenden Grössen bestimmbar ist oder auf Beobachtungen zurückgeht.

Inhaltsverzeichnis.

Fig. 4 The first pages of Neugebauer's unpublished manuscript *Vorlesungen über Babylonische Astronomie Teil I*: *Die Berechnung der Neumonde*, Kopenhagen 1936, pp. I, II and 1 – 30

Fig. 4 (continued)

complete reconstruction of the scheme that was confirmed by other sections of the papyrus. Twelve years later Olaf Schmidt gave a lecture on Egyptian astronomy in the astronomical society; to those who had heard Neugebauer's lectures 12 years earlier, he expressed his regret that he could not explain the Papyrus with the same joy and conviction as he who had deciphered it.[31]

Neugebauer and His Pupil Olaf Schmidt

Olaf Schmidt was Neugebauer's first student. After following Neugebauer's lectures, he also started to read Theodosius, Ptolemy, Autolykos, and other Greek authors under Neugebauer's supervision. This study went on according to the following scheme: during the week, Schmidt would read Theodosius or Ptolemy and write a compilation, richly illustrated with diagrams, of the chapters he had worked through. Once a week he would present his work to his teacher. This meeting took place every Saturday at Neugebauer's home. They discussed the work through the whole morning, and at noon Schmidt was invited to a nice lunch prepared by Neugebauer's wife. The flat was on the upper floor of an apartment house in Hellerup, situated in a huge park with old trees. Neugebauer had been lucky to get a very nice apartment in the newly built houses of the Blidah-park (Fig. 5). According to Schmidt, they enjoyed a beautiful view over the Sound from the dining room.

Schmidt delivered his master thesis on "Theodosius' Sfærik" (Theodosius's Spherics) on November 23, 1937. Neugebauer had supervised the work; but he was

Fig. 5 Photo taken 2009 of some of the houses in the Blidah-park

[31] Schmidt 1950, 128.

Fig. 6 Otto Neugebauer

not a government official but had a professorship sponsored by grants. Therefore, due to formalities, it was Harald Bohr, Schmidt's mentor and teacher in mathematics, who signed in the official documents. Meanwhile Schmidt had become Neugebauer's assistant in Copenhagen. He followed Neugebauer to the United States in the summer of 1939. A stay of 1 year as Neugebauer's research assistant at Brown University was planned; but due to German invasion and occupation of Denmark, the stay ended up becoming 6 years. In 1943 Olaf Schmidt became Doctor of Philosophy in the Brown Mathematics Department, with a thesis *On the Relation between Ancient Mathematics and Spherical Astronomy*. He was thus Neugebauer's first PhD student. Schmidt became an instructor in the Mathematics Department while still continuing his work as Neugebauer's assistant.

Schmidt himself later became a professor of the history of the exact sciences at the Mathematical Institute of Copenhagen University. Here he accepted numerous students to write their thesis on a subject within the history of mathematics or astronomy. And he took over the habit of seeing each single student once a week while working on their thesis. For those of his students with whom he started a closer collaboration, he kept the nice custom of inviting them home for discussions with common lunch on Saturdays.

Altogether, Olaf Schmidt (Fig. 7) visited his friend Neugebauer five times in the USA: from summer 1939 to November 1945 he was alone in Providence; from July 1951 to June 1952 he visited Brown again together with his wife and oldest daughter; the next visit was from January 1966 to June 1966; and the last time was in the fall of 1976, when Schmidt stayed 2 weeks with Neugebauer (Fig. 6) in Princeton.

Fig. 7 Olaf Schmidt

The Neugebauer – Jessen Correspondence

Schmidt did not want anyone to read his correspondence with other scientists, so none of Neugebauer's letters have been kept or catalogued. The letters written to Harald Bohr before the Second World War were burned at the day of the German invasion of Denmark. However, the archives of the mathematical institute contain Neugebauer's correspondence with Børge Jessen. This correspondence consists in 27 letters written by Neugebauer from 1939 to 1978 and the drafts (or copies) of nine letters from Jessen to Neugebauer. Most are written by hand – only a few are typewritten.

The earliest letter is written in German and dates from October 8 1939. Here Neugebauer thanks for Jessen's textbook on geometry, which he and Schmidt had read with great interest. Then he asks for news from Copenhagen, from where he had had no news for many weeks. So he presumes that the shipment of the *Zentralblatt,* sent in August to Jessen, Bohr, and Fenchel in Copenhagen have also not arrived.

During the years 1951–1952, Neugebauer sent a telegram and nine letters to Jessen, all written in English. These are concerned with the early death of

Harald Bohr, with Olaf Schmidt's 1 year stay at Brown, and with the possibility of helping Asger Aaboe (1922–2007) to get a job in the States so that he could become a PhD student of Neugebauer. Several times Neugebauer asks for good photos of Harald Bohr, which he receives and posts in his offices in Brown and Princeton. Three letters from 1956/57 discuss where to publish a second edition of "The Exact Sciences in Antiquity", since Neugebauer "had the impression that Munksgaards are hesitating to bring out a new edition".

The remaining letters are written in the years 1960 and 1969. Here Neugebauer asks for Jessen's personal advice and help in connection with Neugebauer's wish to publish the comprehensive editions of the Astronomical Tables of Al-Khwārizmī and of the Pañcasiddhantikā of Varāhamihīra in the proceedings of *Det Kongelige Danske Videnskabernes Selskab*. Both works were published without delay in the proceedings, and the Academy accepted gratefully the economic support from the Institute for Advanced Studies, as proposed by Neugebauer in the following letter from January 30 1968:

> Dear Børge,
>
> I would very much like to ask you for your private advice in the following matter.
>
> I am just working on a commentary to a famous Sanskrit astronomical work Pañca-Siddhantikā of Varaha Mihira in collaboration with David Pingree of the Oriental Institute in Chicago. My question is the following: I am completely fed up with American publishers and the tyranny of their editors. I therefore would like this book published in Europe, and a publication by the Videnskabernse Selskab would be an ideal solution. I feel, however, that it would be unfair to ask for such a rather expensive publication. I think, however, that I could obtain support from the Institute for Advanced Studies which would contribute a substantial part of the expense. My question is: would the Academy accept such cooperation?

In many of the letters written to Jessen, Neugebauer recalls his good times in Copenhagen the memory of Harald Bohr:

> Countless times I think back to him [Harald Bohr] and to the years in Copenhagen which were among the happiest and most fruitful ones in my life. Unfortunately this lies now all in unreachable distance. [From a letter of Febr. 11 1970.]
>
> *I often think about his calm and wisdom when I see the insanity around us.* [Letter from May 7 1970.]
>
> *So frequently do I think of all his wisdom and kindness which we would need so much in this absurd world.* [Letter from May 26 1970.]

Neugebauer's last letter to Jessen was written January 1, 1978. He apologizes for not writing in Danish (which has deteriorated too much) and thanks Jessen for sending the written version of a lecture "*fra mine læreår*" [from my learning years (which also covered Jessen's time in Göttingen 1929)]. Neugebauer continues:

> How lucky have we been to have known all these excellent persons who have contributed so much to our lives. I often think back in gratitude to my years in Copenhagen and most of all to Harald, the unforgettable friend (Fig. 8).

Before leaving Denmark 1939 Neugebauer had shown his gratitude to Denmark in donating a spectacular cupboard (Fig. 8).

Fig. 8 When Neugebauer left Denmark in 1939, he donated a beautiful old cupboard from the seventeenth century. It stems from southern Germany and is made of pinewood with intarsia of walnut and decorated by locks made of ornamented wrought-iron. It is almost 9 ft high (2.67 m) and is exhibited on a prominent place in the Museum of Decorative Arts of Copenhagen [*Kunstindustrimuseet, Bredgade, København.*] A plate declares it as a gift from Dr. Phil. O. Neugebauer B 105/1939 and gives a long description of the cupboard

References

Becker, H., H.-J. Dahms, and C. Wegeler (eds.). 1998. *Die Universität Göttingens unter dem Nationalsozialismus*, 2nd ed. München: Saur.

Bohr, H., and O. Neugebauer. 1926. Über lineare Differentialgleichungen mit konstanten Koeffizienten und fastperiodischer rechter Seite. *Nachrichten der Gesellschaft der Wissenschaften zu Göttingen, Mathematisch-Physikalische Klasse* 8–22.

Dähnardt, W., and B.S. Nielsen (eds.). 1993. *Exil in Dänemark*. Heide: West-holsteinische Verlagsanstalt Boyens & Co.
Ginzel, F.K. 1909. *Mathematische und Technische Chronologie I*. Leipzig: J. C. Hinrichs'sche Buchhandlung.
Jessen, B. 1986. Matematikere blandt de tysksprogede emigranter. In *På flugt fra Nazismen. Tysksprogede emigranter I Danmark efter 1933*, ed. S. Steffensen, 87–91. København: C. A. Reizels Forlag.
Jessen, B. 1993. Mathematiker unter den deutschsprachigen Emigranten. In *Exil in Dänemark*, ed. W. Dähnhardt and B.S. Nielsen, 127–133. Heide: West-holsteinische Verlagsanstalt Boyens & Co.
Lange, H.O., and O. Neugebauer. 1940. Papyrus Carlsberg Nr. I: Ein hieratisch-demotischer kosmologischer Text. Det kongelige Danske Videnskabernes Selskab. *Historisk- filologiske Skrifter* 1(2).
Neugebauer, O. 1925. Review of *The Rhind Mathematical Papyrus BM 10057 and 10058*, by T.E. Peet. *Mathematisk Tidsskrift* A 66–70.
Neugebauer, O. 1930. Das Mathematische Institut der Universität Göttingen. *Die Naturwissenschaften* 18: 1–4.
Neugebauer, O. 1934a. *Vorlesungen über Geschichte der antiken mathematischen Wissenschaften, Erster Band, Vorgriechische Mathematik*. Berlin: Springer.
Neugebauer, O. 1934b. Über die Rolle der Tabellentexte in der babylonischen Mathematik. Det kongelige Danske Videnskabernes Selskab. *Mathematisk-fysiske Meddelelser* 12: 13.
Neugebauer, O. 1935–1937. *Mathematische Keilschrift-Texte*, 3 vols., *Quellen und Studien zur Geschichte der Mathematik, Astronomie und Physik* A 3. Berlin: Springer.
Neugebauer, O. 1936. Matematiske Kileskrifttekster. *Matematisk Tidsskrift B* 1–9.
Neugebauer, O. 1937. Overlevering af babyloniske matematiske metoder gennem græske skrifter. *Matematisk Tidsskrift B* 17–21.
Neugebauer, O. 1938a. Om babylonsk matematik. *Matematisk Tidsskrift A* 1–10.
Neugebauer, O. 1938b. Review of Lancelot Hogben. *Matematik for millioner, Matematisk Tidsskrift A* 50–57.
Neugebauer, O. 1938–1939. Über eine Methode zur Distanzbestimmung Alexandria – Rom bei Heron. *Det kongelige Danske Videnskabernes Selskab. Historisk-philologiske Meddelelser* 26.2 and 26.7.
Neugebauer, O. 1939. Die Bedeutungslosigkeit der, Sothisperiode' für die älteste ägyptische Chronologie. *Acta Orientalia* 17: 169–195.
Neugebauer, O. 1942. Egyptian planetary texts. *Transactions of the American Philosophical Society* 32(2): 209–250.
Neugebauer, O. 1952. Harald Bohr (1887–1951). *Year Book of The American Philosophical Society* 307–311.
Neugebauer, O., and A. Volten. 1938. Untersuchungen zur antiken Astronomie IV. Ein Demotischer astronomischer Papyrus (Pap. Carlsberg 9). *Quellen und Studien zur Geschichte der Mathematik, Astronomie und Physik B* 4: 383–406.
Pyenson, L. 1995. Inventory as a route to understanding: Sarton, Neugebauer and sources. *History of Science* 33: 253–282.
Ramskov, K. 1995. *Matematikeren Harald Bohr*. Aarhus Universitet, Unpublished Ph.D. dissertation.
Sachs, J., and G.J. Toomer. 1979. Otto Neugebauer, bibliography, 1925–1979. *Centaurus* 22: 257–280.
Schappacher, N. 1998. Das Mathematische Institut der Universität Göttingen 1929–1950. In *Die Universität Göttingens unter dem Nationalsozialismus*, 2nd ed, ed. H. Becker, H.-J. Dahms, and C. Wegeler, 523–551. München: Saur.
Schmidt, O. 1950. Ægyptisk Astronomi. *Nordisk Astronomisk Tidsskrift 1950 NR* 4: 121–134.

Siegmund-Schultze, R. 2009. *Mathematicians fleeing from Nazi Germany*. Princeton: Princeton University Press.
Steffensen, S. (ed.). 1986. *På flugt fra Nazismen. Tysksprogede emigranter I Danmark efter 1933*. København: C. A. Reizels Forlag.
Swerdlow, N.M. 1993. Otto E. Neugebauer (26 May 1899–19 February 1990). *Proceedings of the American Philosophical Society* 137(1): 139–165.

Otto Neugebauer and Ancient Egypt

Jim Ritter

Neugebauer at Göttingen

The Road to Göttingen

On August 21, 1919, Sergeant Otto Neugebauer, of the Austrian Alpine Artillery, left Cassino and the last of the series of Italian prisoner-of-war camps in which he had been interned since the end of the First World War, 9 months earlier, and headed back home to Graz in Austria.[1] He had lived there, since the death of his father before the War, in the house of his uncle and guardian Rudolf Schüssler and was now ready, at the age of nineteen, to resume his life. His uncle had already enrolled him, as a temporary measure and *in absentia*, in the Law Faculty at the Karl Franzen University of Graz for the Summer Semester[2] 1919, but Neugebauer was home in time to begin the Winter Semester 1919/1920 and could change his inscription to the Philosophical Faculty (i.e., Arts and Sciences) with an intended physics major.

[1] Information from Neugebauer's *Tagebuch* (The Shelby White and Leon Levy Archives Center, Institute for Advanced Studies: Otto Neugebauer papers/Box 13). I would like to thank the IAS Archivist, Christine Di Bella, for her generous help.

[2] Austrian, like German, universities operate on a two-semester system: "Winter Semester" and "Summer Semester". Traditionally, (with the corresponding month for Germany in parentheses) the first runs from the beginning of October to the end of February (March) while the second begins in March (April) and ends in September. Actual teaching time is generally:

Winter Semester: beginning of October (mid-October)—end of January (February)
Summer Semester: beginning of March (mid-April)—end of June (July)

J. Ritter (✉)
Institut de mathématiques de Jussieu-Paris Rive Gauche, UMR 7586,
Sorbonne Universités, UPMC-Univ Paris 06, Univ Paris Diderot,
Sorbonne Paris Cité, CNRS, Paris, France
e-mail: jim.ritter@wanadoo.fr

A. Jones et al. (eds.), *A Mathematician's Journeys*, Archimedes 45,
DOI 10.1007/978-3-319-25865-2_4

Besides the standard entrance-level courses in physics (experimental physics, electricity but also geophysics and physical chemistry) and mathematics (calculus and vector analysis) that he took that first semester, he also enrolled in a new course, "Introduction to Relativity Theory", offered by the young instructor (*Privatdozent*) of vector analysis, Heinrich Brell.[3] The subject was Einstein's new theory of gravitation, general relativity, having just been successfully tested by astronomers. This course—and its subject—so interested the young veteran that in the academic year 1920–1921 he took its equally newly-created continuation, "The Mathematical Foundations of Relativity Theory", by the invariant-theory specialist, Roland Weitzenböck. The same subject was certainly discussed in the "Seminar for Theoretical Physics", directed by the principal physics professor at Graz, Michael Radaković, a seminar in which Neugebauer was enrolled every semester after his first. By the summer of 1921, Neugebauer felt confident enough in the field to give a public two-part conference cycle on Hermann Weyl's recent first attempt at a unified field theory, conferences entitled "On the Extension of the General Theory of Relativity by Hermann Weyl" [*Über die Erweiterung der allgemeinen Relativitätstheorie durch Hermann Weyl*].[4] It is of interest to see, at the end of the lecture notes, a diagram created by the budding physicist (Fig. 2a), shown at the conclusion of his second lecture. It is, in fact, a résumé of the new theory, presented in the third edition of Neugebauer's source, Weyl's classic book *Raum Zeit Materie* (Weyl 1919: 251–253). Here Neugebauer has translated Weyl's three pages of discussion and calculation into a single diagram, showing how the Maxwell equations of electrodynamics (Equation 5, lower left) and the gravitational equations related to general relativity (Equation 9, lower right) arise out of a single choice for the action functional $\mathfrak{W}$ (Equation 1, top), and charts the logical connections among them. These classificatory diagrams were as innovative in theoretical physics as they were to become in Neugebauer's later work in the history of mathematics and astronomy, as we shall see. Another premonition of his later attitude is to be seen in the closing lines of the final lecture, where, referring to David Hilbert's views on axiomatization of mature theories,[5] Neugebauer says that "a completed science leads out beyond itself—or back to idealization" (*Eine vollendete Wissenschaft führt über sich selbst hinaus—oder zur Idealisierung zurück*). Though physics here is in his—and Hilbert's—mind, even after moving on to the history of ancient sciences, he will keep, as a central point of his work, the search for organizing principles from which alone, he will feel, any hope of a true understanding of ancient modes of thought can be found.

---

[3] The information about Neugebauer at Graz I owe to the kindness of Prof. Dr. Alois Kernbauer at the University Archives, Institut für Geschichte at the Karl-Franzens-Universität in Graz. For the Graz careers of Brell, Radaković and Weitzenböck see (Aigner 1985). Neugebauer's two public lectures are preserved in (The Shelby White and Leon Levy Archives Center, Institute for Advanced Studies: Otto Neugebauer papers/Box 13 "Vortrag über A.R.T. und Weyl"); the citation is from page 34.

[4] Otto Neugebauer papers, Box 1345, Courtesy of The Shelby White and Leon Levy Archives Center, Institute for Advanced Study, Princeton, NJ, USA.

[5] For Hilbert's view on the role of axiomatization in physics and mathematics, see (Corry 2004).

Fig. 1 Otto Neugebauer's student identification photograph from Ludwig-Maximilian University in Munich, 1922 (Courtesy of the Archiv der Ludwig-Maximilians-Universität, Munich)

Graz had suffered greatly from the War and the subsequent economic crisis and the University had entered into a difficult period in terms of personnel and material; both Brell and Weitzenböck left at the end of the Summer Semester 1921, the first to the Montanistische Hochschule in Leoben[6] and the second to the University of Amsterdam. Though Radaković stayed on until his death in 1934, he was uninterested in the new physics of quantum theory and relativity.[7] Clearly now aiming at further work in the new physics and, more particularly, in relativity, Neugebauer saw that there was no longer a future for him in Graz and, at the end of the Summer Semester, he too decided to move on.

His choice of the Ludwig-Maximilian University in Munich for the Winter Semester of 1921/1922 made sense for a bright young man interested in the new theoretical physics that had revolutionized the face of physics in less than a decade.[8] In 1921, Munich was one of the main centers in Germany, indeed in the world, for this field and boasted some of the most authoritative researchers, grouped around the Institute of Arnold Sommerfeld, author of the 'bible' of the quantum theory of the time, *Atombau und Spekrallinien*.[9] Besides Sommerfeld himself, these included his doctoral student, Wolfgang Pauli, who had already written one of the earliest and

[6] Now the Montanuniversität Leoben for Mining and Metallurgy.

[7] Radaković's research interests centered on classical mechanics and its applications to ballistics and meteorology.

[8] I should like to thank Dr. phil. Claudius Stein for information about Neugebauer's semester at Munich.

[9] For the "Sommerfeld School" of Munich, see (Seth 2010) and (Eckert 2013).

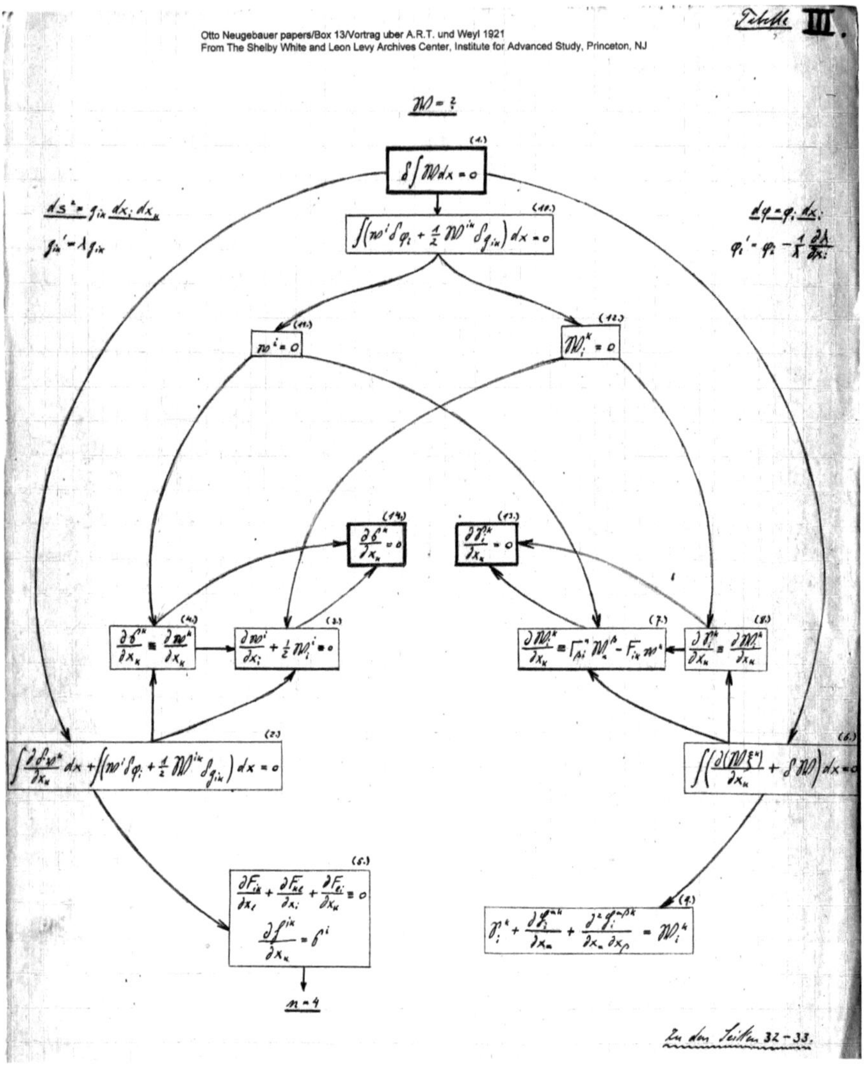

Fig. 2 (**a**) "Tabelle III" from Neugebauer's notes for his lecture on Weyl's unified theory, 1921 (Courtesy of the Shelby White and Leon Levy Archives Center, Institute for Advanced Study, Princeton, New Jersey)

Tabelle I.

'ḥ'-Rechnungen (Übersicht)

R 30
$10 = (\bar{\bar{3}} + \overline{10})\,x$
$x = 10 : (\bar{\bar{3}} + \overline{10})$

M 19
$(1 + \bar{2})\,x + 4 = 10$
$10 - 4 = 6 \quad x = 6 \cdot \bar{\bar{3}}$

M 25
$2x + x = 9$
$2 + 1 = 3$
$x = 9 : 3$

R 25
$x + \bar{2}x = 16$
$x = (16 : 3) \cdot 2$

R 26
$x + \bar{4}x = 15$
$x = (15 : 5) \cdot 4$

R 27
$x + \bar{5}x = 21$
$x = (21 : 6) \cdot 5$

R 24
$x + \bar{7}x = 19$
$x = (19 : 8) \cdot 7$

R 34
$x + \bar{2}x + \bar{4}x = 10$
$x = 10 : (1 + \bar{2} + \bar{4})$

R 32
$x + \bar{3}x + \bar{4}x = 2$
$x = 2 : (1 + \bar{3} + \bar{4})$

R 33
$x + \bar{\bar{3}}x + \bar{2}x + \bar{7}x = 37$
$x = 37 : (1 + \bar{\bar{3}} + \bar{2} + \bar{7})$

R 31
$x + \bar{\bar{3}}x + \bar{2}x + \bar{7}x = 33$
$x = 33 : (1 + \bar{\bar{3}} + \bar{2} + \bar{7})$

R 29 [2])
$\left[\bar{3}\left((x + \bar{\bar{3}}x) + \bar{3}\,(x + \bar{\bar{3}}x)\right) = 10\right]$

K 3 [3])
$x - (\bar{2} + \bar{4})\,x = 5$
$1 - (\bar{2} + \bar{4}) = \bar{4}$
$x = 5 \cdot 4$

R 28
$(x + \bar{\bar{3}}x) - \bar{3}\,(x + \bar{\bar{3}}x) = 10$
$10 : 10 = 1 \quad x = 10 - 1$ [4])

B 1 [5])
$x^2 + \left((\bar{2} + \bar{4})\,x\right)^2 = 100$
$1^2 + (\bar{2} + \bar{4})^2 = 1 + \bar{2} + \overline{16}$
$\sqrt{1 + \bar{2} + \overline{16}} = 1 + \bar{4}, \quad \sqrt{100} = 10$
$x_1 = 10 : (1 + \bar{4}) = 8$
$x_2 = (\bar{2} + \bar{4}) \cdot 8$

Fig. 2 (continued) (**b**) "Tabelle I" from Neugebauer's article on "Arithmetic and Calculational Techniques of the Egyptians" (Reference: Neugebauer 1930a, p. 307)

best surveys of general relativity and its extensions for the authoritative *Encyklopädie der mathematischen Wissenschaften* (Pauli 1921); Gregor Wentzel, Sommerfeld's Assistant, Adolf Kratzer, his former Assistant, while among his students at this time was Werner Heisenberg.[10]

Although Neugebauer had come to Munich to study modern theoretical physics, it was principally general relativity and its extensions that attracted him at this time, as it did so many other bright young scientists. But Munich was now concentrating on quantum theory, rather than relativity. Pauli had become disillusioned with relativity after the September 1920 meeting of the German Association of Natural Scientists and Physicians in Bad Nauheim: "None of the hitherto proposed theories—not even that of Einstein[11]—has so far succeeded in satisfactorily solving the problem of the elementary electrical quantum and it would be advisable to seek a deeper reason for this failure."[12] That deeper reason, Pauli felt, could only come

[10] Sommerfeld's book, first published in 1919, was to see eight thoroughly revised editions over the years. It was translated into English, Russian and French and used extensively as a textbook through the 1920s. Pauli's 237-page article on "Relativitätstheorie" (Pauli 1921), issued that same year as an independent monograph, was also to see a large number of reprints and translations and continues even today to serve as a textbook.

[11] The reference is to Einstein's first unified theory of 1919. For a discussion of this theory, and of Pauli's reaction to it and to that of Weyl, see (Goldstein and Ritter 2003: Section 2).

[12] *Keiner der bisherigen Theorien des Elektrons, auch nicht der Einsteinschen... ist es bisher gelungen, das Problem der elektrischen Elementarquante befriedigend zu lösen, und es liegt nahe,*

from quantum theory. That way lay the future and he persuaded the younger Heisenberg to follow him in abandoning relativity for the quantum. Since this accorded with Sommerfeld's own preference, when Neugebauer arrived a year later he found that Munich was a very exciting place to be indeed—if what one wanted was to do quantum physics.

Neugebauer, as was to become his way, did not follow the general trend. Like the others, he was inspired by Sommerfeld, but, unlike the others, his decision was to abandon physics entirely. As he succinctly put it a few years later, in his autobiographical note for his doctoral thesis: "Stimulated by lectures by Sommerfeld and such a pure mathematical content, I decided to devote myself to mathematics." [*Ich, angeregt durch Vorlesungen von Geheimrat Sommerfeld und solches rein mathematischen Inhaltes, beschloss mich der Mathematik zuzuwenden*].[13] But for mathematics it was Göttingen, not Munich, which was the shining city on the hill, and it was to Göttingen that he went at the end of that single Munich semester.

Mathematics at Göttingen

Otto Neugebauer transferred to the Georg-August University of Göttingen during the Summer Semester of the academic year 1922 when a new mathematics program was now to be in large part centered on the just-founded Mathematical Institute, under the direction of the recently returned Richard Courant, with his own special center of interest, the theory of differential equations.

Although the subjects on which Neugebauer was to be examined for his doctorate were Mathematical Analysis, Geometry and Physics,[14] judging by the courses he took, his own personal mathematical interests at this time lay more in pure mathematics and he took at least five courses with the analytic number theorist, Edmund Landau, one of the few active pure mathematicians at Göttingen.[15] Neugebauer's proverbial energy, as well as his abilities, both mathematical and administrative, were evident from the start. He was chosen by Courant to be his Assistant; first *außerplanmäßige* Assistant in October 1925, replacing Hellmuth Kneser who had been called to Greifswald, then promoted to *planmäßige* Assistant in June 1928, and finally Chief Assistant (*Oberassistent*) at the end of April 1930. He was also chosen to be one of the editorial assistants for the preparation of the first volume of Courant's classic textbook *Methods of Mathematical Physics* (Courant and Hilbert 1923).

---

nach einem tieferen Grund dieses Mißerfolges zu suchen. (Pauli apud Weyl 1920: 650)

[13] Lebenslauf, 22/9/1925, UAG Kur. P.A. Neugebauer, Otto, Band I, Hs N. I would like to express my deepest gratitude to Dr. Ulrich Hunger, at the Universitätsarchiv Göttingen for his most kind assistance.

[14] Bescheinigung, 23/4/1926, UAG Kur. P.A. Neugebauer, Otto, Band 1/II, Bescheinigung 23. April 1926.

[15] For Landau see (Schappacher 1987).

Egyptology at Göttingen

There was another prestigious intellectual center at Göttingen in the early 1920s: the Egyptological Seminar, the Göttinger Seminar für Ägyptologie, housed at this time in a small building, now the Michaelis-Haus. At some point before or after his arrival in Göttingen, Neugebauer had developed an appreciation for the culture of Ancient Egypt.[16] We know that he had read, in German translation[17] (Swerdlow 1993: 140), the most authoritative and influential book on Egyptian history and culture of the opening decades of the twentieth century: *A History of Egypt* by the American Egyptologist James Henry Breasted (1905). The book's section on Egyptian mathematics and astronomy, presumably Neugebauer's first view of the subject, is short and typical of the period's attitude on these subjects. For these reasons it is worth citing in full[18]:

> The science of the time, if we may speak of it as such at all, was such a knowledge of natural conditions as enabled the active men of this age to accomplish those practical tasks with which they were daily confronted. They had much practical acquaintance with astronomy, developed out of that knowledge which had enabled their ancestors to introduce a rational calendar nearly thirteen centuries before the rise of the Old Kingdom. They had already mapped the heavens, identified the more prominent fixed stars, and developed a system of observation with instruments sufficiently accurate to determine the positions of stars for practical purposes; but they had produced no theory of the heavenly bodies as a whole, nor would it ever have occurred to the Egyptian that such an attempt was useful or worth the trouble.
>
> In mathematics all the ordinary arithmetical processes were demanded in the daily transactions of business and government, and had long since come into common use among the scribes. Fractions, however, caused difficulty. The scribes could operate only with those having *one* as the numerator, and all other fractions were of necessity resolved into a series of several, each with *one* as the numerator. The only exception was two thirds, which they had learned to use without so resolving it. Elementary algebraic problems were also solved without difficulty. In geometry they were able to master the simpler problems, though the area of a trapezoid caused some difficulties and errors, while the area of the circle had been determined with close accuracy. The necessity of determining the content of a pile of grain had led to a roughly approximate result in the computation of the content of the hemisphere, and a circular granary to that of the cylinder. But no theoretical problems were discussed, and the whole science attempted only those problems which were continually met in daily life. (Breasted 1905: 100–101)

Non-theoretical, practical and elementary, Egyptian "science" was defined primarily by what it had not achieved. Nothing more needed to be said—and nothing more was said—on the subject in the more than 650 pages of Breasted's book. Only the remark on the rational calendar put in place "more than thirteen centuries before

[16] Interestingly enough, the Ludwig-Maximilians University in Munich was one of the few German universities to have offered courses in both Ancient Egyptian language and in Akkadian (Babylonian). The former, for specialists, were given by Friedrich von Bissing, then in his last semester in Munich and who had primarily archaeological interests. The Akkadian courses were given by Fritz Hommel who, in 1921–1922 offered a seminar on Old Babylonian texts. (LMU 1921). But we can find no trace of any interest expressed by Neugebauer in these areas before Göttingen.

[17] This was a revised version of the book, translated into German by the noted Heidelberg Orientalist Hermann Ranke in 1910 (Breasted 1910).

[18] We cite from the original English. No change was made to this section in the German edition.

the rise" of the earliest period of Egyptian written history seems, in the context, oddly positive and needs perhaps an elucidation, all the more so since, as we shall see, the question played an important role in Neugebauer's own later work. The previous year Eduard Meyer, the leading German ancient historian of that time, had published a monograph on Ancient Egyptian chronology (Meyer 1904) which brought much clarity to what had been a very confused subject. But from Meyer's point of view his most important contribution was elsewhere; the establishment of the exact date of the creation of the Egyptian calendar.[19]

> I do not hesitate to call the introduction of the Egyptian calendar on July 19, 4241 BC the first certain date in the history of the world.

This strangely precise (and much too early) date, accepted almost immediately by the larger Egyptological community, was to dominate—and mislead—Egyptian chronology up to the Second World War and Neugebauer's own work on the question.

At Göttingen, Neugebauer's interest in Egyptology was developed through his close contact with the current holder of the Chair of Egyptology, Kurt Heinrich Sethe, associate (*ausserordentlich*) professor since 1900, and full (*ordentlich*) professor since 1907. Arguably the greatest, and certainly, after his teacher Adolf Erman in Berlin,[20] the most influential Egyptological philologist of the period, Sethe was one of the rare Egyptologists to be interested in and—even rarer—to possess a full command of texts treating numbers and measures. Six years before Neugebauer's arrival in Göttingen, Sethe had published the (still) outstanding work on the subject: *On Numbers and Number Words among the Ancient Egyptians* (Sethe 1916) with the suggestive continuation of the title: *And What Is Be Learned from Them for Other Peoples and Languages: A Contribution to the History of Arithmetic and Language*. The innovative nature of this work was recognized immediately, at least among those few Egyptologists to whom the quantitative was not anathema. Although reviewed only twice[21]—by the German specialist on medical texts, Walter Wreszinski (1917), and by the English Egyptological philologist Battiscombe Gunn (1916)—the praise was laudatory in the extreme. Gunn, who with T. Eric Peet, was the only British Egyptologist to publish significant work on Egyptian mathematics,[22] wrote an unprecedented eight-page review in the major English-language Egyptological journal—and this for an enemy national in the middle of the First World War!—concluding:

[19] *... [S]tehe ich nicht an, die Einführung des ägyptischen Kalenders am 19. Juli 4241 v. Chr. als das erste sichere Datum der Weltgeschichte zu bezeichnen.* (Meyer 1904: 45). We shall return to this question in section "Eduard Meyer and the Sothic period".

[20] Erman, though he never worked himself on Egyptian mathematics, had as students a good number of those who did: Sethe in Germany, Breasted in America, Boris Aleksandrovich Turaev and Vasilij Vasil'evich Struve in the Soviet Union,.... (Erman 1929: 283).

[21] There is also (favorable) mention of the book by Günther Roeder (Roeder 1917: 281) and F. Llewellyn Griffith (Griffith 1917: 273) in their respective reviews of the year in Egyptology.

[22] Besides work on metrological questions, he wrote a 15 page review of Peet's edition of the Rhind papyrus (Gunn 1926) and, with Peet, was the first to analyze problems in detail from the then only partially-published Moscow mathematical papyrus (Gunn and Peet 1929).

> To have had from so great an authority no more than a detailed account of Egyptian numbers and number-words, satisfying the requirements of philology, semantics, and palaeography, would have been a most welcome accession to our science; but the author, owing to the compactness, so to speak, of his subject, has been able to deal with it comparatively, and has step by step adduced illustrative parallels, not only from the civilised peoples of ancient and modern times, but from primitive and illiterate races. In fact he has done what will be done one day for [religion, astronomy, medicine, etc.]. And in this he has marked an epoch: for the first time an Egyptologist of the very front rank has dealt with a part of his science not merely as a special contribution to Egyptology, but with a view to the light it throws upon similar phenomena among other peoples and languages. Because many of the conclusions he arrives at affect Indo-Germanic and Semitic number-lore, his book must be taken into account by whoever wishes in future to examine the origin and evolution of one of the most fascinating and curious fields of universal culture—the art of ciphering. (Gunn 1916: 279)

Neugebauer also was to appreciate this in Sethe's work, the comparative cross-cultural dimension of the history of the mathematical sciences remaining a central theme in his own research.

But Neugebauer had only a comparatively brief direct contact with Sethe. At the end of the Summer Semester of 1923, Sethe left Göttingen for the University of Berlin where he had been called to replace Erman in the chair of Egyptology. His successor at Göttingen, and former student, Hermann Kees, was chosen in large part because of his interest in Egyptian religious texts and beliefs, a subject which had become the center of Sethe's own interests for some years now. So, although Neugebauer kept in epistolary contact with Sethe, he was essentially on his own at Göttingen after 1923 in his interest in Egyptian mathematical texts and numeration.

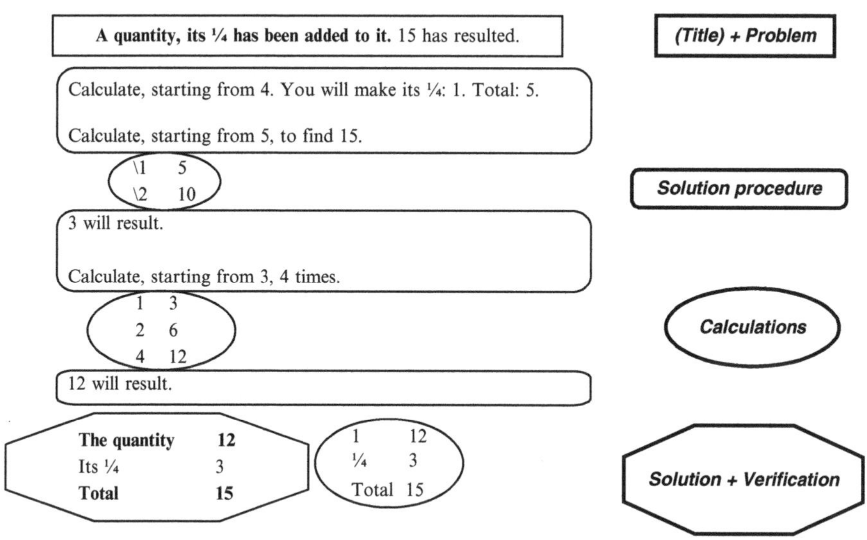

Fig. 3 Translation of Problem 26 of the Rhind Papyrus in the British Museum (BM 10057 + 10058)

The State of Egyptian Mathematical Historiography in 1923

Into what intellectual landscape did Neugebauer wander when he began his initiation into Egyptian language and culture? In particular what was known and understood about Egyptian mathematics by, say, late 1923? One important point needs to made at the outset concerning this domain, one that in large part determines its boundaries and nature and thus the range of possibilities for any research program in the field: the extreme paucity of sources and the elementary nature of their contents.

What mathematical texts were known to exist when Neugebauer began to interest himself in Ancient Egypt is easily resumed:

1 complete papyrus (Rhind papyrus), containing some 80 problems and a few tables, known since its first publication in 1877 (Eisenlohr 1877);
1 partially preserved papyrus (Moscow papyrus) of which only one problem had then been published (Turaev 1917)[23];
6 papyrus fragments, found in 1888–1889 at the town site of Lahun in the Fayum (Lahun fragments), containing the remains of 6 problems and 1 table (Griffith 1898);
2 papyrus fragments of unknown origin, with the remains of 4 problems, held in the Berlin Museum (Berlin fragments) and published at the beginning of the century (Schack-Schackenburg 1900, 1902);
2 wooden exercise tablets, supposedly found in the city of Akhmîm and purchased for the Cairo Museum (Akhmîm tablets), published in 1901 (Daressy 1901: p. 95–96, pl. LXII–LXIV) though not correctly understood until 1923 (Peet 1923b).

Thus from the whole pharaonic period—that is, the beginning of the third millennium BC to the end of the first millennium BC—we possess a total of two papyri, nine small fragments from perhaps three other papyri, a small leather roll, two wooden school tablets, and a potsherd. This was the full extent of known sources in 1923—and, to a great degree, it remains so today. True, the British Museum Leather Roll, containing two copies of a list of calculations, was unrolled in 1927 (Glanville 1927) and the Moscow Papyrus fully published in 1930 (Struve 1930). A few further minor discoveries and publications have followed: a school exercise on an ostracon from the Eighteenth Dynasty tomb of Senenmut at Deir el-Bahri (Hayes 1942: n° 153) and a fragment of papyrus with two incomplete problems (Imhausen and Ritter 2004: 91), found among the others at Lahun in 1890, but for some reason omitted in the original publication. But not another single pharaonic period mathematical source has come to light in all the years since, though post-pharaonic

[23] Though the Moscow papyrus had not yet been published in 1923, photographs of the complete papyrus were in Sethe's possession at Göttingen as part of his work on the Berlin Egyptian Dictionary.

Egyptian mathematical texts (Greek and Roman period texts in demotic writing) began to appear in 1959.

What was then known was the basic structure of Egyptian mathematical texts, that is, their division into tables on the one hand, and problems on the other; this last involving the posing of a problem, its solution, including the carrying out of individual calculations, and finally a verification of the final answer. A typical Egyptian problem is given in translation in Fig. 3, with the various parts of the problem indicated.[24] The solution procedure is expressed as a step-by-step algorithm involving one arithmetic command in each sentence.

Each step can be (though is not necessarily) followed by an explicit calculation effecting the operation. In the example here there is a calculation for each of the last two steps, division, then multiplication. The calculations are always carried out in a two-column arrangement, the division, $15 \div 5$ being carried out as follows:

\1	5
\2	10

The first column is initialized with 1 and the divisor 5 is placed facing it in the second column. The idea is to find entries adding up to the dividend, 15, in the second column, the sum of the corresponding entries in the first column then providing the answer. Various techniques, such as doubling, halving, finding the 2/3, etc. can be applied as required to the columns together.[25] Here one doubling has been carried out to generate the second line and this is sufficient since $5+10=15$ and these two lines are therefore checked (\); the corresponding entries of the first column 1 and 2 will then add up to the answer, 3. Multiplication is similar though inverse in the sense that one factor is placed at the head of the second column, the other factor then being sought in the first column with the corresponding entries of the second column providing the result.

Fig. 4 Translation of the beginning of the beginning of the "2/*n*" table for doubling fractions (from the papyrus UC 32159, Reference: Imhausen and Ritter 2004, p. 95)

2	3	$\overline{\overline{3}}$	2				
	5	$\overline{3}$	1 $\overline{\overline{3}}$	$\overline{15}$	$\overline{3}$		
	7	$\overline{4}$	1 $\overline{2}$ $\overline{4}$	$\overline{28}$	$\overline{4}$		
	9	$\overline{6}$	1 $\overline{2}$	$\overline{18}$	$\overline{2}$		
	11	$\overline{6}$	1 $\overline{\overline{3}}$ $\overline{6}$	$\overline{66}$	$\overline{6}$		
	13	$\overline{8}$	1 $\overline{2}$ $\overline{8}$	$\overline{52}$	$\overline{4}$	$\overline{104}$	$\overline{8}$
	15	$\overline{10}$	1 $\overline{2}$	$\overline{30}$	$\overline{2}$		
	17	$\overline{12}$	1 $\overline{3}$ $\overline{12}$	$\overline{51}$	$\overline{3}$	$\overline{68}$	$\overline{4}$
	19	$\overline{12}$	1 $\overline{2}$ $\overline{12}$	$\overline{76}$	$\overline{4}$	$\overline{114}$	[$\overline{6}$]
	21	$\overline{14}$	1 $\overline{2}$	$\overline{42}$	$\overline{2}$		

[24]The problem is number 26 of the Rhind papyrus, using the standard numbering, first established in its first edition (Eisenlohr 1877). The translation is my own.

[25]The techniques mentioned in most popularizations of Egyptian mathematics are erroneously limited to halving and doubling only. For a more accurate presentation see (Ritter 1995: 50–60).

The tables, used are primarily used as an aid in the carrying out of various steps in the calculations, where these are difficult or time-consuming. An example is given in Fig. 4, the beginning of the so-called 2/*n* table, used for the doubling of odd fractions.[26] Since neither a non-unit fraction (with the exception of 2/3) nor a repeated unit fraction is permitted, two or more distinct fractions must be sought whose sum is the desired doubled fraction. In the table, the numbers are arranged in three or more columns. The first column shows the divisor *n* (in the first entry only it shows both "dividend" 2 and divisor 3). This is followed by columns that alternately shows fractions of the divisor and their values (as a series of unit fractions). As an example, the third line starts with the divisor 7 in the first column, as it is the double of 1/7 that is to be expressed in unit fractions. This is followed in the second column by 1/4 and 1 + 1/2 + 1/4 and in the third by 1/28 and 1/4. This is be understood as 1/4 of 7 is 1 + 1/2 + 1/4, and 1/28 of 7 is 1/4. Since 1 + 1/2 + 1/4 and 1/4 add together to equal 2, the series of unit fractions representing the double of 1/7 is 1/4 + 1/28.

Neugebauer in Copenhagen I

In the spring of 1924, Neugebauer was invited to Copenhagen by the Danish mathematician Harald Bohr, the younger brother of the physicist Niels. This was the first, though as we shall see, far from the last time that the Danish capital was to play a major role in Neugebauer's life.

Bohr had been a frequent visitor to Göttingen in the preceding decade and a half. A quintessential arithmetic-analyst,[27] working on the summability of Dirichlet series and the Riemann zeta function conjecture, he had found a congenial teacher, then colleague in the older Göttingen number-theorist Edmund Landau, with whom he had coauthored a number of articles over the years. But in the year 1923 Bohr had found what was to be his true life's work, a domain of which he was essentially the creator, the theory of "almost periodic functions".[28] He decided to prepare an introduction in German to the new discipline, which was to be published in the Swedish international mathematical journal *Acta Mathematica.* Over the next 3 years, the "introduction" grew into a 258 page classic, published as three separate papers (Bohr 1924, 1925, 1926). Foreseeing the length, Bohr felt that his German,

---

[26] For a discussion of the uses of tables see (Imhausen and Ritter 2004: 95) and the references indicated there. We possess two copies of this table: one occupying almost all of the recto of the Rhind Papyrus (running from the double of 1/5 to the double of 1/101), the other illustrated here, being one of the Lahun fragments (doubling from 1/3 to 1/21).

[27] The term is Salomon Bochner's in (Bochner 1952).

[28] Bohr's original definition of an almost periodic function: A (complex-valued) continuous function f on $\mathbf{R}$ is called *almost periodic* if for any $\varepsilon > 0$, every interval on $\mathbf{R}$ of length greater than a given $l(\varepsilon)$ contains at least one point $\tau(\varepsilon)$ such that $|f(x+\tau) - f(x)| \leq \varepsilon$ for all x. That is to say, an almost periodic function is one which, on a sufficiently long interval, comes arbitrarily close again to any of its already attained values.

though fluent, would need to be vetted by a native speaker, well-versed in mathematics. As Neugebauer had just finished helping Courant with the editorial work on the latter's *Methoden der mathematischen Physik*, the Director of the Institute could now spare his Assistant for a few months to do similar work with Bohr.

Neugebauer's passage to Denmark produced two results, apart from Bohr's first two articles in *Acta*: one was a collaboration with Bohr on a mathematical project in the domain of almost periodic functions that bore fruit some 2 years later in a joint article—Neugebauer's only publication in mathematics—on ordinary linear differential equations with an almost-periodic source term (Bohr and Neugebauer 1926).

The second result was of considerably greater importance for Neugebauer's career. Harald Bohr was the editor (together with Tommy Bonnesen) of the major Danish mathematical journal *Matematisk Tidsskrift B*, a publication of the Danish Mathematical Society.[29] The Society also published *Matematisk Tidsskrift A*, more general and aimed at mathematics teachers. In 1923 Section B of the Journal had received for review a copy of T. Eric Peet's new edition of the Rhind Papyrus but had no reviewer, competent both in mathematics and in Egyptology, in view. Neugebauer's arrival in Copenhagen offered an unlooked-for opportunity that Bohr was quick to seize. As he explained it himself in an introductory footnote to Neugebauer's review:

> When the Journal's Series B accepted the new edition of "Ahmes' Reckoning Book" for review, the young German mathematician O. Neugebauer, who was living in this town and who has been engaged in a detailed study of Ancient Egyptian culture, was so kind as to accept my invitation to write a review for the Journal. Since however the Ancient Egyptian reckoning book presents such an extensive interest to everyone interested in mathematics, the Journal's editors have preferred to print the review in Series A, thereby making it available to a wider circle of readers.[30]

This, Neugebauer's first publication, already reveals what were to become the three leitmotifs of his early interest in ancient mathematics and, more specifically, Egyptian mathematics:

> "Egyptian mathematics was a simple affair ...", at least for us, if we consider the mathematical core of their problems. But against this there stands, at first glance, a dreadfully awkward and cumbersome calculational apparatus. While the basic number system is purely decimal (without place notation), that used for multiplication with integers is *dyadic*....[31] (Neugebauer 1925: 67),

[29] Still publishing, but since 1953 under the title of *Mathematica Scandinavica* and under the editorship of all five of the Mathematical Societies in Scandinavia.

[30] *Da Tidsskriftets Afdeling B modtog den nye Udgave af »Ahmes Regnebog« til Anmeldelse, opholdt sig her i Byen den unge tyske Matematiker Hr. O. Neugebauer der har beskæftiget sig indgaaende med den gamle ægyptiske Kultur, og som var saa elskværdig at imødekomme min Opfordring om at skrive en Anmeldelse til Tidsskriftet. Da den gamle ægyptiske Regnebog jo imidlertid frembyder saa stor Interesse for enhver matematisk interesseret, har Tidsskriftets Redaktioner foretrukket at lade Anmeldelsen trykke i Afdeling A for derved at gøre den tilgængelig en større Læsekreds.* (*apud* Neugebauer 1925: 66 note *).

[31] *"Egyptian mathematics was a simple affair ...", mindestens für uns, wenn wir den mathematischen Kern ihrer Probleme betrachten. Dem steht aber ein auf den ersten Blick erschreckend umständlicher und schwerfälliger Rechenapparat gegenüber. Während das zu Grunde gelegte*

- a concentration on calculational questions,
- a claim, still undeveloped, of the purely additive character of Egyptian mathematics,
- an emphasis on the necessity of avoiding distorting retrospective analyses using contemporary mathematical concepts.

Now Kurt Sethe had, the previous year, already written a review of the new Peet translation (Sethe 1924), and this too in a mathematical journal, the very prestigious *Jahresbericht der Deutschen Mathematiker-Vereinigung* (the *Yearly Report of the German Mathematical Society*). Sethe's article, cited in Neugebauer's own review, focusses on the same main issues as Neugebauer's, that is principally calculational techniques with fractions. Discussions of solution procedures for both men were essentially restricted to comments on the geometric area problems and this choice of focus certainly represents the mainstream of interest in the domain at the time.

The conclusion of this very positive review also puts forward, for the first time in the Neugebauer corpus, what he considered to be the primary qualities of good historical writing:

> In its strictly material way, in its avoidance of all artificial hypotheses, and the sure understanding of historical possibilities, it serves as an excellent guide for the reader.[32] (Neugebauer 1925: 70)

It is in precisely in the name of these values that Neugebauer himself will undertake his reconstruction of the domain of the history of ancient mathematics in the years to follow.

Neugebauer on Egyptian Mathematics and Astronomy

Egyptian Fractions—The Thesis

Neugebauer had arrived at an insight during the preparation of the review of Peet's edition of the Rhind papyrus which he now saw as the central point of a possible thesis. He had remarked in the review, what Peet had not seen—nor any previous commentator—that there was a possible explanation for a strange phenomenon occurring in Problem 48 of that papyrus. This text presents a circle inscribed in a square of side 9 *ḫt* (*khet*), followed by two calculations: the squaring of 9 and the squaring of 8. In view of preceding problems it is clear that here we have the

---

Zahlensystem ein rein dezimales ist (ohne Stellenwert der Zeichen) ist die bei der Multiplikation mit ganzen Zahlen zur Anwendung kommende Methode eine d y a d i s c h e....

The initial English citation is from the book under review (Peet 1923a: Preface). The reference to "dyadic" here is the doubling technique used in explicit calculations.

[32] *In ihrer streng sachlichen Art, in der Vermeidung aller künstlichen Hypothesen und dem sicheren Verständnis für historische Möglichkeiten ist sie geeignet dem Leser ein vorzüglicher Führer zu sein.*

calculation of the area of the two figures; that of the circle being calculated as the equivalent of a square of side 8/9 of the circle's diameter. Now the general practice in mathematical papyri is to use an abstract system of numeration[33] (without metrological units). Here however there is a unique use of the unit of area, *sṯ3t (setjat),* consistently attached to one of the factors. For instance, the calculation for the square reads:

\1	9 *setjat*
2	18 *setjat*
4	36 *setjat*
\8	72 *setjat*

Previous commentators had assumed that the Egyptians, like ourselves, would have conceptualized multiplication as a matter of that type; i.e., the calculation of an area as *length* × *length* = *area*, each factor then having potentially the units of length, only the result being in units of area. Taking the labeling of one of the factors by *setjat* seriously, Neugebauer saw it otherwise, as *pure number* × *area* = *area*:

> with the calculation of the surfaces there comes the clear impression of the original significance of an operation. Determining the surface of a square of side 9 cubits,[34] 9 square cubits will be taken nine times, much more correctly so than any mechanical calculation with "dimensions".[35] (Neugebauer 1925: 69)

The revelation for Neugebauer was that here was clear evidence that the Egyptians did *not* conceptualize as we do, that there was an *additive* basis to their way of thinking that could be of tremendous importance in establishing the evolution of mathematical thought. Such an insight was one of the determining factors in Neugebauer's change of direction in the period 1924–1925. From this point on he turned away from mathematical research and would not become a small fish in a large pond; with the training he had received from the Göttingen mathematicians on one side and Sethe on the other, he would bring a unique combination of gifts to the small but equally significant domain of the history of mathematics and there rout the sloppy thinking and fantastical reconstructions that, as he saw it, had plagued this domain for so long.

The Peet review had illuminated the two legs upon which a thesis could be written: the centrality of the fraction as the core of Egyptian mathematics and the insight that the additive idea lay at the basis of that culture's mathematical thinking. The thesis itself, "The Foundations of Egyptian Fractional Calculations" (*Die Grundlagen der ägyptischen Bruchrechnung*), was finished in the summer of 1926. The colors are announced from the opening page:

---

[33] For this system see (Ritter 2001: 121).

[34] Neugebauer makes an error here, the *setjat* is a square *khet*, not a square cubit *(meḥ).*

[35] *...bei der Berechnung von Flächengrössen zeigt sich die deutliche Empfindung für die ursprüngliche Bedeutung einer Operation. Um die Fläche eines Quadrates von 9 Ellen Seitenlänge zu bestimmen werden 9 Quadratellen neunmal genommen, also viel korrekter als ein mechanisches rechnen mit "Dimensionen".*

> the most important single result of this work is the insight into the exclusively additive foundations of Egyptian mathematics, which gives to the entire further development its specific character.[36] (Neugebauer 1926: 1)

And the larger significance of this is in the opening epigram, taken from the 1869 Tübingen Inaugural Lecture of a mathematician and historian of mathematics whom Neugebauer viewed with respect, Hermann Hankel:

> Whoever knows the history of mathematics and has an eye open for the typical character of an epoch cannot overlook the influence that a period and a tradition have exercised on the development of mathematical science. Were I allowed to document these facts in detail here, then you Gentlemen would recognize in the state of mathematics of any epoch the reflection of all the traditions which characterize that period. For even mathematics too is a science which is carried on by *men*, and every period, like every people, has only *one* spirit.[37] (Hankel 1869: 25)

Neugebauer saw in the additive structure of Egyptian mathematics just that culture-specific trait which characterizes the Egyptian spirit and would provide an explanation for their "dreadfully awkward" choice of calculational techniques.

Radical changes had come to history in recent times, Neugebauer argued in the opening lines of his thesis:

> It is not only Greek science that succumbed to the magic that a millennial past had cast over all Egyptian thought; modern [historical] science too has had to gradually learn to approach things "in an unprejudiced manner", and so to understand how they came to be. Next to the demand not to consider all phases of a process as simultaneous and equivalent for our understanding, there is another: so far as possible to guard against the uncritical transfer of common modern concepts and experiences to ancient conditions.[38] (Neugebauer 1926: 1)

With the demand for an "unprejudiced manner" to determine truly how things "came to be", one detects an echo of the famous motto of the positivist school in history, put forward by the great German historian Leopold von Ranke in his first book:

---

[36] *Das wichtigste prinzipielle Ergebnis der vorliegenden Arbeit ist die Einsicht in die ausschließlich additive Grundlage der ägyptischen Mathematik, welche der gesamten weiteren Entwicklung ihr spezifisches Gepräge gibt.*

[37] *Wer die Geschichte der Mathematik kennt und ein offenes Auge für den typischen Charakter einer Zeit hat, kann den Einfluss nicht übersehen, den Zeitcharakter und Volkseigenthümlichkeit auf die Entwicklung der mathematischen Wissenschaft ausgeübt haben. Wäre es mir erlaubt, diese Thatsache hier ausführlich zu begründen, so würden Sie, hochverehrte Herren, in dem Zustande der Mathematik in jeder Epoche den Reflex aller der Eigenthümlichkeiten erkennen, welche jene Zeit charakterisiren. Es ist eben Mathematik auch eine Wissenschaft, die von M e n s c h e n betrieben wird, und jede Zeit, sowie jedes Volk hat nur E i n e n Geist.*

Though Neugebauer cites only the last sentence, he clearly has in mind the whole paragraph.

[38] *Nicht nur die griechische Wissenschaft ist dem Zauber erlegen, den eine tausendjährige Vergangenheit über alles ägyptische Denken gebreitet hatte; auch die moderne Wissenschaft hat erst allmählich lernen müssen, „vorurteilslos" an die Dinge heranzutreten und sie so zu verstehen, wie sie geworden sind. Neben die Forderung, nicht alle Phasen eines Prozesses wie Gleichzeitiges und für unser Verständnis Gleichwertiges zu betrachten, tritt die andere, sich. soweit als irgend möglich davor zu hüten, uns geläufige moderne Begriffe und Anschauungen auf antike Verhältnisse kritiklos zu übertragen.*

> To history has been assigned the office of judging the past, of instructing the present for the benefit of future ages. To such high offices this work dare not aspire: it wants only to show what actually happened—*wie es eigentlich gewesen.*[39] (Ranke 1824: Vorrede)

If Neugebauer's history is informed by the positivist movement, his view of mathematics is firmly in the foundationalist tradition of the Göttingen school; the unification of mathematics under the aegis of set theory and the securing of its foundation through formalized logic.[40] This had already been the case in his work on Weyl's unified field theory and marked, for Neugebauer, the radical changes in recent mathematics:

> The mathematics of the last century too has experienced a major change; its "arithmetization"[41] has made great strides and investigations of its logical foundations have reached a decisive stage. Both directions have sharpened the capacity to single out the conceptual core of mathematical theorems and operations. Clearly history too must try to recognize the relation in which the concepts which are original to a given historical development stand to those concepts which, to modern ways of thinking, must have occupied this place from a purely logical point of view.[42] (Neugebauer 1926: 1)

The core that Neugebauer had detected is the additive structure, which he saw as essential for the development of mathematics.

The implication of all this was clear for Neugebauer: the time had come to rethink the history of mathematics in the light of these two revolutions in modern thought. Previous histories of mathematics had suffered from their unawareness of how this modern historical rigor, largely philological in origin, together with a steadfast refusal of anachronism, had outdated their traditional approach. This was particularly visible for Neugebauer in the few standard studies on ancient, particularly Egyptian, mathematics:

> I need only point to the arbitrary constructions of a M[oritz] Cantor or a [Friedrich] Hultsch. The critical and careful view of the historian of mathematics has, on this point, not succeeded in keeping up with contemporary philological work."[43] (Neugebauer 1926: 1)

---

[39] *Man hat der Historie das Amt, die Vergangenheit zu richten, die Mitwelt zum Nutzen zukünftiger Jahre zu belehren, beigemessen; so hoher Ämter unterwindet sich gegenwärtiger Versuch nicht: er will bloß zeigen, wie es eigentlich gewesen.*

[40] For mathematics at Göttingen see (Rowe 2004).

[41] For the arithmetization program in mathematics see (Petri and Schappacher 2007) and (Jahnke and Otte 1981).

[42] *Auch die Mathematik der letzten Jahrhunderte hat eine große Wandlung erfahren; ihre „Arithmetisierung" hat große Fortschritte gemacht und die Untersuchungen über ihre logischen Grundlagen sind in ein entscheidendes Stadium getreten. Beide Richtungen haben den Blick dafür geschärft, den begrifflichen Kern mathematischer Sätze und Operationen herauszuschälen. Es ist klar, daß auch die Geschichte gerade der Anfänge der Mathematik danach streben muß, das Verhältnis zu erkennen, in dem die Begriffe, die in der gegebenen geschichtlichen Entwicklung die ursprünglichen sind, zu jenen Begriffen stehen, die nach modernen Anschauungen diesen Platz in rein logischer Hinsicht einnehmen müssten.*

[43] *...ich brauche etwa nur auf die willkürlichen Konstruktionen von M. Cantor oder Hultsch hinzuweisen. Kritik und Sorgfalt der Historiker der Mathematik haben es in diesem Punkte nicht vermocht, mit der gleichzeitigen philologischen Arbeit Schritt zu halten.*

Neugebauer's view of the correct way of proceeding is clear from the way the thesis is organized. The first chapter of the thesis is far from being an empirical summary of Egyptian calculational practices, gathering together observations of the kind he made on Problem 48. It starts with a chapter on the "Conceptual Foundations of Egyptian Mathematics" where as he stated in a later article,

> pure linguistic and psychological considerations, … showed that the original number concept is not limited to '*natural*' integers, but includes, as equally legitimate elements, precisely the "natural" fractions, while the 'algorithmic' fractions appear first as the inevitable result of a true *calculational technique* ('division')….
>
> The linguistic-psychological motivation naturally reaches deeper than the so-to-speak empirical, that inferred from the calculational formalism….[44] (Neugebauer 1930a: 336 n. 115)

For Neugebauer this additive spirit infuses all calculational operations of the Ancient Egyptians. Addition, subtraction, and multiplication, of course, but also division which is nothing other than a multiplicative test and thus ultimately an addition as well. All efforts to reconstruct anachronistically a multiplicative thinking for the Egyptians are thus necessarily condemned to failure—this criticism was particularly aimed at previous attempts by Eisenlohr, Cantor and Hultsch to find a prime-number concept in the Rhind papyrus.

Turning then to the numbers appearing in Egyptian calculations, Neugebauer distinguishes two types: natural [*natürlich*] and algorithmic [*algorithmisch*]. The former includes the integers and a small number of (unit) fractions used in everyday life, such as 1/2, 1/3, 1/4,…, 1/8. These comprise the fundamental core of Egyptian numbers. All other fractions make up the algorithmic class and constitute the only *extension* of the number concept developed in Ancient Egypt.[45]

The second part of the thesis is dedicated to a detailed analysis of the calculations with fractions appearing in the Rhind papyrus. He shows how the various techniques, particularly the use of "auxiliaries" [*Hilfszahlen*], a name given to a particular calculational technique involving the addition of fractions, flow naturally from such an additive spirit.

This dismissal of the work of two of the leading historians of mathematics of the period (e.g., Cantor 1894 and Hultsch 1895) is not just the typical iconoclastic enthusiasm of the young doctoral student but remained a constant in Neugebauer's attitude to those who held views he considered insufficient or outdated. See section "The Neugebauer style" below. It has to be placed in the context of a general polemic against Cantor's cultural history, see (Lützen and Purkert 1993).

[44] *…rein sprachliche und psychologische Überlegungen … zeigten, daß sich der ursprüngliche Zahlbegriff nicht auf den der* „natürlichen" g a n z e n Zahlen *beschränkt, sondern als gleichberechtigte Elemente eben die „natürlichen" Brüche mit umfaßt, während die „algorithmischen" Brüche erst als zwangsläufiges Resultat einer wirklichen* Rechentechnik *erscheinen („Division")….*
Die sprachlich-psychologische Motivierung reicht natürlich tiefer als die sozusagen empirische aus dem Rechenformalismus erschlossene…

[45] For fractions, this corresponds more or less to the general distinction introduced in (Benoit et al. 1992: 11) between "special-status fractions" and "*quantièmes*", but there it is applied to other cultures as well and based on a distinction in the *written* forms of the elements of the two classes (special signs for the special-status fraction vs. systematic constructability for the *quantièmes*).

An example is given in Fig. 4, taken from the verification in Problem 34 of the Rhind papyrus. The transcription incidentally shows two major notational innovations introduced for the first time into the study of ancient mathematics by Neugebauer in his thesis: 1° the representation of fractions 1/*n* by a bar over the integer *n* (and a double bar over 3 for 2/3). and 2° the use of boldface type to represent numbers and words written in red ink in the Egyptian texts. Both conventions are now universally adopted.

The calculation shows a multiplication of 5 1/2 1/7 1/14 by 1 1/2 1/4, knowing that the answer should be equal to 10. Starting with the first factor and producing the second factor by successive halving, the problem is now to show that the sum of all the fractions in the second column is indeed equal to 10. The scribe has summed the integers and the halves and quarters to produce 9 1/2 1/8. The complement needed to reach 10, i.e. 1/4 1/8, must now be shown to be equal to the sum of the remaining six fractions from column 2. To do this the scribe chooses the smallest fraction, here 1/56 and puts down the integer 1 in red ink underneath (the "auxiliary"); the other fractions are then attributed numbers on the basis of their relation to 1/56. The two fractions 1/28 for example are given the auxiliary 2 since they are the double of 1/56. The sum of the auxiliaries are added, their sum here is 21. Finally to show their equality with 1/4 1/8, the corresponding auxiliaries of these last, namely 14 and 7 respectively are added together. Since their sum is also 21 the scribe has determined that the sum of the six remaining fractions are indeed equal to 1/2 1/4, and thus that the result of the multiplication is indeed 10.

But the main emphasis in the second section of the thesis is on an attempt to use the additive principle to determine the origin of the Rhind 2/*n* table; i.e., finding the algorithm that would have been used by the Egyptians to decompose fractions of the form 2/*n* for odd *n* between 5 and 101 into sums of unit fractions. Neugebauer's failure to find a single unique method for the construction of the table, a failure common to all approaches to this question both before him and since, leads him to postulate a complicated developmental history, in which several stages, separated in time and not well integrated one with the other, would explain the difficulties.

To finally establish his thesis on the non-existence of any multiplicative concept in Egyptian mathematics, Neugebauer has to confront the commonly appearing term Egyptian *sep*, translated as 'times', as in *sep* 2 ('twice'), *sep* 3 ('three times'), etc. and as such, generally used in the expression for the operation of multiplication: *waḥ-tep m* p *sep* q('Calculate, starting with *p*, *q* times') as we have seen in our example above (section "The state of Egyptian mathematical historiography in 1923"). In the thesis he suggests a way of dealing with this and refers the reader to a forthcoming article (Neugebauer 1926: 5). This article duly appeared the following year (Neugebauer 1927); a short note in which, taking a cue from Sethe's discussion of the word *sep* (Sethe 1916: 46), Neugebauer points out that in the syntactic structure of phrases involving *sep* and a number, *sep* is treated as an object, so saying *sep* 5 "five times" in Egyptian is no more multiplicative than saying "five apples" [my example]. Thus for Neugebauer the last objection to a completely additive conception of number is answered.

The thesis, defended on 21 April 1926 at Göttingen, had been reviewed by the two Egyptologists, Kurt Sethe (now in Berlin) and Hermann Kees, and by Richard Courant, who wrote the résumé of the opinions expressed.[46] In his summary of the reports by the Egyptologists, he stressed the recognition on their part that the thesis marked a turning point from the Egyptological perspective:

> Both of the reports offered from our colleagues, SETHE in Berlin and KEES, show that the accompanying work of Herr NEUGEBAUER represents a quite decisive advance in the unriddling of an important question and deserves high recognition as an achievement.[47] (Courant apud UAG Math. Net. Prüf. Neugebauer, Otto: 19 April 1926: 1)

But when Courant turns to the mathematical point of view, he sounds distinctly less enthusiastic—at least about the subject matter.

> From a mathematical point of view, it is naturally not a question of deep or difficult arithmetic problems; rather one can compare the solved exercises [of the Rhind papyrus] to the disentanglement of a complicated spectrum and the disclosure of the number-theoretic regularities contained therein.[48] (Courant apud UAG Math. Net. Prüf. Neugebauer, Otto: 19 April 1926: 1)

In closing, Courant takes up a motivation, shared by Neugebauer himself, that of "reclaiming" a central role for contemporary mathematicians in the comprehension and appreciation of past mathematics:

> It is not the first time that a mathematician has achieved something essential for Egyptology; the beginning of scientific Egyptology, as is well known, is closely linked to the names of some great French mathematicians. I am especially happy that again today a scholar, whom I know to be a mature mathematician, takes up the tradition.[49] (Courant apud UAG Math. Net. Prüf. Neugebauer, Otto: 19 April 1926: 2)

---

[46] UAG Math. Nat. Prüf. Neugebauer, Otto: 19 April 1926.

[47] *Die bei den von Kollegen S e t h e in Berlin und Kollegen K e e s erstatteten Gutachten zeigen, dass die beiliegende Arbeit von Herrn N e u g e b a u e r vom Standpunkte der Ägyptologie aus einen ganz entschiedenen Fortschritt in der Enträtselung einer wichtigen Frage bedeutet und als Leistung hohe Anerkennung verdient.*

[48] *In mathematischer Hinsicht handelt es sich natürlich nicht um tiefe oder schwierige arithmetische Probleme, vielmehr kann man die gelöste Aufgabe mit der Entwirrung eines komplizierten Spektrums und der Aufdeckung der darin enthaltenen zahlentheoretischen Gesetzmässigkeiten vergleichen.*

The reference here is to the then current semi-empirical methods of analyzing atomic spectra by using the "old" Bohr-Sommerfeld quantum theory to derive numerical relations among the various lines of atomic spectra. The Göttingen physics department, with Max Born and his student Friedrich Hund, was then at the forefront of just such an approach; see (Hund 1927).

[49] *Es ist nicht das erste Mal, dass ein Mathematiker wesentliches für die Ägyptologie geleistet hat; die Anfänge der wissenschaftlichen Ägyptologie sind ja bekanntlich eng mit den Namen einiger grosser französischer Mathematiker verbunden. Es freut mich ganz besonders, dass heute wieder ein mir als reifer Mathematiker bekannter Gelehrter die Tradition aufnimmt.*

"French mathematicians" refers to those mathematicians and engineers who participated in Napoleon's Egyptian Expedition or in the edition of the monumental *Description de l'Égypte* which codified its scientific results: Gaspard Monge, Jean-Joseph Fourier, Edmé François Jomard, and Pierre Simon Girard.

One hears here an echo of the "nostrification" for which Göttingen was well-known: claiming for mathematics (and where possible, Göttingen mathematics) that which had been done in other domains.[50]

Neugebauer's received his doctorate with his thesis adjudged "excellent (*ausgezeichnet*)" and his oral defense "very good (*sehr gut*)".[51] For the completion of the Sethean program there remained now only a comparative study of the Mesopotamian material, a good, quick Habilitation subject. After a summer partially passed in the South of France with mathematician friends from Göttingen, Hans Hopf and the Russian Pavel Sergeyevich Alexandrov,[52] Neugebauer turned to the study of Sumerian with Anton Deimel in Rome.

Egyptian Mathematics—The **Quellen und Studien** *Articles*

The period following Neugebauer's successful thesis defense marked a period of particularly intense intellectual and organizational activity in his life. Besides his research work, he continued as Courant's Assistant, but also developed important ties with the Berlin scientific publisher, Julius Springer, with whom he would in the following years develop and act as editor for no less than four important scientific series. In 1929, with the Kiel historians of Greek mathematics, Otto Toeplitz and Julius Stenzel, he founded the "Sources and Studies on the History of Mathematics, Astronomy and Physics" (*Quellen und Studien zur Geschichte der Mathematik, Astronomie und Physik*) series on the history of science.[53] Two years later, he became the founding editor of the major review journal for mathematics, the "Journal for Mathematics and Its Neighboring Areas" (*Zentralblatt für Mathematik und ihre Grenzgebiete*)[54] and in 1932, its associated monograph series "Results in Mathematics and Its Neighboring Areas" (*Ergebnisse der Mathematik und ihrer Grenzgebiete*). Finally in 1933, together with the engineer Wilhelm Flügge, he created the *Zentralblatt für Mechanik*.

But Egypt was not forgotten. Neugebauer continued his work on the unpublished Moscow mathematical papyrus, to which he had access thanks to Sethe's photographs and the Russian contacts of his friend Alexandrov. In 1928, he spent time in Leningrad with Vasilij Vasil'evich Struve, who was to edit the papyrus as the first

---

[50] For the example of such an annexation in the case of Minkowski and special relativity, see (Walter 1999).

[51] UAG Kur. P.A. Neugebauer, Otto: Band 1/II, Bescheinigung 23. April 1926.

[52] See (Bečvářová and Netuka 2010: 16).

[53] This consisted of two parts, series A to publish editions of ancient texts—the first volume was Struve's edition of the Moscow papyrus (Struve 1930)—and series B for research articles in the subject.

[54] Now called *Zentralblatt MATH*, this was created to replace the venerable but ailing *Jahrbuch über die Fortschritte der Mathematik*. For the history of these two journals see (Reinhard Siegmund-Schultze 1993).

volume of *Quellen und Studien A*. Most importantly, he was working on a pair of long articles, some 120 pages in all, on Egyptian mathematics to be published in the first volume of *Quellen und Studien B* and which would turn out to be, in a sense, his final word on Egyptian mathematics.

At the end of the introduction to the first of these articles, which basically resumes and extends his thesis, Neugebauer acknowledges his sources of inspiration.[55] The first, as we have seen already, is Sethe, to whom the article is dedicated "in admiration and gratitude":

> I must in conclusion recall three books which have been of particular influence... The first is SETHE's *On Numbers and Number Words...*, which in particular first opened up for me the possibility of historical ways of looking at concepts of number and fraction which were not founded on mere "intuition".[56] (Neugebauer 1930a: 303)

The remaining two influences on the young Neugebauer in these first publications are the historian of Egyptian art Heinrich Schäfer and the French anthropologist Lucien Lévy-Bruhl. As Neugebauer goes on to explain:

> Then [follows] H. SCHÄFER's *On Egyptian Art, Particularly Drawing...*, especially through its principal insights into the Egyptian conception of "perspective" style, and finally LÉVY-BRUHL's *How Natives Think*, which SCHÄFER's work completes in a much more general manner.[57] (Neugebauer 1930a: 303)

It was probably Sethe who introduced Neugebauer to the seminal work of Schäfer, then director of the Egyptian Museum in Berlin, as Sethe and Schäfer were longtime friends, having been students together under Erman in Berlin.[58] And as

---

[55] Though in large part composed in the autumn of 1928, Neugebauer states in a note (Neugebauer 1930a: 301) at the beginning of the article that as he cites the Moscow Mathematical Papyrus, he had wanted to wait until the official publication of that text in (Struve 1930).

[56] *Ich muß zum Abschluß dreier Werke gedenken, die von bestimmendem Einfluß auf die eigentlichen Grundlagen meiner Anschauungen geworden sind .... Das erste S e t h e s „Von Zahlen und Zahlworten"..., das mir überhaupt erst die Möglichkeit einer nicht auf bloße „Intuition" gegründeten geschichtlichen Betrachtungsweise des Zahl- und Bruchbegriffs erschlossen hat.*

[57] *Dann H. S c h ä f e r s „Von ägyptischer Kunst besonders der Zeichenkunst"..., vor allem durch seine prinzipiellen Einsichten über die ägyptische Auffassung der „perspektivischen" Darstellungsweise und schließlich L é v y-B r u h l s „Fonctions mentales dans les sociétés inférieures"..., das S c h ä f e r s Werk in ganz allgemeiner Hinsicht ergänzt.*

The earliest, and best, English-language review of the version of Schäfer's book used by Neugebauer (Schäfer 1919) is to be found in (Davies 1921), with a detailed summary of the first edition. Only a much later, considerably revised (first by the author, then by the editor and the translator) version was translated into English in 1974 under the title *Principles of Egyptian Art*. Lévy-Bruhl's now very perjorative-sounding French title (Fonctions mentales dans les sociétés inférieures) is here rendered into English by that under which the authorized English-language version of the book was published in 1925.

[58] As Erman later recalled about them: "And within the individual generations [of my students] harmony and friendship ruled, and I can only picture many of my students in their youth, like... Sethe and Schäfer, as pairs." [*Auch innerhalb der einzelnen Generationen herrschte Eintracht und Freundschaft, und manche meiner Schüler wie...Sethe und Schäfer kann ich mir in ihrer Jugend nur als Paare denken.*] (Erman 1929: 283).

Neugebauer himself points out, Schäfer led him to Lévy-Bruhl and the 'primitive mind'.

The impact of these formative influences can be seen in the conclusions that Neugebauer himself put forth as central to his project.

> I would like to set up a sentence of Lévy-Bruhl's as almost a leitmotiv for the setting for pre-Greek, and particularly Egyptian, mathematics: "Their mentality does not lend itself well to the operations familiar to us; but through means which are proper to itself, it knows how to obtain, up to a certain point, the same results."[59] (Neugebauer 1930a: 303)

In his thesis Neugebauer had emphasized two aspects of the importance of Sethe's 1916 groundbreaking work, the possibility of achieving a level of *rigor* in philological analysis comparable to that in mathematics and the introduction of a true *historicity* of number concepts. In the 1928–1930 paper, he was equally influenced by another aspect, the importance of cross-cultural *comparativeanalysis* to tease out possible laws of mathematical development. In the pursuit of this third point of the program, he went to Rome to study Assyriology with Anton Deimel in view of doing in that field what he had begun in the Egyptological domain, but with the essentially important addition of a potentially much larger corpus of unpublished texts to edit and study. On 12 November 1927 he obtained his *Habilitation* with a thesis on the origins of the sexagesimal system in Mesopotamia and was launched on the massive undertaking of the *Mathematische Keilschrifttexte* edition.

If his thesis had been an attempt to found new ideas about the nature of mathematical activity on a historical basis, the new pair of articles on the subject seek to analyze the totality of known Egyptian mathematics, divided into "Arithmetic and Calculation of the Egyptians" [*Arithmetik und Rechentechnik der Ägypter*] (Neugebauer 1930a) and "The Geometry of Egyptian Mathematical Texts" [*Die Geometrie der ägyptischen mathematischen Texte*] (Neugebauer 1931). Such a division for him now however is purely conventional, denoting merely a separation into domains of *application* of one and the same fundamental Egyptian—but also pre- and early Greek—mathematical nucleus, a "unitary developmental field for the mastery of calculational (that is, 'arithmetic-algebraic') problems" [*einheitliches Entwicklungsfeld zur Beherrschung rechnerischer (d. h. „arithmetisch-algebraischer") Aufgaben*] (Neugebauer 1930a: 302).

The main thrust of his work is now to unveil the organizational principles at work in the Egyptian mathematical texts. The addition of "algebraic" to "arithmetic" in the above quote is symptomatic of a certain shift in his thinking since the thesis. The organizing principle that he will adopt in these two articles is algebraic and thematic. Thus, for example, the *ʿḥʿ (aḥa)* problems (those using this term meaning

[59] *Einen Satz L é v y—B r u h l s möchte ich geradezu als Leitmotiv für die Einstellung zur vorgriechischen, insbesondere ägyptischen Mathematik, hinstellen: „Leur mentalité se prête mal aux opérations qui nous sont familières; mais, par des procédés qui lui sont propres, elle sait obtenir, jusqu'à un certain point, les mêmes résultats".*

The citation from Lévy-Bruhl is to be found at (Lévy-Bruhl 1922: 205), in the chapter entitled "The pre-logical mentality and its relation to numeration".

'quantity' in their formulation, like Problem 26 of the Rhind papyrus, Fig. 3), with examples in the Rhind, Moscow, 'Kahun' (Lahun) and Berlin texts, represent a particular method, essentially algebraic in nature,[60] and expressible in modern form as a linear equation $x+ax+bx+c=d$, where a and b are fractions, and c and d integers. As is common with Neugebauer, at least since his work in Graz on unified theories, recognizing the correct underlying organizational principle allows a classification of the material. The Neugebauer classificatory diagram for the *aḥa* problems is reproduced here as Fig. 2b, organized, not by the specific form of the equation, but by the "consistent application" [*konsequent Anwendung*] (Neugebauer 1930a: 308) of the solution method.

Egyptian 'geometric' problems as laid out in the second article (Neugebauer 1931) are in fact metrological texts and thus akin to the arithmetic problems treated in the first article. Neugebauer's organizational principle in their treatment however is by geometric form: figures, surfaces, volumes grouped under their particular forms, as was traditional in their treatment by previous authors. The solution methods being straightforward for the most part, they all serve as a means of classification. But Neugebauer's net is thrown wide and he includes (published) material not previously brought together in this context, including a late second millennium literary school text (Papyrus Anastasi I), a Ptolemic-period papyrus in Greek that he felt reflected earlier Egyptian practice, and architectural diagrams taken over from Schäfer's work. *Exhaustivity*, the marshaling of *all* available evidence, was another trademark of Neugebauer which marked already his early Egyptian work.

These two articles were to be the last substantive work that Neugebauer ever published on the subject of Egyptian mathematics. When he spoke of it later, either in his *Vorgriechische Mathematik* (Neugebauer 1934) or in *The Exact Sciences in Antiquity* (Neugebauer 1957) or in a small number of occasional book reviews, it was never a question of more than a selection of the points he had already made in full before 1931.

The Return to Egypt—Neugebauer and Egyptian Astronomy

On January 26, 1932 Otto Neugebauer was named Associate (*außerordentlicher*) Professor[61] and seemed set to create a world center for the study of ancient science at Göttingen. Almost exactly a year later, Adolf Hitler was named Chancellor of Germany and on April 7, 1933 his government promulgated the "Law for the Restoration of the Civil Service" [*Gesetz zur Wiederherstellung des Berufsbeamtentums*], banning from public employment, and thus from employment at any German university, all non-Aryans and political opponents of Nazism. Neugebauer, being neither a Jew nor a Communist, was thus not directly affected

---

[60] Neugebauer specifically opposes the interpretation of these problems as examples of a 'false position' method, common in earlier work (Neugebauer 1930a: 309).

[61] UAG Kur. P.A. Neugebauer, Otto: Band III, Document 11.

but the writing was clearly on the wall and, with the idea of awaiting a more favorable climate in Germany, he activated his Copenhagen contacts. Harald Bohr obtained for him an offer of a position at the University of Copenhagen. On October 18, 1933 he asked for a 1-year leave of absence from his position at the University to be effective from the beginning of 1934. This was granted on November 1, 1933, with the condition that he relinquish his Göttingen salary for this period.[62] He was to renew this formality every year until May 28, 1936 when, it being clear that far from abating, the Nazi pressure on academic life was increasing, he announced that he would no longer ask for an extension of his leave of absence and thus resigned from his position at the University of Göttingen.[63]

Once settled in Copenhagen for the Second Semester of 1933/34, Neugebauer taught a course on "pre-Greek (i.e., Egyptian and Babylonian) mathematics",[64] which, written up under the same title (Neugebauer 1934), would become the first volume of a planned trilogy: *Lectures on the History of the Ancient Mathematical Sciences* [*Vorlesungen über Geschichte der antiken mathematischen Wissenschaften*], with a second volume planned on pre-Greek mathematical astronomy and a third on Greek science. The treatment of Egyptian mathematical texts in the first volume is essentially a résumé of the two *Quellen und Studien* articles and will constitute the last contributions of Neugebauer to Egyptian mathematics, aside from the later reworking of the same material in his popular *The Exact Sciences in Antiquity* (Neugebauer 1957).

With the second volume of the *Vorlesungen* in view, Neugebauer had been turning his attention more and more to ancient astronomical texts. Naturally he was principally interested in the abundant and advanced Babylonian material. However he did find the time, in conjunction with a colleague, Aksel Volten, of the Egyptological Institute of the University of Copenhagen and specialist in Demotic,[65] to publish, in what was to be the last fascicule of the last volume of *Quellen und Studien*, a Roman period Egyptian astronomical text, papyrus Carlsberg 9, found in a temple library in Tebtunis and dating from around 144 AD (Neugebauer and Volten 1938).[66] Unlike most Egyptian astronomical papyri which are mythological in content it deals with a 25-year lunar cycle, serving to connect the religious lunar calendar with the civil solar one, presented in the form of tables generated by a numerical subtraction algorithm. Though it contains a list of the zodiacal signs, a borrowing

---

[62] Note from Schnoering at the Ministry to Kurator Valentiner. UAG Kur. P.A. Neugebauer, Otto: Band I, Document II 16.

[63] Letter from Kurator Valentiner to Neugebauer. UAG Kur. P.A. Neugebauer, Otto, Band I, Document II 17.

[64] "Über vorgriechische Mathematik", course notes in The Shelby White and Leon Levy Archives Center, Institute for Advanced Studies: Otto Neugebauer papers.

[65] Demotic refers to both a very late stage (seventh century BC to fifth century AD) of the ancient Egyptian language, and to its corresponding written form, an extremely cursive form of hieratic.

[66] Succinctly reedited in (Neugebauer and Parker 1969: Text vol. p. 220–225, Plate vol. p. 65).

from Hellenistic astronomy, it was recognized by Neugebauer as stemming from an indigenous Egyptian tradition, a rarity among extant late Egyptian sources.[67]

One other late text, from the same Tebtunis temple library, connects up to an even older native Egyptian tradition: papyrus Carlsberg 1, that Neugebauer and the director of the Egyptological Institute, Hans Ostenfeld Lange, published as a monograph a little over a year later (Lange and Neugebauer 1940).[68] The papyrus contains a hieratic text with a demotic commentary; large parts of this text exist also in hieroglyphic form on the ceilings of the royal tombs of Seti I and Ramesses IV, dating back to the thirteenth and twelfth centuries BC respectively. This time, the context is mythological: the sky goddess Nut and the appearance and disappearance of the stars. Though all aspects of the text were treated in the study, it was clear that Neugebauer's main interest in it lay in "Chapter E" the marking of the hours of any night by the culmination (in the earlier versions, the rising) of one of a group of thirty-six constellations, known today by the Greek name of "decans".

Though the *Vorlesungen* were never completed,[69] the material, in an enormously expanded form, was published in various forms over the following years by Neugebauer. In the Egyptian case this was in the shape of a dozen articles culminating and subsumed in the monumental three volumes of *Egyptian Astronomical Texts*, written with his Egyptological partner at Brown University, Richard A. Parker (Neugebauer and Parker 1960, 1964, 1969).[70]

The Neugebauer Style

Once the final decision to abandon mathematics (as a research but not as a social and intellectual milieu) had been taken, Neugebauer's future agenda was fixed. There were to be two main tasks: one negative, clearing the dead wood of outworn ideas and overly speculative theories, often accepted simply because the necessary compelling arguments and marshaling of evidence had not been carried out; the other positive, replacing these with accurate, justifiable—and justified—explanations, using the intensive rigor of the new philology and the new mathematics and the extensive breadth of Sethe's cross-cultural comparative method. We have already seen a number of examples of the positive side of this approach in the case of Egypt. The negative, critical aspect was to create for Neugebauer an image of a

[67] The original of the text dates back to the fourth century BC (Parker 1950: 24–29).

[68] See now (Neugebauer and Parker 1960: 36–94, pls. 36–42).

[69] Swerdlow (Swerdlow 1993: 147) argues plausibly that Neugebauer no longer felt that its original program of presenting Egypt and especially Mesopotamia as simply prolegomena to Greek science was in any way adequate or pertinent.

[70] Unlike his work on Egyptian and Babylonian mathematics or on Babylonian astronomy, Neugebauer never published alone a major article or monograph on Egyptian astronomy; all are cosigned with an Egyptologist, Volten, Lange, Parker. It is unclear if and to what extent he ever learned Demotic in which the majority of the astronomical texts were written.

redoubtable polemicist, always ready to take on other prominent historians, dead—like Moritz Cantor or Eduard Mayer—or alive. To understand the dynamic of Neugebauer's interventions into questions of ancient science, we shall review here two examples of the Neugebauer interventionist style and of its reception in the professional milieux which dealt with Ancient Egyptian mathematics and astronomy.

Eduard Meyer and the Sothic Period

Since the star Sirius constitutes one of the decans, that marking, in principal, the first day of the year, that is, the beginning of the first month of the Inundation season, Neugebauer's studies of the papyri Carlsberg 1 and Carlsberg 9 led him directly to the nature of the relationship between the Egyptian civil and lunar calendars; and this in turn led to the question of the 'Sothic cycle' so important for Eduard Mayer's Egyptian chronology.

Eduard Meyer, who, as we have seen (section "Egyptology at Göttingen"), believed that he had fixed the first certain date in Antiquity, July 19, 4241 BC, had presented Egyptology with a major problem. His argument turned on the so-called "Sothic cycle", the relation between the Egyptian civil year of 365 days and the astronomically observed annual heliacal rising of the bright star Sothis (our Sirius). There is a discrepancy of ¼ of a day per year between the Egyptian civil calendar of 365 days and the time for the Earth to make one complete revolution around the Sun, and thus $4 \times 365 = 1460$ years between coincidences of the astronomical heliacal rising of Sirius and any fixed date of the civil calendar. Based on the few dated mentions of the gap between the Egyptian New Year and the rising date of Sirius in Egyptian texts, Meyer assumed that the Egyptians must have fixed their civil calendar on a date when the rising of the Nile (the first day of the first month of the Inundation Season, our July 19) and the rising date of Sirius coincided. Since these occurrences are separated by 1460 years, the only acceptable candidates for that event were 2781 BC or 4241 BC. The former could be ruled out since it was known that the civil calendar was in use before the Old Kingdom (during which the first date falls) and thus the minimal date for the introduction was 4241 BC. Since Mayer assumed that such a fixing of the calendar year required repeated and accurate observations of the rising of Sirius, he had to postulate the existence of a corps of trained astronomers in Egypt at a time a full millennium earlier than previously accepted dates for the invention of writing and the creation of a centralized state. This in turn put pressure on Egyptologists to push back all historical dates in Egypt to fill the thousand-year void, an essentially impossible task.

In his work on ancient Egyptian astronomy, Neugebauer had shown how the astronomical papyri's cycles could have been discerned and the tables constructed—by means of numerical schema deriving from dead reckoning over long periods of time, with no need for precise astronomical observation.

Installed in Copenhagen, Neugebauer had had his interest in Egypt revitalized through the influence of two men; the Egyptologist Aksel Volten whom we have already met as Neugebauer's co-author of the edition of the papyrus Carlsberg 9 and an old friend from Göttingen, the Croatian-born mathematician Willy Feller.[71] Feller too had been an Assistent of Courant from 1925 to 1928, obtaining his doctorate with the latter in the same year as Neugebauer, then had gone as a *Privatdozent* to the University of Kiel. Refusing to take the Nazi oath in 1933, Feller had come to Copenhagen for a year, working with Herbert Busemann on differential equations and becoming friendly with Harald Bohr—and of course meeting up again with Neugebauer when the latter arrived in 1934. Offered a professorship at the University of Stockholm in the Winter Semester of that year, Feller nonetheless kept in close touch with his Copenhagen friends and colleagues. In 1939, when Neugebauer left the no-longer safe haven of Copenhagen for Brown University in Providence, Rhode Island, Feller too was offered a professorship there and became the executive editor of Mathematical Reviews when his friend founded that American replacement for his Zentralblatt, from which Neugebauer and numerous other anti-Nazi editors had resigned in 1938. It was in discussions with Feller, who shared a strong interest in Ancient Egypt, that the idea of taking on Meyer's isolated date arose and it was Feller who pushed Neugebauer into publication in 1939:

> I would like to emphasize that the core of this work came out of general discussions with Dr. W. Feller of Stockholm. I must also thank Dr. Feller for the bibliography on the Nile as well as the checking of the manuscript and the proofs.[72] (Neugebauer 1939a: 169)

In Neugebauer's typical manner, the article is divided into two parts, the negative and the positive. First, under the rather menacing epigram "Lasciate ogni speranza",[73] he attacks the very possibility that at the date proposed by Meyer, some thousand years before the invention of writing in Egypt and the establishment of a State, there could have existed a body of astronomers capable of the observations and calculations necessary for the construction of such a calendar. Furthermore, in only 8 years, the calendar would have been already 2 days out of synchronization, a fact hardly to be missed by such a hypothetical body of scientists. In short the Sothic cycle and the 365-day civil year were originally two independent systems for the Egyptians.

Having disposed of Meyer's correlation of the civil and Sothic year, Neugebauer passed to the replacement theory. He pointed out that dead reckoning, i.e., counting the number of days between successive inundations, would have permitted the establishment of a 365 day civil year by simply averaging over this interval for some reasonable time, say 50 years—the variability of the inundation would have become visible only over some much longer period of time during which the Sothic cycle could have been recognized as providing a tighter link with the rising of the Nile.

---

[71] For Feller (born Vilibald Srećko Feller) see (Birnbaum et al. 1970).

[72] *Ich möchte hervorheben, daß der Kern dieser Arbeit in gemeinsamer Diskussion mit Dr. W. Feller, Stockholm, entstanden ist. Ebenso habe ich Herrn Dr. Feller für die Literaturnachweise über den Nil sowie für Durchsicht von Manuskript und Korrekturen zu danken.*

[73] *Lasciate ogni speranza voi ch'intrate!*, "Abandon all hope, you who enter!" is the final line of the inscription Dante places at the entrance to Hell in his *Divina Commedia* (canto 3, line 9).

Only thereafter did the question of the relative lapse between the Sothic and the civil date become of interest to Egyptians.

The article caused an immediate stir in the Egyptological community. Jean Capart, the eminent Belgian Egyptologist and historian of art, immediately contacted Neugebauer in his capacity as editor of the Belgian journal *Chronique d'Égypte*, requesting a shortened translation into French. This was very quickly forthcoming under the title of "La période sothique" and Capart prefaced it in enthusiastic terms:

> Professor O. Neugebauer has had the kindness to sum up for *Chronique d'Égypte* his recently published study on the "Sothic period". At the beginning of this article he had written, concerning those, I suppose, whose opinion he combats: *Lasciate ogni speranza*. He might equally well have written: *Chronologia egyptiaca liberata*.[74] For Professor Neugebauer's proof liberates Egyptian historical studies from the Procrustean bed to which Ed. Meyer had felt it necessary to bind them. (Capart *apud* Neugebauer 1939b: 258)

The "first certain date in the history of the world" had vanished. It was not only Capart who was excited; the German Egyptologist Alexander Scharff, giving a talk in the summer of 1939 to the Bavarian Academy of Sciences, began by stating:

> We can never be sufficiently grateful to the ... mathematician O. Neugebauer for a physically small but intellectually momentous work, in which, with compelling conclusions of amazing simplicity, he abolishes the astronomical foundations of the entire edifice of Egyptian chronology that Ed. Meyer had once constructed with such enormous astuteness.[75] (Scharff 1939: 3)

But it was Neugebauer's negative argument that had really swept the field. His proposed replacement did not in itself fix any particular period as a candidate for the introduction of the calendar since it was only Meyer's hypothesis of the existence of an organized body of professional astronomers in fifth millennium Egypt that Neugebauer opposed. In fact, Neugebauer was rather tempted to simply leave the date there where Meyer had placed it, around 4200 BC. It was the Egyptologists who saw the liberation as one of being able to significantly lower the date of the introduction of the calendar; indeed this was the major point of Scharff's lecture, cited above, and by the time Neugebauer was settled in the US at Brown he had finally become convinced (Neugebauer 1942: 401 n. 17) that it was likely to have been introduced during the period of State formation at the beginning of the third millennium BC.[76]

---

[74] Latin for "Egyptian chronology liberated".

[75] *Wir können dem ... Mathematiker O. Neugebauer gar nicht dankbar genug sein für eine umfänglich kleine, inhaltlich bedeutungsschwere Arbeit, in der er mit zwingenden Schlüssen von verblüffender Einfachheit die astronomischen Grundlagen des ganzen einst von Ed. Meyer mit gewaltigem Scharfsinn erdachten Gebäudes der ägyptischen Chronologie aufhebt.*

[76] That no "clearing of the ground' is ever truly established once and for all is made clear by the fact that some celebrations of the centennial of Meyer's publication in 2004 trot out the full hypothesis, initial date and all, e.g., (Zulian 2004), where, interestingly, Neugebauer is completely ignored.

Kurt Vogel and the Eye of Horus

Though Neugebauer had been the first trained mathematician[77] to turn to a professional study of Ancient Egyptian mathematics, he was not to remain so long. Kurt Vogel,[78] 11 years older than Neugebauer, had also studied mathematics and physics seriously—at Erlangen and Göttingen—and, after his service in the War, had become a teacher of mathematics at the Maximilians-Gymnasium in Munich, where indeed he remained until his retirement. He too had a long-standing interest in Ancient Egypt and had also learned the language while at Göttingen. In 1927 he decided to go back for a doctorate under the University of Munich Egyptologist Wilhelm Spiegelberg, with a view to working in the area of Egyptian mathematics. His 1929 thesis (Vogel 1929a) was entitled "The Foundations of Egyptian Arithmetic in Relation to the 2:*n* Table of the Rhind Papyrus" [*Die Grundlagen der ägyptischen Arithmetik in ihrem Zusammenhang mit der 2:n-Tabelle des Papyrus Rhind*]. At the University, his work attracted the patronage of mathematician Constantin Carathéodory and of the historian of mathematics, Heinrich Wieleitner. The latter was associated with the venerable *Archive for the History of Mathematics, the Sciences and Technology* [*Archiv für Geschichte der Mathematik, der Naturwissenschaften und der Technik*] and opened its pages to Vogel. Over the following 3 years he published some ten articles on Egyptian mathematics before turning his attention to Greek and Babylonian subjects.[79]

The similarities to Neugebauer's own history, the choice of thesis topic, the striking similarities in range of subjects and even titles—Vogel published his 1958–1959 synthesis of ancient mathematics under the title *Vorgriechische Mathematik* (Vogel 1958/1959)—bear witness to what had developed as a fierce rivalry that persisted between the two over a long period. From their respective fortresses of *Quellen und Studien* and *Archiv*, they traded negative reviews of each other's work—and of that of each other's protégés—with occasional forays into enemy territory. Upon invitation by the editors, Neugebauer published his (essentially very critical) essay review of Vogel's thesis in the pages of *Archiv* (Neugebauer 1930b), while Vogel published a string of reviews in Neugebauer's *Zentralblatt für Mathematik*. When the London Leather Roll was finally unrolled and published in 1927 (Glanville 1927), the first (though minor) addition to the Egyptian mathematical corpus in over a quarter of a century, both men rushed into print (Vogel 1929b; Neugebauer 1929) with predictably opposed interpretations.

One typical example of this opposition can be seen at work in the exchange, a year later, over the question of the so-called 'Horus-eye fractions' in Egyptian mathematics.[80] In 1911, the Berlin Egyptologist Georg Möller was preparing the second

---

[77] Though T. Eric Peet had majored in mathematics as an Oxford undergraduate, he had never practiced it, having turned to archeology immediately after graduation.

[78] What follows is based on (Mahoney and Schneider 1986) and (Folkerts 1983).

[79] Most of these have been reprinted in (Folkerts 1988).

[80] A detailed discussion of these "fractions" and their current status can be found in (Ritter 2003).

volume of his great hieratic sign-list, that is the list of ordinary Egyptian written signs and since it was generally believed that hieratic signs were in all cases cursive forms of pictographic originals, Möller had organized his sign-list by hieroglyphic form, organized by class of objects represented. In particular he believed he had shown that a series of seven signs representing the dimidiated fractions—1/2, 1/4, 1/8, …, 1/64—of the basic unit of volume, the *ḥeqat*, used to measure grain came originally from component parts of the Eye of Horus, a frequently encountered religious symbol in Ancient Egypt (Möller 1911). This had become received opinion by 1930 when Neugebauer looked into the matter (Neugebauer 1930c). On the negative side, he first passed in review the mathematical uses of these signs in conjunction with ordinary fractions in the mathematical texts and presented an argument that they were not fractions but integer multiples of a known smaller capacity unit, the *ra*. This is a direct outcome of the additive basis of Egyptian mathematics, for in such a system, rather than fractional parts of a unity—1 *ḥeqat* measure—it is more natural to think of dyadic parts of the *ḥeqat*'s equivalent of 320 *ra*, i.e., 160, 80, 40, 20, 10, 5. Then he assembled arguments to show that the *ḥeqat*-"Horus-eye fractions" system was originally separate from and independent of the *ra* capacity measure system. Finally, turning to the form of the hieroglyphic signs, the sole basis for Möller's identification, he marshaled all the then-available evidence to show that, at least for 1/16 and 1/32, this identification simply did not stand up to scrutiny.[81] In its place Neugebauer argued for the signs in question coming from an abbreviated form of the writing of their equivalents, 20 *ra* and 10 *ra*, thus arguing for a secondary origin of the Horus-eye fractions as integral multiples of a smaller basic unit.

The response was immediate—and from Vogel. In the very next issue of the same journal with his "On the Question of the Parts of the Bushel" (Vogel 1930), he countered with a series of arguments against Neugebauer's refusal to see the Horus-eye signs as fractions on grounds largely already presented in his own work and in reviews of those of his rival, while Neugebauer in turn, in "Once More, the Parts of the Bushel" (Neugebauer 1932) brought out his earlier criticisms of Vogel and reiterated his own position. And there the "exchange" ended.

In this case, unlike the result of the Neugebauer—Meyer "debate", nobody listened to either Neugebauer's critiques of Möller's hypothesis nor to his counterproposal for the origin of the supposed 'Horus-eye fractions'. Möller's hypothesis was simply repeated in virtually every reference to Ancient Egyptian capacity measurement for over 70 years. There is, in fact, only one single mention of Neugebauer's argument in either the Egyptological or in the history of science literature in all this time, a passing reference at the end of an article on the religious aspects of the Eye of Horus (Müller-Wollermann 1986).

Looking at the ways in which Neugebauer was heard—or ignored—by the Egyptological community in his attempts to address them directly on contentious issues points to the limitations of Neugebauer's program to bring rigor to the field

[81] In his edition of the Rhind papyrus, Peet had already expressed skepticism about Möller's hieroglyphic equivalents for these capacity measures (Peet 1923a: 25–26).

of Egyptology. He was listened to when he offered something that was already on the agenda of the Egyptologists; in the case of Meyer's chronology, this was the unease they felt with an overlong chronology and here Neugebauer's critique was welcomed enthusiastically, though not his comfort with Meyer's early date in general. On the other hand, Möller's suggestion of a religious origin for metrological measures fit well with the general theological and humanistic interests of the Egyptologists and Neugebauer's (and even Peet's) objections found no traction and evoked not the slightest interest.[82]

The Essential Tension

That interest in fundamental questions that drew Otto Neugebauer as a young veteran first to unified theories based on the new theory of relativity, then to pure mathematics, led him finally to a search for the founding principles of mathematical and logical thought in humanity's earliest preserved records. In the early twentieth century, in terms of known texts, that could only mean Ancient Egypt. But more than that, Neugebauer also had a personal interest in that culture. He is of course well-known for his later harsh judgement that

> Egypt provides us with the exceptional case of a highly sophisticated civilization which flourished for many centuries without making a single contribution to the development of the exact sciences. (Neugebauer 1975: II 559)

Yet he would still write, many years after he had abandoned Egypt as a research domain, that

> of all the civilizations of antiquity, the Egyptian seems to have been the most pleasant.... There is probably no other country in the ancient world where cultivated life could be maintained through so many centuries in peace and security. (Neugebauer 1957: 71)

But it was not only the culture that had attracted him to the Nile Valley. The contact at Göttingen with the one Egyptologist who most represented the attempt to bring rigor to the study of Egyptology determined in large part not only Neugebauer's decision to leave pure mathematics for a study of its history but also the manner of interpellating the ancient texts. An appreciation of the cultural context, a sensitivity to the foundational questions lying behind calculational techniques and notational conventions, an appreciation of the possibilities of classificatory and structural approaches: in all this Neugebauer was a product of Göttingen, not only of Courant and Hilbert but of Sethe too.

If we ask to what extent and how Neugebauer influenced the domain of Egyptological research, the reply is more nuanced. Unlike the situation in Assyriology, there are no generic references to Neugebauer's work in either the Egyptological nor the in history of Egyptian science communities today, no standard references to his publications. Recognition of his work on fractions is limited,

[82] For more detail on this question see (Ritter 2003).

at most, to citing the corresponding ten pages of *The Exact Sciences in Antiquity* as a general source for the novice, with no mention made of the general historical and foundational motivation behind it. It was not that he was either ignored or unappreciated by the Egyptologists of the period. T. Eric Peet for example, in a 1934 critical review of a book on Egyptian art, bemoaned the absence in it of an appreciation of the new and exciting work being done in the field.

> Much new light has recently been thrown on Egyptian drawing and sculpture, and on Egyptian mathematics, by the ability of some writers, notably Schafer in the case of drawing and sculpture, and Gunn, Neugebauer, Vogel,...in the case of mathematics, to clear their minds effectively of the modern point of view, to go back behind the Greek and to see the subject as the Egyptians saw it. (Peet 1934: 120)

That Peet was able to appreciate this in Neugebauer is all the more striking since many at the time—and even more so later—saw in his use of elementary algebraic symbolism in the analysis of Egyptian mathematics not a mere notational convenience, the organizing principles for him being the solution methods employed, but the actual ancient conceptual framework. Ironically this made Neugebauer seem to be yet another anachronistic modernizer, a position against which he fought so hard in his early days.

Even when he was listened to by Egyptologists in the 1920s and 1930s he was always seen as essentially an outsider, "der Mathematiker" in Scharff's words cited earlier. This was not entirely a misreading; there was always an essential tension in Neugebauer's work. As Noel Swerdlow very finely observed in his obituary of his colleague and friend (Swerdlow 1993), there were always two aspects to Neugebauer: the historian sensitive to the specificities of given historical and cultural contexts and the practicing mathematician, concerned with the rigor of analysis and the essential unity of all mathematics. From the beginning Neugebauer attempted to synthesize these two aspects: " It has been my endeavor," he explained in the Introduction to his thesis, "so far as it was in my power, to emphasize here both tendencies—those of the historical as well as of the mathematical sciences." [*Es war mein Bestreben, beide Tendenzen, sowohl die der historischen wie der mathematischen Wissenschaften, soweit es in meinen Kräften stand, hier zur Geltung zu bringen.* Neugebauer 1926: 1]. It was not that he had any illusions about the modern mathematical interest of the extant Egyptian material but he could hold at bay within himself the impatience of the professional mathematician because he was interested in other, more fundamental questions about the nature of all mathematical cognition, questions that the Ancient Egyptian material like any other could supply if approached in the right manner. But the paucity of the Egyptian material compared to the Mesopotamian, and its more elementary nature meant that the early "desire to investigate the logico-conceptual foundations of mathematics of one of the most interesting people in Antiquity" [*die logisch-begrifflichenGrundlagen der Mathematik eines der interessantesten Völker des Altertums*, Neugebauer 1926: *Vorwort*] gave way to the sheer intellectual challenge of the sophisticated application of arithmetic methods and their application to physical phenomena in

Mesopotamian mathematical astronomy. And to *this* Neugebauer, Egypt had nothing more to offer.

References

Aigner, A. 1985. *Das Fach Mathematik an der Universität Graz*, Publikationen au dem Archiv der Universität Graz 15. Graz: Akademische Druck und Verlagsanstalt.

Bečvářová, M., and I. Netuka. 2010. *Jarník's notes of the lecture course* Punktmengen und reelle Funktionen *by P. S. Aleksandrov (Göttingen 1928)*. Prague: Matfyzpress.

Benoit, P., K. Chemla, and J. Ritter. 1992. *Histoire de fractions, fractions d'histoire*, Science networks 10. Basel: Birkhäuser.

Birnbaum, Z., et al. 1970. William Feller, 1906–1970. *The Annals of Mathematical Statistics* 41: IV–XIII.

Bochner, S. 1952. Harald Bohr. *Bulletin of the American Mathematical Society* 58: 72–75.

Bohr, H. 1924. Zur Theorie der fastperiodischen Funktionen. I. Eine Verallgemeinerung der Theorie der Fourierreihen. *Acta Mathematica* 45: 29–127.

Bohr, H. 1925. Zur Theorie der fastperiodischen Funktionen. II. Zusammenhang der fastperiodischen Funktionen mit Funktionen von unendlich vielen Variabeln; gleichmässige Approximation durch trigonometrische Summen. *Acta Mathematica* 46: 101–214.

Bohr, H. 1926. Zur Theorie der fastperiodischen Funktionen. III. Dirichletentwicklung analytischer Funktionen. *Acta Mathematica* 47: 237–281.

Bohr, H., and O. Neugebauer. 1926. Über lineare Differentialgleichungen mit konstanten Koeffizienten und fastperiodischer rechter Seite. *Nachrichten von der Gesellschaft der Wissenschaften zu Göttingen. Mathematisch-Physikalische Klasse* 1926: 8–22.

Breasted, J.H. 1905. *A history of Egypt: From the earliest times to the Persian conquest*. New York: Charles Scribner's Sons.

Breasted, J.H. 1910. *Geschichte Ägyptens*, German translation by Hermann Ranke. Berlin: Karl Curtius.

Cantor, M. 1894. *Vorlesungen über Geschichte der Mathematik. I. Von den ältesten Zeiten bis zum Jahre 1200 n. Chr.* 2nd ed. Leipzig: Teubner.

Corry, L. 2004. *Hilbert and the axiomatization of physics (1898–1918): From Grundlagen der Geometrie to Grundlagen der Physik*. Dordrecht: Kluwer.

Courant, R., and D. Hilbert. 1923. *Methoden der mathematischen Physik*, vol. 1. Berlin: Springer.

Daressy, G. 1901. *Ostraca (CGC 25001–25385)* (Catalogue générale des Antiquités égyptiennes du Musée du Caire). Cairo: Institut Français d'Archéologie Orientale.

Davies, N. de G. 1921. [Review of Schâfer, *Von ägyptischer Kunst*]. *The Journal of Egyptian Archæology* 7: 222–228.

Eckert, M. 2013. *Arnold Sommereld: Science, life and turbulent times 1868–1951*. New York: Springer.

Eisenlohr, A. 1877. *Ein mathematisches Handbuch der alten Ägypter (Papyrus Rhind des British Museums) übersetzt und erklärt*, 2 vols. Leipzig: J. C. Hinrichs.

Erman, A. 1929. *Mein Werden und mein Wirken: Erinnerungen eines alten Berliner Gelehrten*. Leipzig: Quelle & Meyer.

Folkerts, M. 1983. Kurt Vogel: Biographie und Bibliographie. *Historia Mathematica* 10: 261–273.

Folkerts, M. ed. 1988. *Kurt Vogel: Kleinere Schriften zur Geschichte der Mathematik*, 2 vols. Wiesbaden: Franz Steiner.

Glanville, S.R.K. 1927. The mathematical leather roll in the British Museum. *The Journal of Egyptian Archaeology* 13: 232–239.

Goldstein, C., and J. Ritter. 2003. The varieties of unity: Soundings in unified theories 1920–1930. In *Revisiting the foundations of relativistic physics*, ed. A. Ashtekar et al., 93–149. Dordrecht: Kluwer.

Griffith, F.Ll. 1898. *Hieratic papyri from Kahun and Gurob*, 2 vols. London: Bernard Quaritch.

Griffith, F.Ll. 1917. Bibliography 1916–1917: Ancient Egypt. *The Journal of Egyptian Archaeology* 4: 261–279.

Gunn, B. 1916. [Review of Sethe, *Zahlen und Zahlworten*]. *The Journal of Egyptian Archæology* 3: 279–286.

Gunn, B. 1926. [Review of Peet, *The Rhind Mathematical Papyrus*]. *The Journal of Egyptian Archæology* 12: 123–137.

Gunn, B., and T.E. Peet. 1929. Four mathematical problems from the Moscow mathematical papyrus. *The Journal of Egyptian Archaeology* 15: 167–185.

Hankel, H. 1869. *Die Entwicklung der Mathematik in den letzten Jahrhunderten*. Tübingen: Füs'sche Sortimentsbuchhandlung.

Hayes, W.C. 1942. *Ostraka and name stones from the tomb of Sen-Mut (No. 71) at Thebes*. New York: Metropolitan Museum of Art.

Hultsch, F. 1895. *Die Elemente der ägyptische Teilungsrechnung*. In *Abhhandlungen der philologisch-hlstorischen Classe der Königlich Sächsischen Gesellschaft der Wissenschaften zu Leipzig 17*. Leipzig: Hirzel.

Hund, F. 1927. *Linienspektren und periodisches System der Elemente*, Structur der Materie in Einzeldarstellungen 4. Berlin: Springer.

Imhausen, A., and J. Ritter. 2004. Mathematical fragments. In *The UCL Lahun papyri: Religious, literary, legal, mathematical and medical*, British archaeological reports 1209, ed. M. Collier and S. Quirke, 71–96. Oxford: Archaeopress.

Jahnke, H.N., and M. Otte. 1981. Origins of the program of "arithmetization of mathematics". In *Social history of nineteenth-century mathematics*, ed. H. Mehretne, H. Bos, and I. Schneider, 21–49. Boston: Birkhäuser.

Lange, H.O., and O. Neugebauer. 1940. *Papyrus Carlsberg No. I: Ein hieratisch-demotischer kosmologischer Text*, Det Kgl. Danske Videnskabernes Selskab, Historisk-filologiske Skrifter, vol. I, no. 2. Copenhagen: Ejnar Munksgaard.

Lévy-Bruhl, L. 1922. *Les Fonctions mentales dans les societes inferieures*, 6th edn. Paris: Félix Alcan [First edition: 1910].

Ludwig-Maximilians-Universität München. 1921. *Verzeichnis der Vorlesungen: Winter Halb-Jahr 1921/1922*. Munich: Universitätsverlag.

Lützen, J., and W. Purkert. 1993. Conflicting tendencies in the historiography of mathematics: M. Cantor and H. G. Zeuthen. In *The history of modern mathematics*, Images, ideas and communities, vol. 3, ed. E. Knobloch and D. Rowe, 1–42. San Diego: Academic Press.

Mahoney, M.S., and I. Schneider. 1986. Eloge: Kurt Vogel, 30 September 1888–27 October 1985. *Isis* 77: 667–669.

Meyer, E. 1904. *Ägyptische Chronologie*. In *Abhandlungen der Königlich Preussischen Akademie der Wissenschaften aus dem Jahre 1904. Philosophisch-historische Classe*. Abhandlung I. Berlin: Königliche Akademie der Wissenschaften in commission bei G. Reimer.

Möller, G. 1911. Die Zeichen für die Bruchteile des Hohlmaß und das Uzatauge. *Zeitschrift fiir Agyptische Sprache und Altertumskunde* 48: 99–101.

Müller-Wollermann, R. 1986. Udjatauge. *Lexikon der Ägyptologie* 6: 824–826.

Neugebauer, O. 1925. [Review of Peet, *The Rhind Mathematical Papyrus*]. *Matematisk Tidsskrift A* 1925: 66–70.

Neugebauer, O. 1926. *Die Grundlagen der ägyptischen Bruchrechnung*. Berlin: Julius Springer.

Neugebauer, O. 1927. Über die Konstruktion von *sp* "Mal" im mathematischen Papyrus Rhind. *Zeitschrft für Ägyptische Sprache und Altertumskunde* 62: 61–62.

Neugebauer, O. 1929. Zur agyptischen Bruchrechnung. *Zeitschrift für Ägyptische Sprache und Altertumskunde* 64: 44–48.

Neugebauer, O. 1930a. Arithmetik und Rechentechnik der Ägypter. *Quellen und Studien B* 1: 301–380.

Neugebauer, O. 1930b. Die Grundlagen der ägyptischen Arithmetik: Bemerkungen zu einem Buch dieses Titels von Dr. K. Vogel. *Archiv für Geschichte der Mathematik, der Naturwissenschaften und der Technik* 4(13): 92–99.

Neugebauer, O. 1930c. Über den Scheffel und seine Teile. *Zeitschrift fiir Agyptische Sprache und Altertumskunde* 65: 42–48.

Neugebauer, O. 1931. Die Geometrie der agyptischen mathematischen Texte. *Quellen und Studien* B1: 413–451.

Neugebauer, O. 1932. Nochmals die Scheffelteile. *Zeitschrift für Ägyptische Sprache und Altertumskunde* 68: 122–123.

Neugebauer, O. 1934. *Vorlesungen über Geschichte der antiken mathematischen Wissenschaften. Erster. Band: Vorgriechische Mathematik*, Die Grundlagen der mathematischen Wissenschaften in Einzeldarstellungen 43. Berlin: Springer.

Neugebauer, O. 1939a. Die Bedeutungslosigkeit der 'Sothisperiode' für die älteste ägyptische Chronologie. *Acta Orientalia* 17: 169–195.

Neugebauer, O. 1939b. La Période sothiaque. *Chronique d'Égypte* 28: 258–262.

Neugebauer, O. 1942. The origin of the Egyptian calendar. *Journal of Near Eastern Studies* 1: 396–403.

Neugebauer, O. 1957. *The exact sciences in antiquity*, 2nd ed. Providence: Brown University Press, [First edition: Princeton: Princeton University Press, 1952].

Neugebauer, O. 1975. *A history of ancient mathematical astronomy*, 3 vols. Berlin: Springer.

Neugebauer, O., and R.A. Parker. 1960. *Egyptian astronomical texts I. The early decans*. London: Lund Humphries for Brown University Press.

Neugebauer, O., and R.A. Parker. 1964. *Egyptian astronomical texts II. The Ramesside star clocks*. London: Lund Humphries for Brown University Press.

Neugebauer, O., and R.A. Parker. 1969. *Egyptian astronomical texts III. Decans, planets, constellations and zodiacs*, 2 vols. London: Lund Humphries for Brown University Press.

Neugebauer, O., and A. Volten. 1938. Untersuchungen zur antiken Astronomie IV. Ein demotischer astronomischer Papyrus (Pap. Carlsberg 9). *Quellen und Studien* B4: 383–406.

Parker, R.A. 1950. *The calendars of ancient Egypt*, Studies in ancient oriental civilization 26. Chicago: Chicago University Press.

Pauli, W. 1921. "Relativitätstheorie" in *Encyklopädie der mathematischen Wissenschaften mit Einschluss ihrer Anwendungen, Band V, Teil 2*, 539–775. Leipzig: Teubner.

Peet, T.E. 1923a. *The Rhind mathematical papyrus: British Museum 10057 and 10058*. London: Hodder & Stoughton for Liverpool University Press.

Peet, T.E. 1923b. Arithmetic in the Middle Kingdom. *Journal of Egyptian Archaeology* 9: 91–95.

Peet, T.E. 1934. [Review of Williams, *The Decoration of the Tomb of Per-neb*]. *The Journal of Egyptian Archæology* 20: 119–120.

Petrie, B., and N. Schappacher. 2007. On arithmetization. In *The shaping of arithmetic after C. F. Gauss's* Disquisitiones Arithmeticæ, ed. C. Goldstein, N. Schappacher, and J. Schwermer, 343–374. Berlin: Springer.

Ritter, J. 1995. Measure for measure: Mathematics in Egypt and Mesopotamia. In *A history of scientific thought: Elements of a history of science*, ed. M. Serres, 44–72. Oxford: Blackwell Publishers.

Ritter, J. 2001. Egyptian mathematics. In *Mathematics across cultures: The history of non-western mathematics*, ed. H. Selin, 115–136. Dordrecht: Kluwer.

Ritter, J. 2003. Closing the Eye of Horus: The rise and fall of Horus-eye fractions. In *Under one sky (AOAT 297)*, ed. J. Steele and A. Imhausen, 298–323. Münster: Ugarit-Verlag.

Roeder, G. 1917. Ägyptologie (1916). *Zeitschrift der Deutschen Morganländischen Gesellschaft* 71: 272–295.

Rowe, D. 2004. Making mathematics in an oral culture: Göttingen in the Era of Klein and Hilbert. *Science in Context* 17: 85–129.

Schack-Schackenburg, H. 1900. Der Berliner Papyrus 6619. *Zeitschrift für ägyptische Sprache und Altertumskunde* 38: 135–141.

Schack-Schackenburg, H. 1902. Das kleinere Fragment des Berliner Papyrus 6619. *Zeitschrift für ägyptische Sprache und Altertumskunde* 40: 65–66.

Schäfer, H. 1919. *Von ägyptischer Kunst, besonders der Zeichenkunst: Ein Einführung in die Betrachtung ägyptischer Kunstwerke*, 2 vols. Leipzig: J. C. Hinrichs.

Schappacher, N. 1987. Das Mathematische Institut der Universität Göttingen 1929–1950. In *Die Universitât Göttingen unter dem Nationalsozialismus*, ed. H. Becker, H.-J. Dahms, and C. Wegeler, 345–373. Munich: Saur.

Scharff, A. 1939. Die Bedeutungslosigkeit des sogenannten ältesten Datums der Weltgeschichte und einige sich daraus ergebende Folgerungen für die ägyptische Geschichte und Archäologie. *Historische Zeitschrift* 141: 3–32.

Seth, S. 2010. *Crafting the quantum: Arnold Sommerfeld and the practice of theory, 1890–1926*. Cambridge: MIT Press.

Sethe, K. 1916. *Von Zahlen und Zahlworten bei den alten Ägyptern und was für andere Völker und Sprachen daraus zu lernen ist: Ein Beitrag zur Geschichte von Rechenkunst und Sprache*. Straßburg: Karl J. Trübner.

Sethe, K. 1924. [Review of Peet: *The Rhind Mathematical Papyrus*]. *Jahresbericht der Deutschen Mathematiker-Vereinigung* 33: 139–143.

Siegmund-Schultze, R. 1993. *Mathematische Berichterstattung in Hitlerdeutschland: der Niedergang des "Jahrbuchs über die Fortschritte der Mathematik"*. Göttingen: Vandenhoeck & Ruprecht.

Struve, V.V. 1930. *Mathematischer Papyrus des Staatlichen Museums der schönen Kunst in Moskau, Quellen und Studien A 1*. Berlin: Springer.

Swerdlow, N.M. 1993. Otto E. Neugebauer (26 May 1899–19 February 1990). *Proceedings of the American Philosophical Society* 137: 138–165.

Turaev [Touraeff] B.A. 1917. The volume of the truncated pyramid in Egyptian mathematics. *Ancient Egypt* 1917: 100–102.

Vogel, K. 1929a. *Die Grundlagen der ägyptischen Arithmetik in ihrem Zusammenhang mit der 2:n-Tabelle des Papyrus Rhind*. Munich: Beckstein.

Vogel, K. 1929b. Erweitert die Lederrolle unsere Kenntnis ägyptischer Mathematik? *Archiv fur Geschichte der Mathematik, der Naturwissenschaften und der Technik* 11: 386–407.

Vogel, K. 1930. Zur Frage der Scheffelteile. *Zeitschrift für ägyptische Sprache und Altertumskunde* 66: 33–35.

Vogel, K. 1958/1959. *Vorgriechische Mathematik. Teil I: Vorgeschichte und Ägypten; Teil II: Die Mathematik der Babylonier*, Mathematische Studienhefte, Heft 1–2, 2 vols. Hannover: Schöningh.

von Ranke, L. 1824. *Geschichte der romanischen und germanischen Völker von 1494 bis 1514*. Berlin: Georg Reimer.

Walter, S. 1999. Minkowski, mathematicians, and the mathematical theory of relativity. In *The exploding worlds of general relativity*, Einstein studies, vol. 7, ed. H. Goenner, J. Renn, J. Ritter, and T. Sauer. Boston: Birkhäuser.

Weyl, H. 1919. *Raum·Zeit·Materie*, 3rd ed. Berlin: Springer.

Weyl, H. 1920. Elektrizität und Gravitation. *Physikalische Zeitschrift* 21: 949–951.

Wreszinski, B. 1917. [Review of Sethe, *Zahlen und Zahlworten*]. *Orientalistische Literaturzeitung* 20: 18–22.

Zulian, M. 2004. La hipótesis sotíaca de Eduard Meyer: una revisión a 100 años de su publicación. *Antiguo Oriente* 2: 75–83.

As the Outsider Walked in the Historiography of Mesopotamian Mathematics Until Neugebauer

Jens Høyrup

The Background

In an obituary of Jules Oppert (Heuzey 1906: 73*f*) we find the following[1]:

> With Jules Oppert disappears the last and the most famous representative of what one may call the creation epoch Assyriology. When he entered the scene, Assyriological science had existed but a few years. The decipherment of the Persian texts, inaugurated by Grotefend in the beginning of the last century, had opened the way; the proper nouns common to the two Persian and Assyrian versions of the trilingual Achaemenid inscriptions provided a firm base for the determination of the value of a certain number of signs; Rawlinson recognized the polyphonic character of the Assyrian system, and Hincks justly defended the syllabic principle against Sauley. After a few works on Old Persian, Oppert brought his main effort to the Assyrian inscriptions. After having been entrusted together with Fresnel with a mission to the Babylonian area, he published in 1859, after his return, the second volume (actually the first in date) of his Expédition en Méslopotamie [*sic*] in which, by means of recently discovered sign collections or syllabaries, he established the principal rules of decipherment. This volume, Oppert's masterpiece, constitutes a turning point; it put an end to the gropings and established Assyriology definitively.

Similarly, Samuel Noah Kramer (1963: 15) states that

> Rawlinson, Hincks and Oppert – cuneiform's "holy" triad – non only put Old Persian on firm ground, but also launched Akkadian and Sumerian on the course to decipherment.

Kramer's whole description of the process of decipherment of the three languages (pp. 11–26) shows the importance during the initial phase of knowledge derived from classical and Hebrew sources (often very approximative knowledge,

[1] My translation, as everywhere in what follows when no translator is indicated.

J. Høyrup (✉)
Section for Philosophy and Science Studies, Roskilde University, Roskilde, Denmark
e-mail: jenshoyrup@gmail.com; http://www.ruc.dk/~jensh

A. Jones et al. (eds.), *A Mathematician's Journeys*, Archimedes 45,
DOI 10.1007/978-3-319-25865-2_5

as it turned out, except for the Hebrew *language* and *terminology*) and of bi- and trilingual texts.[2]

So much concerning the conditions for the beginning of cuneiform scholarship. The conditions for initial work on matters connected to cuneiform *mathematics* (understood broadly, as numero-metrological practice) are reflected slightly later in Heuzey's obituary:

> Oppert's scientific activity pointed in very different directions: historical and religious texts, (Sumero-Assyrian) bilingual and purely Sumerian texts, juridical and divinatory texts, Persian and neo-Susian texts, there is almost no branch of the vast literature of cuneiform inscriptions he has not explored. The most particular questions – juridical, metrological, chronological – attracted his curiosity [...].

Administrative, economical and historiographic documents were indeed not only a main source for metrology; reversely they could only be understood to the full once the pertinent metrology itself was understood, for which reason they were also the main motive for understanding numeration and metrology.

This is illustrated by the earliest discovery of sexagesimal counting. In connection with work on calendaric material, Edward Hincks (1854a: 232) describes a tablet ("K 90") containing "an estimate of the magnitude of the illuminated portion of the lunar disk on each of the 30 days of the month"[3] without going into the question how its numbers were written; in a parallel publication (1854b) "On the Assyrian Mythology" concerned with the numbers attached to the gods he refers to the "use of the different numbers to express sixty times what they would most naturally do" and bases this claim on the numbers on the tablet just mentioned, where 240 is written iv (Hincks uses Roman numerals to render the cuneiform numbers), and where "iii.xxviii, iii.xii, ii.lvi, ii.xl, etc." stand for "208, 192, 176, 160, etc.". Henry Rawlinson's contribution to the topic in (1855) (already communicated to Hincks when the second paper of the latter was in print, in December 1854) consists of a long footnote (pp. 217–221) within an article on "The Early History of Babylonia", in which he states that the values ascribed by Berossos (ed. Cory 1832: 32) to σάρος (*šār*), νήρος (*nēru*) and σώσσος (*šūši*), namely respectively 3600, 600 and 60 years, are "abundantly proved by the monuments" (p. 217), giving as further confirmation an extract of "a table of squares, which extends in due order from 1 to 60"(pp. 218–219), in which the place-value character of the notation is obvious but only claimed indirectly by Rawlinson.[4]

---

[2] A more detailed description of the process, confirming this picture, is found in (Sayce 1908: 7–35). Even more detailed is (Fossey 1907: 102–244).

[3] Archibald Henry Sayce, when returning to the text (now identified as K 490) in (1887: 337–340), reinterprets the topic as a table of lunar longitudes.

[4] That Rawlinson is anyhow also interested in the mathematics *per se* and not only as a means for chronology (after all, he was interested in *everything* Assyro-Babylonian) is however revealed by what comes next in the note, namely that "while I am now discussing the notation of the Babylonians, I may as well give the phonetic reading of the numbers, as they are found in the Assyrian vocabularies".

Oppert wrote a number of major papers on metrology (1872, 1885, 1894; etc.), which confirm the picture. The sources are archaeological measurements combined with evidence contained in written sources (mostly indicating concrete measures rather than dealing with metrology) and comparison with other metrologies.[5] The first of these papers draws, inter alia, on the "Esagila tablet", a copy from 229 BCE of an earlier text and described by Marvin Powell (1982: 107) as

> a key document for Babylonian metrology, because it 1) describes in metrological terms a monument that has been explored and carefully measured, 2) links the standard system of mensuration with the Kassite system, 3) makes it possible to identify the standard cubit with the NB [Neo-Babylonian/JH] cubit, and 5) enables us to calculate the absolute length of these units as well as the area of the iku used in both Sumerian-OB [Old Babylonian/JH] and in Kassite-Early NB documents

– which means that it fits precisely into the general pattern of Oppert's and contemporary work on metrologies, even though the full exploitation of the document was not possible at a moment when the Esagila complex had not yet been excavated, and when relative chronologies preceding the neo-Assyrian epoch were still not firmly established.[6]

Over the following five decades, work with this focus was pursued by a number of scholars – beyond Oppert also Vincent Scheil, François Thureau-Dangin, Herman Hilprecht, Franz Heinrich Weißbach, Arthur Ungnad, François-Maurice Allotte de la Fuÿe, Louis Delaporte, Ernst Weidner and others.[7] The outcome was a fair understanding of the many different metrologies (Thureau-Dangin 1909, 1921) including brick metrology (Scheil 1915b); of the place-value system and the function of tables of reciprocals (Scheil 1915a, 1916)[8]; and of techniques for area determination (Allotte de la Fuÿe 1915) – all (as far as allowed by available sources) in contexts extending from the mid-third (occasionally the outgoing fourth) to the late first millennium BCE.

Hilprecht's discussion of "multiplication and division tables" (1906) deserves special mention. It made available an important text group, but also cast long shad-

[5] Since Mesopotamian metrology varies much more over the epochs than Oppert had imagined, it is obvious that the comparative method led him astray as often as to the goal. The task may be claimed only to have been brought to a really satisfactory end by Marvin Powell (1990).

[6] This is well illustrated by the chapter "History and Chronology [of Chaldaea]" in the second edition of George Rawlinson's *Five Great Monarchies of the Ancient Eastern World* from (1871: I, 149–179). The author can still do no better than his brother Henry had done in (1855) – all we find is a critically reflective combination of Berossos and Genesis, with a few ruler names from various cities inserted as if they were part of one single dynasty.

This was soon to change. In (1885: 317–790), Fritz Hommel was able to locate everything from Gudea onward in correct order; absolute chronologies before Hammurapi were still constructed from late Babylonian fancies (Hommel locates Sargon around 3800 BCE and Ur-Nammu around 3500 and lets the Ur, Larsa and Isin dynasties (whose actual total duration was c. 350 years) last from c. 3500 until c. 2000 BCE – pp. 167*f*).

[7] See (Friberg 1982: 3–27).

[8] Actually, Scheil's understanding was not broadly accepted: Meissner (1920: II, 387) from 1925 does not know about sexagesimal fractions. Meissner also mixes up the place-value and the absolute system.

ows: not understanding sexagesimal fractions and thus wishing all numbers occurring in the tables to be integers, he interpreted the table of reciprocals as a table of division of 12,960,000 – a number he then finds (p. 29) in an interpretation of Plato's *Republic* VIII, 546B–D (the notoriously obscure passage about the "nuptial number"). That allowed him to confirm a statement he quotes from Carl Bezold on p. 34:

> Mathematics was with the Babylonians, as far as we now know, first of all in the service of astronomy and the latter again in the service of a pseudo-science, astrology, which probably arose in Mesopotamia, spread from there and was inherited by the gnostic writings and the Middle Ages [...].

In this way, Babylonian mathematical thought was made much more numerological and linked much more intimately to esoteric wisdom than warranted.[9]

The Earliest "Properly Mathematical" Texts

All these insights built on the combination of archaeological measurement (of building structures and of metrological standards) with various kinds of written documents and (with gradually dwindling importance) comparative studies. None of this material except some tables of multiplication, reciprocals and powers belonged to genres which were soon to be considered as "properly mathematical" texts.[10]

A few such texts were published during the years 1900–1928. In 1900, hand copies without transliteration of the two extensive Old Babylonian problem collections BM 85194 and BM 85210 appeared in CT IX. In (1906: pl. 15), Hilprecht copied hand copies of two more, identifying (p. 62) the contents of one as "Divisors of 12,960,000 and their quotients in geometrical progression" and the other as "arithmetical calculations". However, since the CT-texts could be judged by Ernst Weidner (1916: 257) to be "the most difficult handed down in cuneiform" and Hilprecht's word problem containing "arithmetical calculations" is actually even more abstruse, it is barely a wonder that no attempt was made to approach them for long.[11] In (1916: 258), Weidner announced to have lately "had the occasion to copy a whole sequence of similar texts", and he gave a transliteration and an attempted

[9] Esoteric numerology certainly left many traces in Mesopotamian sources – but not in sources normally counted as "mathematical"; the only exception is a late Babylonian metrological table starting with the sacred numbers of the gods (W 23273, see (Friberg 1993: 400)). Apart from that, even the text corpus produced by the Late Babylonian and Seleucid priestly environment kept the two interests strictly separate.

[10] I disregard the "metrological tables", which were not yet understood as mediators between the various metrologies and the place value system. I also disregard mathematical astronomy, where the extension of the place value system to fractions had been understood better (Epping 1889: 9*f*), (Kugler 1900: 12, 14), without this understanding being generalized, cf. (Scheil 1915a: 196).

[11] Weidner mentions as the only exception "an occasional notice" by Hommel in a *Beilage* to the *Münchener Neuesten Nachrichten* 1908, Aug. 27, Nr. 49, p. 459, which I have not seen. He says

translation of two sections from one of them, the tablet VAT 6598; Weidner's contribution was immediately followed up by Heinrich Zimmern (1916) and Ungnad (1916), both of whom improved the understanding of the text and the terminology in general, drawing on the same text and on the texts published in CT IX, from which Ungnad transliterated and translated a short extract in (1916) and another one in (1918).

The next step was C. J. Gadd's publication (1922) of a first fragment of BM 15285, a text about the subdivision of a square into smaller squares, smaller triangles, etc. This text was quite different from those published previously, but a few terms were shared, which confirmed readings proposed by Weidner and Zimmern.

Also in (1922: pl. LXI–LXII), Thureau-Dangin published hand copies of AO 6484, a major Seleucid problem text, but without seeing more in it than "arithmetical operations".

Finally, Carl Frank published six mathematical texts from the Strasbourg collection in (1928), with transliteration and attempted translation.

By then, however, the study of cuneiform mathematical texts had also been taken up at Neugebauer's Göttingen seminar. In 1985 Kurt Vogel told me about the immense astonishment when one morning Hans-Siegfried Schuster related that he had discovered solutions of second-degree equations in a cuneiform text. Vogel did not date the event, but it must have taken place in late 1928 or (most likely, see below, note 36) very early 1929.

Confronted Readings

Before we shift our attention to this new phase, we may look at what had been achieved – and what not yet – up to then by confronting Weidner's interpretation and the commentaries it called forth with that of Neugebauer of the same text in MKT.[12] Some of the differences, we should be aware, come from the fact that Neugebauer's transliterations follow the conventions of Thureau-Dangin's *Syllabaire accadien*, which was only published in (1926).

This is Weidner's transliteration and translation from (1916: 258*f*) (left) and Neugebauer's treatment of the same text in MKT (I, 280, 282) (right) (Fig. 1)[13]:

nothing about its substance being in any way important, only that it interprets the final clause *ne-pé-šum* of problems as "quod erat demonstrandum".

[12] This is certainly "whiggish" historiography – and it has to be, if our aim is to locate Neugebauer's achievement in its historical context.

[13] Here and in what follows, when quoting transcriptions and transliterations (also of single words and signs), I follow the conventions of the respective originals. When speaking "from the outside", on my own, I follow modern conventions. Since the delimitation is not always clear, some inconsistencies may well have resulted.

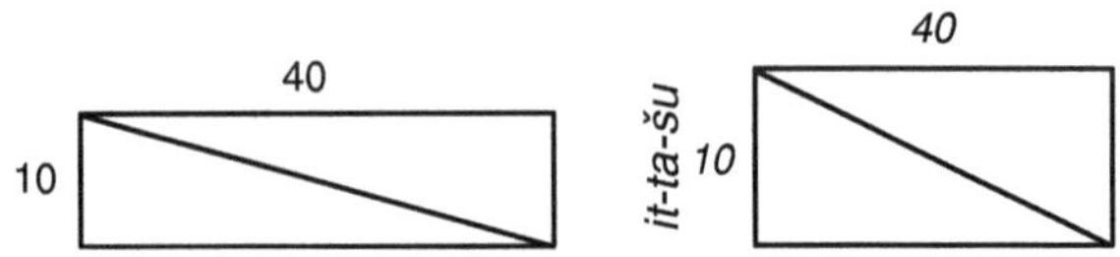

Fig. 1 The drawing on the tablet, as rendered by Weidner (*left*) and Neugebauer (*right*)

1	*2 ú da 40 ú šir zi-li-ip-tu-šu en-nam za-e 10 sag* 2 Ellen (?) Seite (?), 40 Ellen Tiefe (?). Seine Diagonale berechne du. 10 (ist) die Höhe	1	2 kùš dagal 40 kùš sukud *ṣí-li-ip-ta-šu* en-nam za-e 10 sag 2 Ellen Weite, 0;40 Ellen(sic) Höhe, Seine Diagonale (ist) was? Du? 0;10, die Breite
2	*šá-ne 1 40 ta-mar ka-bi-rum 1 40 a-na 40 ú šir i-ši-ma* als Quadrat 1 40 erhältst du. Die Quadratfläche 1 40 auf 40 Ellen Tiefe (?) ist sie,	2	*šu-tam-ḫir* 1,40 *ta-mar qà-qá-rum* 1,40 *a-na* 40 kùš sukud *i-ši-ma* (?) quadriere. 0;1,40 siehst Du (als) Fläche. 0:1,40 mit 0;40 Ellen(sic) Höhe multipliziere und (?)
3	*1 6 40 ta-mar a-na tab-ba 2 13 20 ta-mar a-na 40 ú šir* 1 6 40 erhältst du. Zu verdoppeln, 2 13 20 erhältst du. Zu 40 Ellen Tiefe (?)	3	1 6 40 *ta-mar a-na* tab-ba 2,13,20 *ta-mar a-na* 40 kùš sukud 0;1,6,40 siehst Du. Mit ⟨2⟩ verdopple. 0;2,13,20 siehst Du. Zu 0;40 Ellen (sic) Höhe
4	*daḫ-ḫa 42 13 20 zi-li-ip-to ta-mar ne-pi-šum* hinzufügen, 42 13 20 als Diagonale erhälts du. (Also) ist es gemacht worden.	4	daḫ-ḫa. 42,13,20 *ṣí-li-ip-ta ta-mar ne-pé-šum* addiere. 0;42,13,20 (als) Diagonale siehst Du. Verfahren.

As already seen by Weidner (1916: 359) (the diagram, indeed, leaves litte doubt) the text contains a "calculation of the diagonal of a rectangle whose sides are given". If the given sides are *a* and *b*, Weidner states the diagonal to be $a+\frac{2a\cdot b^2}{3600}$, whereas Neugebauer gives $a+2a\cdot b^2$ [14]; Weidner's divisor 3600 is a symptom that he writes at a moment when he has certainly more or less understood the use of the place

[14] Neugebauer tries to make sense of this impossible formula by interpreting it as an approximation to $a+\frac{2a\cdot b^2}{2a^2+b^2}$; (Neugebauer 1931a: 95–99) explains the origin of the guess, which he finds in the music theory of Nicomachos and Iamblichos – classical Antiquity remained a resource when other arguments were not available.

Difficulties in the handling of the sexagesimal system may be the reason that Weidner did not discover that the formula – adding a length and a volume – is impossible because a change of measuring unit would not change the two addends by the same factor (this is the gist of "dimension analysis").

value system even for fractions but still writes in a spirit untouched by this understanding.[15]

Some of the other differences between the two transliterations hinge on different ways to render the same cuneiform character even though it is understood in the same way, as can be seen from the translations. For instance, Weidner has Ú, the sign name, where Neugebauer has kùš, the Sumerian reading of the sign when meaning a cubit ("Elle"), as it had been identified in the meantime.[16] Such changes are immaterial for our present concern.

Somewhat more pertinent is the disagreement in the first line concerning DA/dagal. These are different signs but rather similar in the Old Babylonian period.[17] Weidner's mistake illustrates the difficulty of reading a cuneiform text whose genre and terminology is as yet unknown. Fortunately for him, the two words are more or less synonymous according to his dictionary (Delitzsch 1914: 130*f*), respectively "side" and "breadth".

Most significant are the cases where Weidner, as he states himself, had to guess at a meaning from the context – the context presented by the present text as well as that of the CT-IX texts, which Weidner had evidently studied intensely without getting to a point where he could make coherent sense of them.

This starts with ŠIR (now UZU = šir$_4$), which again is similar to the sign read by Neugebauer (SUKUD, meaning "height"[18]); since the sign is often found in CT IX, Weidner concludes that it must refer to a dimension, and he finds in Rudolph Brünnow's list from (1889: 200 #4558) that it may stand for *naqbu*, "depth".[19] This seems to make sense, after which the interpretation of *ziliptum* "follows by itself from the context". Neugebauer's spelling *ṣi-li-ip-ta* corresponds to modern orthography, but even he is not able to connect the word to the verb *ṣalāpum*, whose sense "cross out" was not yet established – at least still not in (Bezold 1926: 113, 238).

[15] In detail: Weidner supposes the dimensions of the rectangle to be 10 and 40, even though the initial "2 cubits" should make him understand that 10 stands for 10′, and 40 in consequence (if the calculations are to be meaningful) for 40′ – both corresponding to the unit nindan (1 nindan = 12 cubits); instead he wonders (col. 259) what these 2 cubits may be. Weidner therefore supposes the product to be 4000, about which he says that "it is written in cuneiform as 1 6 40, i.e., 1 (3600)+6 (·60)+40. But this number can also be understood as $1+6(\cdot\frac{1}{60})+40(\cdot\frac{1}{3600}) = \frac{4000}{3600} = 1{,}11$".

A small remark on notations: the ′–″ notation was used (and possibly introduced) by Louis Delaporte in (1911:132) (′ and ″ only); Scheil (1916: 139), immediately followed by Ungnad (1916: 366), uses ′, ″ and °, as does later Thureau-Dangin. Strangely, Neugebauer believed in (1932a: 221) that the °-′-″ notation had been created "recently" by Thureau-Dangin (similarly MKT I, p. vii n. 5); I have not noticed references from his hand to (Scheil 1915a), but he had referred to Ungnad (1916) on several occasions – e.g., (Neugebauer 1928: 45 n.3). Neugebauer's own notation goes back at least to (1929: 68, 71).

[16] See, e.g., (Thureau-Dangin 1921: 133).

[17] For such similarity I rely on (Labat 1963).

[18] This reading goes back to (Zimmern 1916: 323).

[19] Now *nagbu*, interpreted "spring, fountain, underground water" ((CAD XI, 108), cf. (AHw 710)). The error was pointed out by Ungnad (1916: 363), who also proposed the reading sukud, "height".

The next word *en-nam*, thus Weidner with many references to CT IX, "must mean 'calculate'". Neugebauer's "was" corresponds to what he had observed in (1932b: 8) – that en.nam stands where other texts have the interrogative particle *mīnûm*.[20]

Weidner's reading of *za-e* as "you" is correct, and conserved in MKT. However, Weidner connects it to his preceding presumed imperative; it was Ungnad (1916: 363*f*) who pointed out that this word, here and often in the CT-IX texts, marks the beginning of the calculation. Ungnad does not feel sure that a Sumerian za.e, "you", is meant, and as we see Neugebauer adopts his doubt.[21]

šá-ne is interpreted by Weidner as "square" simply because 1 40 is the square of 10; he confesses not to be able to explain it further; the correct reading of the sign group as *šu-tam-ḫir*, adopted by Neugebauer, was suggested by Zimmern (1916: 322*f*) and explained as the "imperative of a [verb] *šutamḫuru*, 'to raise to square' (literally let stand against, let correspond to each other)".

The ensuing *ka-bi-rum* is interpreted (reasonably, if only the reading had been correct) as "breadth", and Weidner then supposes that it refers to the square understood as a "broad rectangle". The proper reading (as given in MKT) means "ground" (in mathematical texts the basis of a prismatic volume).

Weidner does not comment upon his interpretation of *ta-mar* in line 2 (and again in line 3) as "erhältst du", but it is obviously derived from the context. Neugebauer's philologically correct reading "you see" goes back to Ungnad (1916: 364).

In the end of line 2, Weidner understands *i-ši-ma* as "it is", which forces him to understand *a-na* (translated "auf") as a multiplication (without specifying that this is what he does).[22] Zimmern (1916: 322) and Ungnad (1916: 364) point out that *i-ši* is the imperative of *našûm*, "to raise", and that this term (always "raising *to*", which explains *a-na*) is used repeatedly for multiplication in CT IX; this understanding (but not the translation) recurs in MKT.

In the next line, the interpretation of *tab-ba* as doubling is correct, and goes back to (Delitzsch 1914: 152); only Neugebauer's familiarity with a much larger range of texts allows him to see that the scribe has omitted a number 2 – yet even he, trapped by the interpretation of the operation as just multiplication, does not see that *ana* should be taken in its ordinary sense "to" (doubling "until twice").

Also the interpretation of *daḫ-ḫa* as addition is correct.[23] The derivation of the closing phrase *ne-pí-šum* from the verb *epēšum*, "to do"/"to proceed" is correct too, even though the actual grammatical interpretation is mistaken, as pointed out by

[20] In (1929: 88), Neugebauer still accepted Weidner's interpretation. Arguments that a verbal imperative was most unlikely and an alternative orthography for *mīnûm* unsupported by other evidence were first given by Thureau-Dangin (1931: 195*f*); the idea that it is a (pseudo-) Sumerogram for that word was first hinted at by Albert Schott, see (Neugebauer 1932b: 8 n. 18).

[21] No longer needed, since other texts have the Akkadian *atta*.

[22] I wonder whether Weidner was led to this conclusion by numerical necessity alone (1 6 40 being indeed the product of 40 and 1 40) or by the parallel use of ἐπί in Greek mathematics.

[23] Unfortunately, Weidner's commentary equates this Sumerian word with *eṣēpu*, building on a hint in Delitzsch (1914: 134); Delitzsch's supportive examples are conjugated forms of *waṣābum*, also the actual equivalent in Old Babylonian mathematical texts.

Zimmern (1916: 322) and Ungnad (1916: 364); they both correct to "Verfahren", "way to proceed", as taken over by Neugebauer.

Weidner's article deals not only with this but also with another section of the same tablet, in which a different approximation to the same diagonal is found, namely $a+\frac{b^2}{2a}$ – much better, both by being meaningful and by being more precise even with the actual numbers and unit. On the tablet, this section comes first, and with hindsight it seems a reasonable assumption that the second method (the one Weidner presents first) is a second approximation gone awry.[24]

In connection with the first approximation, Weidner makes only two new terminological observations, one wrong and one slightly problematic. Firstly, he translates the passage *½ 2 30 dùg-bi 1 15 ta-mar* as "Die Hälfte von 2 30, als seinen Quotienten 1 15 erhältst du", believing from his inspection of the CT-IX texts (probably from parallels to the present passage) that DÙG stands for the result of a division, and taking *bi* to be the Sumerian possessive suffix (thus "its quotient"). Zimmern (1916: 322) corrected this Sumerographic reading, replacing it by phonetic Akkadian *ḫi-pí* "break" (*viz* "break off ½") – cf. also (Ungnad 1916: 364 n.5) (the signs read by Weidner and Zimmern are the same).

Secondly, Weidner (1916: 261) states that "*igi-dú-a* with a number enclosed between *igi* and *dú* means substantially [*sachlich*] that the ensuing higher power of 60 is divided by the enclosed number". Zimmern (1916: 324) specifies that *igi* must be understood as "part", and *dú* (du_8 since Thureau-Dangin's *Syllabaire*) as "to split", while Ungnad (1916: 366) suggests an interpretation that comes close to the determination of the reciprocal of the enclosed number[25] – clearly the understanding of the Old Babylonian calculators, as was soon to be known with certainty, whereas that of the Ur III inventors had probably been the corresponding fraction of 60 (cf. (Scheil 1915a) and (Steinkeller 1979)) – closer indeed to Weidner's understanding without being identical.

Beyond the attempted "substantial" and philological interpretation of the mathematics of the text, Weidner (1916: 259) also speaks about its purpose:

> Oriental science was never undertaken for its own sake but was always science with a purpose [*Tendenzwissenschaft*], and therefore the present piece of text was of course not written down by the Akkadian in order to show how right triangles were calculated in his times, but it must have had a very real background. It is probably the calculation of an architect or a surveyor, who has then executed his task in agreement with the calculation.

Later, as a commentary to the only approximate character of the calculations, Weidner continues thus:

> However, if we take into account, as already pointed out, that this is nothing but applied mathematics in the service of the architect and the surveyor, then we arrive at a milder judgement. We know sufficiently well, indeed, that these gentlemen do not always insist on maximal precision in their work.

---

[24] A possible interpretation is offered in (Høyrup 2002: 271*f*).

[25] He does not use the term "reciprocal" but speaks about the operation of dividing 1 by the number in question.

The insight that the text might be a school problem had to wait.

Beyond objections and direct commentaries to Weidner, some further important observations concerning the terminology are made by Zimmern and Ungnad. Zimmern (1916: 323) notices that two different terms express addition, *daḫ-ḫa*[26] and UL.GAR. Ungnad (1916: 367) points out that there are also two ways to express subtraction, the operation BA.ZI (Akkadian *nasāḫum*, "to tear out") and the observation that one entity exceeds (DIR) another one by so and so much; he also mentions KIL.KIL (NIGIN) (col. 366*f*) as a term for squaring and recalls the already known use of ÍB.DI (íb.si$_8$) for "square root".

Both also end their articles by hoping for new texts and new insights in the area. Apart from a transliteration and translation of another problem from CT IX (namely, BM 85194, obv. III, 23–30) produced by Ungnad in (1918), it lasted quite a while before this wish was fulfilled.

As already mentioned, the next text to be discussed was a large fragment of BM 15285, a text about the subdivision of a square in various smaller figures (Gadd 1922). It contains drawing of these together with verbal descriptions, and even though drawing and description are only conserved together in a few cases, Gadd was able to make new observations (1922: 151) on the terminology – not least to identify *mitḫartum* with a square, which, as he states, agrees perfectly with Zimmern's reading *šutamḫurum*, and to show that ÍB.DI (íb.si$_8$) was used ideographically for *mitḫartum*. He was also able to confirm the interpretation of *kippatum*[27] as "ring"/"circle", as derived already by Thureau-Dangin on the basis of non-mathematical texts, and to read SAG.KAK as "triangle".

As equally mentioned, in the same year Thureau-Dangin published a hand copy of AO 6484, a fairly long mathematical text from the Seleucid era, no. 33 of 58 texts from the collections of the Louvre and the Musée du Cinquantenaire. However, all he has to say about it is that it contains "arithmetical operations (fragmentary tablet). Probably from the first half of the second Seleucid century". In spite of his interest in anything that had to do with mathematics, evident since his astute analysis of a field plan from Ur III in (1897), he did not return to the text in the following years, which can probably be taken as evidence that he understood no more than what he had already said in 1922.

Then we come to Frank's edition from (1928)[28] of 50 texts from the Strasbourg collection, six of which were mathematical. Frank offers hand copies, translitera-

---

[26] To this he links Akkadian *ruddûm* instead of *waṣābum* – a mistake in the context of the mathematical texts, as it was to turn out when more of these became known.

[27] Gadd says *kibbatum*, but that orthography has already disappeared in (Bezold 1926: 147).

[28] According to what is written on p. 6, the hand copies were made in 1914, after which Frank had no more access to the tablets; he only received his old copies and notes in 1925, after which he could resume working on them. Actually, what Frank received through the mediation of a friend were only draft hand copies; what he had originally prepared for an edition arrived too late (Waschow 1932a: 211), cf. (Thureau-Dangin 1934).

tion and German translation of some of the texts, partial transliteration mixed with explanatory translation of the others – and a short general commentary (pp. 19*f*). In this commentary it is stated that

> the following texts, like those close to them in CT IX, cannot yet be understood in all details. More intensive work than what is intended here, and indeed on all "mathematical" texts, would in itself be most welcome.

The quotes around "mathematical", however, point back to an important insight, entirely missed in 1916: that these texts are *Rechenaufgaben*, that is, school texts.

Neugebauer Enters the Game

Very soon – as a matter of fact almost immediately – more intensive work was indeed taken up. Neugebauer had already published a paper in (1927) about the origin of the sexagesimal system, and in (1928) a short note from his hand pointed out that the approximation discussed second in (Weidner 1916) might be meant as an approximation to the exact value predicted by the Pythagorean theorem; he also suggested that both Greek geometry and the Indian śulba-sūtras might have borrowed from the Babylonians. The watershed was (Neugebauer 1929), appearing in the first issue of the *Quellen und Studien* B[29] and dealing with the mathematical Strasbourg texts. How much had happened can be illustrated by a confrontation of Frank's text of no. 10 with Neugebauer's new translation (1929: 67*f*) and transliteration (MKT I, pp. 259*f*) (the article from 1929 brings a translation only and a handful of notes correcting Franks transliteration). The figure in Neugebauer's first line is taken from his transliteration, but corresponds to what is found in his translation.

1 Oben Zahlen: 1, 3; 783, 1377. 1 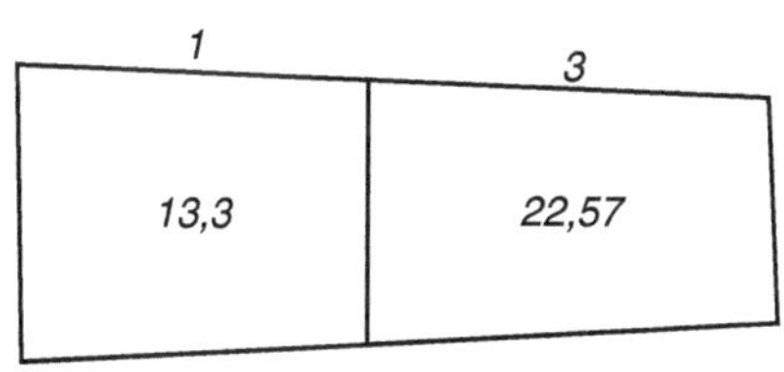

[29] Full title *Quellen und Studien zur Geschichte der Mathematik, Astronomie und Physik*. Abteilung B: *Studien*.

2	sag-gi-gud(!) (so wohl, nicht bi) *ina libbi* 2 íd-meš 783 a-šà(g) [sa]g(?) ein Viereck (*ummatu*), darinnen 2 'Flüsse', 738 das erste Feld,	2	SAG-KI-GUD *i-na libbi* 2 id-meš 13,3 a-šà an Ein Viereck, darinnen zwei Flüsse, 13,3 (=783) die obere Fläche,
3	1377 a-šà(g) *šanū*^{u} … 3 (?) gál uš-ki … 1377 das zweite Feld … $^{1}/_{3}$ untere Länge …	3	22,57 a-šà ⟨ki-⟩2 i[gi] 3 gál uš ki *i-n*[*a*] 22,57 (=1377) die zweite Fläche [und] ein Drittel der unteren Länge für
4	uš-an-na-ta sag(!)-an-na *eli* RI dirig-a von der oberen Länge an die obere Breite größer als RI	4	uš an-na *ša* sag an-na u-gù RI dirig die obere Länge, die obere Breite größer als die Trennungslinie
5	*ù* RI eli sag-ki-ta dirig gar-gar … igi (?) und RI größer als die untere Breite …	5	*ú* RI u-gù sag ki-ta dirig gar-gar ⸢36⸣ und die Trennungslinie größer als die untere Breite, zusammen
6	uš-ne-ne sag-meš *ù* RI en-nam Die Längen, Breiten und RI berechne	6	uš-ne-ne sag-meš *ù* RI en-nam Die Längen, Breiten und die Trennungslinie berechne
7	za-e ak-da-zu-de 1 *ù* 3 ḫ̆e-ga[r] Wenn du dabei (so) verfährst: 1 und 3 (seien) angesetzt(?);	7	za-e ki-da-zu-dè 1 *ù* 3 ḫ̆é-gar Du verfährst so: 1 und 3 lege (?)
8	1 *ù* 3 gar-gar 4 igi 4 dù-ma 15 1 u. 3 addiert = 4; (60) durch 4 dividiert = 15;	8	1 *ù* 3 gar-gar 4 igi 4 du$_{8}$-ma 15 1 und 3 zusammen (ist) 4. Das Reziproke von 4 (ist) 0;15 (=1/4) und
9	15 *a-na* 36 nim 540 in-se 540 *a-n*[*a*] 15 auf 36 erhöht gibt 540; 540	9	15 *a-na* 36 nim 9 in-sum 9 *a-na* 0;15 (=1/4) mit 36 erhöht gibt 9. 9 mit
10	1 nim 540 in-se 540 *a-na* 3 nim 1620 auf 1 erhöht gibt 540; 540 auf 3 erhöht = 1620;	10	1 nim 9 in-sum 9 *a-na* 3 nim 27 1 erhöht gibt 9. 9 mit 3 erhöht 27.
11	540-ta sag-an-na *eli* RI dirig um 540 ist die obere Breitseite größer als RI;	11	9 *ša* sag an-na u-gù RI dirig Um 9 ist die obere Breite über die Trennungslinie größer,
12	1620 ta (?) RI *eli* sag-ki-ta dirig um 1620(?) ist RI größer als die untere Breitseite.	12	27 *ša* RI u-gù sag ki-ta dirig um 27 ist die Trennungslinie gegen die untere Breite größer:

13 igi 1 dù 1 *a-na* 783 nim Divisor 1. 1 auf 783 erhöht	13 igi 1 du$_8$ 1 *a-na* 13,3 nim Das Reziproke von 1 ist 1. Mit 13,3 (=783) erhöht
14 783 in-se igi 3 dù 20 *a-na* gibt 783. Divisor 3 (d. h. 60: 3) 20 auf	14 13,3 in-sum igi 3 du$_8$ 20 *a-na* gibt 13,3(=783). Das Reziproke von 3 (ist) 0;20 (=1/3). Mit
15 1377 nim 27540 in-se 1377 erhöht macht 27540.	15 22,57 nim 7,39 in-sum 22,57 (=1377) erhöht gibt 7,39 (=459).
Rs.	Rs.
1 783 *eli* 459 en-nam dirig 783 ist größer als 459: berechne die Differenz.	1 13,3 u-gù 7,39 en-nam dirig 13,3 (=783) gegen 7,39 (=459) berechne den Überschuß.
2 324 dirig 1 *ù* 3 gar-gar 4 324 ist die Differenz. 1 und 3 addiert:=4;	2 5,24 dirig 1 *ù* 3 gar-gar 4 5,24 (=324) ist der Überschuß. 1 und 3 zusammen (ist) 4.
3 bar(!) (= *mišil*) 4 (!) QU 2 igi 2 dù 30 *a-na* 324 die Hälfte von 4 geteilt: 2; (60) durch 2 dividiert = 30, auf 324	3 1/2 4 gaz 2 igi 2 du$_8$ 30 a-na 5,24 Halbiere 4 (das ist) 2. Das Reziproke von 2 (ist) 0;30 (=1/2). Mit 5,24(=324)
4 9720 in-(se)-*ma* nu- GIR 9720 nu-dù gibt 9720, nicht …; 9720 nicht teilbar	4 2,42 in⟨-sum⟩-ma nu-GÌR 2,42 nu-du$_8$ gibt 2,42 (=162), nicht ..... 2,42 (=162) nicht teilbar
5 en-nam *a-na* 9720 ḫ̮e-gar *ša* 540 in-se berechne. Zu 9720 soll gelegt werden, 'daß, was 540 gibt'.	5 en-nam *a-na* 2,42 ḫ̮é-gar *ša* 9 in-sum Berechne mit 2,42 (=162) gelegt, was 9 gibt.
6 200 ḫ̮e-gar igi 200 dù 18 in-še 200 sei gelegt, durch 200 dividiert gibt 18(?);	6 3,20 ḫ̮é-gar igi 3,20 du$_8$ 18 in-sum 0;03,20 (=1/18) gelegt. Das Reziproke von 0;03,20 (=1/18) gibt 18.
7 18 *a-na* 1 nim 18 uš-an(!?)-na 18 18 auf 1 erhöht: 18 die obere Langseite; 18	7 18 *a-na* 1 nim 18 uš an-na 18 18 auf 1 erhöht (ist) 18. Die obere Länge (ist) 18
8 *a-na* 3 nim 54 uš-ki us-ki-ta auf 3 erhöht: 54 die untere Langseite; von (?) der unteren Langseite	8 *a-na* 3 nim 54 uš ki {uš ki-ta} Mit 3 erhöht: 54 (ist) die untere Länge von der [oberen] Länge aus

9	*mišil*(!) 36 sag(?)-ne 18 (statt 17!) *a-na* 72 nim die Hälfte von 36 die Breiten(?) 18(!), auf 72 erhöht	9	½ 36 gaz ne 17(sic) *a-na* 1,12 nim Halbiere die Breite 36. 18 mit 1,12 (=72)
10	1296 *i-na* 36 a-šà(g) dù 864 1296; durch 36 Felder(?) teilbar;	10	21,36 *i-na* 36 a-šà du_8 14,24 (ist) 21,36 (=1296). Von 36,00 (=2160) subtrahiert (ist) 14,24 (=864)
11	igi 72 ba-dù 50 *a-na* 864 nim 864 durch 72 teilbar; 50 auf 864 erhöht	11	igi 1,12 uš du_8 50 *a-na* 14,24 nim Das Reziproke von 1,12 (=72), der Länge, ist 0;00,50 (=1/72). Mit 14,24 (=864) erhöht
12	43 200 (!) in-se 22 4*a-na* 26(!) daḫ-ḫi-*ma* 48 GAB(?) gibt 43200 (!); 22 zu 26 (!) hinzugefügt = 48, teilbar (?),	12	12 in-sum 12 *a-na* 36 daḫ-ma 48 gibt 12. 12 mit 36 addiere. 48 [ist es.]
13	48 sag-an-na 12 *a-na* 27 daḫ 48 obere Breitseite; 12 zu 27 hinzugefügt	13	48 sag an-na 12 *a-na* 27 daḫ 48 die obere Breite, 12 mit 27 addiert:
14	39 RI 12 sag-ki-ta in-se 39 RI, gibt 12 von der unteren Breitseite aus.	14	39 RI 12 sag ki-ta in-sum 39, die Trennungslinie, von 12, der unteren Breite, gibt es

The most striking difference between the two translations is probably that Neugebauer conserves the sexagesimal place value notation (though still, probably as help to readers not accustomed to it, translating parenthetically into decimal notation). This is in any case the reason he gives to make a revised translation instead of just copying Frank, and we see indeed that Frank time and again locates the numbers in a wrong sexagesimal order of magnitude, which did not facilitate his understanding of this very complicated procedure.[30] Once this was corrected, Neugebauer was also able to correct a number of Frank's readings – but this was, if we are to believe his words, at least in the main a secondary effect of getting the numbers right.[31]

Of particular importance was Neugebauer's insight that igi *n* should be understood as the reciprocal of *n*. As we have seen, this almost coincides with what Ungnad had said in 1916 (but not fully, cf. below, note 43). However, Neugebauer's

[30] In MKT I, p. 263, Neugebauer characterizes it as *umwegig,* "roundabout". A possible understanding of the underlying idea, based on a proposal by Jöran Friberg, is in (Høyrup 2002: 241–244). The procedure itself was perfectly understood by Neugebauer.

[31] The interpretation of RI as "Trennungslinie", the parallel transversal dividing the trapezoidal quadrangle into two strips, is probably an exception to this rule; according to p. 70, n. 14 it was due to V. V. Struve.

explanation was much more transparent, and from now on it was generally accepted.[32]

From 1929 to 1935 there were few important but a number of less decisive changes in Neugebauer's translation. In obv. 2, the quadrangle becomes a trapezium, and the rivers become strips – but both in agreement with the commentary from 1929, there is no change in the interpretation. Obv. 4 becomes clearer, "die obere Länge. Was die obere Breite über die Trennungslinie hinausgeht", and the beginning of obv. 5 and a number of similar passages are modified correspondingly. In obv. 5 and elsewhere, "zusammen" becomes "addiert, and in obv. 6 and elsewhere the imperative "berechne" for en.nam becomes "was", in agreement with the understanding of this term as a logogram for *mīnûm*. In obv. 7, "Du verfährst so" becomes "Du bei deinem Verfahren", in better agreement with the Sumerian expression and indeed a perfect translation of the corresponding Akkadian phrase *atta ina epēšika*, with which Neugebauer was now familiar; further, "lege" becomes "mögest du nehmen", in better agreement with the precative prefix ḫe but less close to the semantics of ĝar; similarly elsewhere. In obv. 8, "Das Reziproke von 4 (ist) 0;15" becomes "Das Reziproke von 4 gebildet und 0;15 (ist es)"; this at least renders the presence of a verb du_8, even though it semantics ("split"/"detach", correctly described by Zimmern, cf. above) is not respected[33] (nor the imperative found in parallel syllabic texts); similarly elsewhere. In obv. 9, a change from literal to "substantial" translation takes place, and "erhöht" becomes "multipliziert". In rev. 3, on the other hand, "halbiert" becomes "abgebrochen" – here, the "substantial" translation is replaced by a literal one. Rev. 5 becomes "Was mit 2,42 sollst du nehmen, das 9 gibt", both clearer and closer to the original (apart from the semantics of ĝar) than the 1929 version. In rev. 8, MKT understands that the repetition in the end is a dittography, and the attempted repair from 1929 disappears. In rev. 9, "subtrahiert" becomes "brich ab", an attempted return from "substantial" to literal translation – not quite unobjectionable, "brich ab" is used in the preceding line and elsewhere for gaz/*ḫepûm*, while du_8 elsewhere designates the "detaching" or "splitting off" of a reciprocal (rendered "substantially" in MKT by "gebildet").[34] In rev. 12, "mit 36 addiere" becomes "zu 36 addiere", which fits the preposition *ana* better but still conflates the symmetrical operation ĝar.ĝar, connected with *u* ("and"), and the asymmetric operation daḫ, connected with *ana*; similar rev. 13.

In the programmatic statement for MKT (I, p. viii) it is said that "in principle, the translation is obviously literal",[35] but then explains why this principle cannot always

[32] As I have experienced several times, this does not mean that today's Assyriologists are generally familiar with the place value system. Indeed, unless they work on astronomical texts or mathematical school texts (very few do), they never see it in use.

[33] dù, we remember, had become du_8 in Thureau-Dangin's reform.

[34] Footnote 5a in (Neugebauer 1930a: 122) reveals that "subtrahiert" was chosen originally because Neugebauer had mistakenly believed to improve Frank's reading *a-šà(g) dù* by changing it into *uṣuḫ*. The same note shows that Neugebauer is now perfectly aware that the correct literal translation would be "abgespalten"; we may perhaps presume the deviating translation in MKT to be nothing but a slip.

[35] "Die Übersetzung ist selbstverständlich im Prinzip eine wörtliche".

be respected – a dilemma every translator knows all too well. As we see, the 1929 version followed the same rule – but not in the same way; sometimes, MKT becomes more literal than the early translation, sometimes less. For the purpose of understanding what Neugebauer saw as the mathematical structure of the texts, this was immaterial.

The 1929-article also dealt with Frank's text no. 8, in front of which Frank had given up, offering no transliteration and only translation of small isolated bits. The text is indeed very difficult, firstly because it is badly broken, secondly because it gives only problem statements (fortunately illustrated by diagrams) but no indication of the procedure. Also fortunately they can be arranged in groups that belong together. Taking advantage of this, Neugebauer succeeded in reconstructing the problems, and showed that they presuppose the ability to solve mixed second-degree equations; in the final paragraph (pp. 79*f*) he summarized the outcome of the analysis:

> One may legitimately say that the present text presents us with a fair piece of Babylonian mathematics that enriches our all too meagre knowledge of this field with essential features. Quite apart from the use of formulas for triangle and trapezium we see that complex linear equation *systems* were drawn up and solved, and that the Babylonians drew up systematically problems of *quadratic* character and certainly also knew to solve them – all of it with a computational technique that is wholly equivalent to ours. When this was the situation already in Old Babylonian times, in future one will have to learn to look at the later development with different eyes.

In a note added after the proofs were finished (that is, in March 1929), Neugebauer points out that the solution of a problem from CT IX (namely BM 85194, rev. II, 7–21) shows how to solve quadratic equations, and acknowledges the decisive contribution of Schuster.[36] Schuster himself published an analysis of the second-degree *igûm-igibûm* problems from the Seleucid text AO 6484 in the following issue of *Quellen und Studien* B (1930).

In the first issue, Neugebauer and Struve (1929) had published an article purportedly dealing with the Babylonian treatment of the geometry of the circle, actually also with the truncated cone as well as with other configurations that allowed Neugebauer to establish UR.DAM as a term for the height in a plane or solid figure.[37] Apart from establishing which mathematical insights, method and "formulae"[38] were used in the texts, this and subsequent publications in *Quellen und Studien* B[39] (and one in Weidner's *Archiv für Orientforschung*, namely (Waschow 1932b)) thus established the meaning of a number of technical terms while giving more precision

---

[36] This is why Schuster's discovery should probably be dated in early 1929.

[37] Actually, verbal forms of *warādum* ("to descend") are involved, but for the immediate technical purpose this was not decisive, as observed by Thureau-Dangin (1932b: 80) in the note where he made the grammatical analysis of the term.

[38] In the sense of "standard schemes" – no symbolic writing was of course intended as far as the Babylonians were concerned.

[39] For instance (Schuster 1930), (Neugebauer 1932b), (Waschow 1932a).

to earlier proposals or putting them on a firmer ground. Thureau-Dangin (1931: 195) was thus mistaken when believing in a kind of division of labour, where he was going to take care of terminology and grammar and Neugebauer of the substance.[40] As it turned out, he was also mistaken on his own account, from (1932a) onward he too was to take up both aspects of the texts – and in a note from (1933a: 310), Neugebauer could justly point out that a philological disagreement between the two was due to a "substantial" disagreement about the construction of a fortification.

Beyond mathematical substance and terminology, Neugebauer and the other contributors to *Quellen und Studien* also elucidated the historical setting of the mathematical texts, to the very limited extent it could at all be done at the moment.[41]

In (1932b: 6*f*), Neugebauer made a first (fully adequate) division of the Old Babylonian material into two groups, represented respectively by the Strasbourg texts and the CT IX-texts. He further correctly suggests that the former are slightly older and the latter slightly younger, and even (probably also correct, see (Goetze 1945: 149)) that the Strasbourg texts are from Uruk, and that AO 8862, though not properly a member of the Strasbourg group, is still likely to be related to it.

Negatively, Neugebauer points out in the conclusion of the same paper (p. 24) that the Old Babylonian mathematical texts are wholly unconnected to astronomy, and that they go far beyond the practical concerns of surveying and accounting. This was a rebuttal of opinions held by many Assyriologists at the time, expressed for instance by Bruno Meissner (1920: II, 380) – cf. also (Weidner 1916: 259) as quoted above, and Hilprecht quoting Bezold in note 7. Already in (1929: 73) Neugebauer had pointed out that the Strasbourg problems were constructed in such a way that they produced neat solutions – which implies that they were *constructed*, and thus that they were school texts and not a surveyor's working notes. This had already been understood by Frank (1928: 19) (cf. above), but Frank had underplayed his insight even more than Neugebauer did here.

A last insight into the cultural embedding of Babylonian mathematics – in this case, of the Seleucid period – was due to Schuster. In (1930: 194) he observes that the colophon of AO 6484, like that of other tablets published in (Thureau-Dangin 1922), shows it to have been written by "a representative of a large family of priests known since long from other texts from the Seleucid epoch".

---

[40] "[...] les études d'O. Neugebauer, qui ont pour objet plutôt le fond que la forme des textes, apportent au philologue d'utiles données".

[41] This limitation was emphatically pointed out by Neugebauer in (1932b: 24). In (1934a: 204), he was perhaps even more emphatic when pointing out that "we still know practically nothing about how Babylonian mathematics was situated within the overall cultural framework".

We may take it as an expression of the same explicitly agnostic attitude that Neugebauer never spoke of Babylonian "mathematicians". We may recognize *mathematics* in the texts, but nothing was known about the social role of their authors, in particular, whether any social role or identity (even a part-time role or an aspect of identity) would allow this characterization.

The Sexagesimal System

The understanding of the sexagesimal place value system was mentioned several times above, but some aspects of it deserve separate discussion.

I shall not go into the speculations of Thureau-Dangin, Neugebauer and others concerning its origin: before the metrological and numerical notations of the proto-literate period were deciphered,[42] all such attempts had to remain speculations – some of them sensible, some of them definitely not sensible, but never more than speculations.

Until this point, (Ungnad 1917) was not mentioned, even though this publication was often referred to during the critical years. It was important both for the information it gave and for the problematic traces it left.

Ungnad discussed (pp. 41*f*) the boasting of Assurbanipal that he was able to "*u-pa-ṭar I.GI A.DU.E it-gu-ru-tí*". He pointed out that *itguru* (<*egērum*, "to twist", "to be(come) twisted/confused/...") meant "complicated", and took *paṭārum* to mean "solve" (that is, solve problems). Since A.DU can be read a.rá, a term familiar from tables of multiplication as well as lexical lists, it had to mean "multiplication"; finally, concerning I.GI, a phonetic writing of igi, he claimed with reference to (Hilprecht 1906: 21*ff*) that *[x] IGI y GÁL.BI = z* means that *[x]: y = z*, that is, that the term refers to division – which of course seemed to make beautiful company with the multiplication.[43] On the whole, Assurbanipal was thus supposed to have boasted that he was able to solve complicated problems of division and multiplication. This interpretation of the quotation was still repeated by Adam Falkenstein in (1953: 126), whereas (Fincke 2003: 111) "straightens" it into "I solved complicated mathematical problems".

In (1929), Neugebauer had already translated igi as "reciprocal". However, in an editorial note to Schuster's analysis of what is now known as *igûm-igibûm*-problems (Schuster 1930: 196 n.1), he cites Ungnad for the insight that igi may mean reciprocal, but also (in the Assurbanipal-passage) "division" simply/*schlechthin*. Written with sign names, the two unknown quantities dealt with in the problems are ŠI and ŠI.BU.Ú. ŠI may also be IGI, for which reason Schuster called them *igû* and *šipû*. Inspired by Ungnad Neugebauer now feels tempted to translate the former term "divisor", and since the two quantities turn out to be each other's reciprocals, this seems to him to suggest that the latter term should be translated "multiplier". Since *šipû* could not in any way be connected to a known term for multiplication, he ended up by opting for *Nenner* ("denominator") and *Zähler* ("numerator"), though characterizing the choice as "disputable".

When MKT was published, ŠI.BU.Ú had become *igi-bu-ù*. Yet Neugebauer still uses the same "disputable" translation of the two terms, in the absence of more adequate words; he is quite aware and explains (MKT I, p. 349) that they constitute a pair of reciprocal numbers (already Schuster had assumed that this was meant by

[42] That is, until (Friberg 1978; 1979) and the definitive analysis in (Damerow and Englund 1987).

[43] Ungnad's failure to take his own article from (1916) into account indicates that he had not yet fully realized that igi designates the reciprocal, not a quotient in general – cf. Neugebauer's remark in (1930b: 187 n. 8).

the text). The first to recognize that the two terms are Akkadianized forms of Sumerian igi and igi.bi, "igi" and "its igi" was Thureau-Dangin (1933: 183*f*).[44] This insight was then taken over in MCT (p. 130) by Neugebauer and Abraham Sachs.

However, the story did not end here. In H. Goetsch's "Die Algebra der Babylonier" (1968: 83), a problem supposedly dealing with *Nenner* and *Zähler* is quoted – but without Neugebauer's explanation that these names are used in the absence of better alternatives. Nor is it revealed that they stand for a pair of reciprocals – perhaps because this is told by Neugebauer in connection with a different problem.

More rectilinear was the progress in the understanding of how the sexagesimal place value system works. In (1930b: 188–193), Neugebauer described the *system* constituted by tables of reciprocals and multiplication (not yet being aware that this system is Old Babylonian and thus does not concern the large Seleucid table of reciprocals AO 6456) – in particular that those numbers that occur as multiplicands (Neugebauer's *Kopfzahlen*) are those that turn up as reciprocals,[45] the multiplicand 7 being the only exception – in Neugebauer's later terminology, today in general use, *regular numbers*. In (Neugebauer 1931b), these results were presented in a more systematic way and on the basis of a larger text material; but now the irrelevance of the Seleucid material was recognized.

The two articles develop the idea that the system of tables was originally meant as a way to express general fractions,[46] and only accidentally became a system based on place value – in particular due to the presence of the table with multiplicand 7, because of which the tables contained everything needed for any multiplication. This idea (as well as the idea that creation of the place value system was inspired by metrology, which Neugebauer had maintained since (1927)) was made possible by neglect of the fact that more than a millennium of sexagesimal absolute value counting precedes the place value notation.[47] Given the apparently very sudden implementation during Ur III (a process of which no hints were known in the 1930s), an only accidental development is now implausible, and the development from weight metrology (now known to be created much later than the absolute sexagesimal system) impossible.[48]

Two more articles in *Quellen und Studien* B deal with the place value system: (Neugebauer 1931c) is a mathematical analysis centred upon the notion of regular/

---

[44] In (1932a: 52), Thureau-Dangin still speaks of *igû* and *šibû*.

[45] This observation had already been made by Hilprecht (1906: 21), but did not make much sense in his context of "Plato's number".

[46] This idea could possibly explain his otherwise not obvious translation of igi/igi.bi as "Nenner"/"Zähler".

[47] The absolute sexagesimal system is described in (Thureau-Dangin 1898: 81*f*). That it goes back to the fourth millennium BCE was not known in 1898, nor in 1930, but in any case it precedes every hint of use of the place value notation by many centuries; besides, the original curviform character of its signs shows them to belong with the earliest phase of writing.

Neugebauer does discuss the absolute system in (1927: 8–13), but mixing it up with speculations that thwart his understanding.

[48] Since they are peripheral to my topic, I shall not document these claims, just refer to (Powell 1976) as a seminal publication.

irregular numbers, (Neugebauer 1932c) proposes how AO 6456, the big Seleucid table of reciprocals, might have been constructed (suggesting also that the same method was used for the Old Babylonian standard table with its 30 or fewer entries). None of them are of specific interest for the present investigation.

Neugebauer's contributions concerned the internal structure of the place value system. Thureau-Dangin's *Esquisse d'une historie du système sexagésimal* from (1932a) is very different in approach. It deals with other aspects of Sumerian and Akkadian numeration too, including spoken numerals as well as the absolute sexagesimal system and the absolute notations for fractions, and shows that metrological systems, though compatible with sexagesimality, cannot be the starting point of the sexagesimal system, whether place-value or absolute (while recognizing on p. 33 that the use of gin in the generalized sense of a sixtieth is probably borrowed from metrology). It also points out very explicitly that the place value system was introduced as an *instrument de calcul* (p. 51). This publication can thus be considered a culmination and completion of the development of the preceding eight decades, and gives much more insight into the overall numerical culture of ancient Mesopotamia than Neugebauer's papers on the topic from 1930 to 1932. However, even though a strongly revised version appeared in English translation in *Osiris* in (1939), and even though it also reveals its author's broad knowledge of relevant aspects of the mathematics of other pre-Modern cultures (from ancient Egypt and Greece to Fibonacci and Stevin), this study never had much impact on the historiography of mathematics.

Neugebauer's Project

The preceding two sections concerned what Neugebauer did concretely to the understanding of Babylonian mathematics. This, however, was part of a programme, which is expressed in the inaugural statements from the first issue of *Quellen und Studien* B (Neugebauer et al. 1929: 1–2). Here we read:

> Through the title *Quellen und Studien* we want to express that we see in the constant reference to original sources the necessary condition for every serious historical research. It shall thus be our first aim to make accessible *sources*, that is, to offer them inasfar as possible in a form which not only may meet the demands of modern philology but also, through translation and commentary, will enable the non-philologist to check for himself the words of the original in any moment. To fulfil the legitimate requests of *both* groups, philologists and mathematicians, will only be possible if we succeed in producing close collaboration between them. To open the road for that will be one of the main purposes of our undertaking.

The *Quellen und Studien* were to appear in two sections:

> One, A, *Quellen*, will contain the actual large editions, containing the text in its original language, a philological apparatus and as literal a translation as possible, which makes the text as accessible also for the non-philologist as can be done. [...]. The issues of section B,

> *Studien*, will collect articles that are closely or less closely associated with the material that can be drawn from the sources.
>
> The *Quellen und Studien* will offer contributions to the history of mathematics. However, they do not address specialists of the history of science alone. They will certainly propose their material in a form which may *also* be useful for the specialist. But beyond that they address all those who feel that mathematics and mathematical thought are not only concerns of a particular science but profoundly connected to the totality of our culture and its historical development, and that a bridge can be found between the so-called *Geisteswissenschaften* and the apparently so ahistorical "exact sciences". Our final aim is to participate in the building of such a bridge.

Unfortunately, as Neugebauer had to observe in (1934a: 204), "we still know practically nothing about how Babylonian mathematics was situated within the overall cultural framework" – cf. above, note 41. The bridge he was able to build was thus one between mathematics and highly technical Assyriological philology. No doubt, even the general educated public (not to speak of historians of mathematics) would find the latter field much more arcane than the former.

Another kind of programmatic statement is found in (Neugebauer 1933b: 316*f*)[49] – a kind of elaboration of the negative conclusions of (1932b: 24). Neugebauer starts by summing up polemically the picture of Mesopotamian mathematics that had been derived from field plans and tables: "the level of purely empirical mensuration, loaded with all kinds of number-mystical ballast" – "chaldaeic wisdom" which was then supposed (cf. above, note 7) to be

> continued in Pythagorean wisdom, from which by pure miracle exact Platonic mathematics grew out: indeed a miracle, this almost unmediated transition from Pythagorean number mysticism to a rigorous theory of irrational numbers operating with the class separation of "Dedekind's cut".

Thanks to "the work of Junge, Vogt, E. Sachs, Frank and others", he goes on, this construction had been deprived of one of its main pillars, the Pythagoras legend. The destruction of the other main pillar, the belief in purely empirical and numerological Babylonian mathematics, was now to be accompanied by the introduction of a new understanding of Old Babylonian mathematics: not at all in the style of [Greek Euclidean] geometry but rather of "pure formal-algebraic character". In terms of a later epoch, the programm is thus anti-Orientalist, anti-new-age in spirit. In 1933, readers may have observed implicit anti-Spenglerianism.

Neugebauer still published a number of articles on Babylonian mathematics in *Quellen und Studien* B and other journals during the next few years.[50] In Vol. **4** of *Quellen und Studien* B from 1937 to 1938, however, he has five articles on ancient astronomy but nothing more on non-astronomical mathematics, Mesopotamian or otherwise. By then we may say that his work on Babylonian mathematics had come to an end, apart from the volume he prepared with Sachs in 1945 (MCT), which can

[49] The main theme of this article is the link between, on one hand, tables of cubes and cube roots (known since Rawlinson) and a recently discovered tabulation of n^3+n^2, on the other the third-degree problems of a text now known as BM 8200+VAT6599.

[50] Among these, I shall mention in particular (Neugebauer 1934b), the first description of the mathematical series texts.

be seen as a mandatory supplement to MKT, necessitated or at least invited by the new texts to which he had got access by then. Neither the discovery and publication of a number of texts from Eshnunna (Baqir 1950a, b, 1951, 1962) nor the problematic edition of the mathematical Susa texts (nor E. M. Bruins's venomous slander) ever provoked him to take up the topic again.

MKT is thus at the same time a marvellous culmination and a farewell.[51] Whatever programmatic statement we find here may therefore be considered definitive.

Actually, we find very little. MKT appeared in three volumes in *Quellen und Studien* A in 1935–1937; with due respect for Struve's edition of the Moscow Papyrus (1930), it was certainly the weightiest contribution to this section. As we remember, section A was to "contain the actual large editions, containing the text in its original language, a philological apparatus and as literal a translation as possible, which makes the text as accessible also for the non-philologist as can be done". In agreement with this description of the section, the *Vorwort* of vol. I (p. v) starts by stating that the purpose of the work "from the very beginning [in 1929] was to procure a complete collection of all mathematical cuneiform texts", and that this aim had been achieved in the sense that probably no essential published material had escaped notice, while all unpublished material to which Neugebauer had had access had been included.

As mentioned above, there is also a programmatic statement (p. viii) that "in principle, the translation is obviously literal", but that this principle cannot always be respected. But that is all.

In the end of vol. III (pp. 76–80) we then find a *Rückblick*, a retrospect on the three volumes. It mostly contains tentative conclusions and delineations of open questions, but one passage (p. 79) confirms the apparently restrictive programme formulated in the *Vorwort*:

> It does not belong among the tasks that I have proposed for myself in this edition to develop the consequences which can be drawn from this text material. I have outlined them within a broader framework in my *Vorlesungen* (Neugebauer 1934a/JH), and sketched the connections to Greek mathematics in a work "Zur geometrischen Algebra" (Neugebauer 1936/JH); I hope to finish in the not too distant future a detailed investigation of all questions pertaining to the history of terminologies [which never appeared/JH].

Still, the following page – apart from indexes and reproductions of tablets the final page of the work – draws some general conclusions. These pertain not least to the nature of and conditions for the development of early mathematics (p. 80):

[51] It can hence be considered a paradox that Assyriologists, after the appearance of MKT, tended to put aside any tablet containing too many numbers in place-value notation as "at matter for Neugebauer" (as formulated to me with regret by Hans Nissen at one of the Berlin workshops on "Concept Formation in Mesopotamian Mathematics" in the 1980s). As we have seen, the fathers and giants of Assyriology, from Hicks, Rawlinson and Oppert to Thureau-Dangin, considered anything mathematical as very important. Even after the revival of active work on Mesopotamian mathematics during the last three decades and many new insights, an *Encyclopedia of Ancient History* planned by Blackwell and Wiley in 2009 suggested 500 words for "Mathematics, Mesopotamian" – the same as was dedicated to Mesopotamian hairstyles (I succeeded in raising the limit to 700 words).

> Since our knowledge of these things is of relatively recent date, and current datings had to be pushed considerably, there is an obvious danger to overestimate the mathematics of the Babylonians. In order to somehow gloss over the lack of a basis in sources, many familiar books change elementary mathematical things into "propositions" and "discoveries" that must be ascribed to great men. It seems to me that we should not stamp the Babylonians as such discoverers. What is often overlooked and cannot be sufficiently emphasized is the terrible difficulty and slowness of the development of the very simplest fundamental mathematical concepts, first of all of a genuine computational technique. This, however, is not the achievement of a single person; it can only be understood within a historical process, inextricably attached to the emergence of a whole culture.

So, the broader programme of the *Quellen und Studien* had not been forgotten – only the limits imposed by available sources (and by the lack of relevant sources) prevented Neugebauer from filling it out.

Why Neugebauer, Why Göttingen?

As we have seen, many outstanding Assyriologists had been interested in everything mathematical they could get their hands on. Gradually, they had come to understand the many different metrologies well (except those of the proto-literate period). Assyriologists' attempts to understand the two CT IX texts and the Strasbourg texts had yielded important insights into the mathematical terminology; actually, most of the basic vocabulary for mathematical operations was already acceptably well understood in 1928, thanks to Weidner, Zimmern, Ungnad and Frank. When it came to understanding such texts, however, progress was blocked.

On the other hand, once the breakthrough had been effectuated by Neugebauer, Struve and Schuster, even Thureau-Dangin was able to participate in the new development. What was so special, we may ask, about Neugebauer and his Göttingen circle, which allowed the opening of a road which even the most eminent of Assyriologists had not been able to find on his own?

Other Assyriologists may have been blocked by their expectation that the Babylonians could have engaged in nothing but "empirically based" practical calculation. As cited above, this was the opinion expressed by Meissner in his survey from 1925. Some, like Frank, may have been stopped by the habit to translate all numbers into Arabic numerals, sometimes mistaking orders of magnitude (and, in general, by not understanding to the full the floating-point nature of the sexagesimal place-value notation). Many will surely have had a mathematical training that did not suffice as support for the mathematical fantasy required for the task.

None of this is valid for Thureau-Dangin, except perhaps the low expectations concerning the level of the Babylonians[52] – nor indeed for Allotte de la Fuÿe, ancient

[52] His characterization of AO 6484 as "arithmetical operations" might suggest exactly such low expectations.

polytechnicien who, though born in 1844, was still quite active, but mostly interested in third-millennium documents.[53]

Assyriologists, including Thureau-Dangin, were of course interested in many other topics than mathematics. As long as nothing beyond practical calculation was expected to exist, they may simply not have looked for it; once it was known there was something to be found, that situation may have changed. After all, however, it only changed radically in the case of Thureau-Dangin, as illustrated by Wolfram von Soden's case. Von Soden was certainly interested in mathematics: he wrote extensive and thorough reviews of MKT in (1937), of TMB in (1939) and of TMS in (1964); I also experienced his interest in the mathematical area personally in correspondences I had with him during the 1980s. He was even (as far as I am aware of) the first to suspect publicly that the picture of Babylonian mathematics constructed by Neugebauer and Thureau-Dangin was too modernizing (von Soden 1937: 189–191), which might well have spurred him to pursue this particular interest. Apart from the reviews, though, only two publications from his industrious hand deal with mathematics as such[54]: an analysis of a number of problem texts from Eshnunna from (1952), and a collaborative work (Gundlach and von Soden 1963) treating a problem text from Eshnunna and one from Susa. Even in Thureau-Dangin's case his *full* concentration of matters mathematical only lasted some 5 years, from 1932 to 1936. This can be seen in his "Notes assyriologiques", containing miscellaneous observations on the material he worked on: during these years, almost everything in these notes concerns mathematics and its applications; before 1932 and after 1936, that is not the case.

In any case, Neugebauer and his collaborators initiated a breakthrough where nobody else had succeeded. It is important not to leave out from this observation the collaborators and participants in the seminar: the contributions of Schuster, Struve, Heinz Waschow and Schott, all fully trained and active Assyriologists, are very visible in *Quellen und Studien* B, and explicitly acknowledged by Neugebauer in MKT and elsewhere.[55]

---

[53] In 1930 he published an article on protoliterate (Jemdet Nasr) metrology and mensuration, in 1932 another one on AO 6456, the Seleucid table of reciprocals. On his mathematical interest and competence, see (de Genouillac 1939).

[54] I disregard publications where general *a priori* ideas about the nature of Mesopotamian mathematics enter as part of a broader argument, such as (von Soden 1936).

[55] It may perhaps be adequate to recapitulate some elements of what these four Assyriologists did later in connection with Mesopotamian mathematics.

Schuster published oft-cited works on Sumero-Babylonian bilingual texts in 1938 and on Hatto-Hittite bilinguals in 1974 and 2002; he lived until 2002, but seems not to have worked on mathematical texts after 1930.

Struve, as curator of the cuneiform collection of the Ermitage in Leningrad, analyzed its corpus of Ur III accounts, which induced him to draw a very grim picture of the social system that implemented the place value system in its social engineering (Struve 1934) – a picture that has now been amply confirmed by Robert Englund (1990). He lived until 1965 but seems never to have published more on "mathematics proper".

Neugebauer's personal stamina and competence may have been decisive – we are dealing with the statistics of very small numbers, where personalities count for very much and become primary facts allowing no full explanation from or reduction to general factors. But it was probably important for this stamina and competence to come into play, *both* that Neugebauer himself was not primarily an Assyriologist but a historian of mathematics (in the dichotomy of (Neugebauer et al. 1929), not primarily "a philologist" but "a mathematician"), *and* that he was able to enter into close collaboration with and inspire a number of Assyriologists.[56] As a non-Assyriologist, he could concentrate (at least until 1937) on Babylonian *mathematics* alone, and thereby come to know the totality of the corpus much better than anybody had done before 1929; The preface of MKT I (p. v) lists how few higher-level texts were at all known by then. However, his deep respect for sources, as reflected in the programme for *Quellen und Studien*, caused him to seek philological collaboration and advice, and kept him free of the danger of rational reconstruction based on what the Babylonians *might* have done, if only they had been more or less Greek or more or less modern mathematicians. In Leopold von Ranke's words (in the sense von Ranke really used them in 1824 (von Ranke 1885: vii), against lazy invention and too hasty generalization), Neugebauer's proudly modest aim was to find out *wie es eigentlich gewesen*.

Correspondingly, it was probably decisive for the way in which Thureau-Dangin could contribute when the parallel work of the two began, that *his* starting point was that of the philologist, a reader and interpreter of texts, also in his approach to the history of mathematics (where he was much more akin to for instance Kurt Vogel than to Neugebauer). Even his aim is covered by von Ranke's maxim.

Waschow prepared an edition of the important Seleucid problem text BM 34568, published in MKT III (pp. 14–22). In his dissertation from (1936), an edition of letters from the Kassite period, he states in the (unpaginated) CV that he had entered active army service in 1934 and was at the moment serving as a non-commissioned anti-aircraft officer while intending to continue scholarly activity in parallel. In 1938 he published *4000 Jahre Kampf um die Mauer*, about siege techniques since Old Babylonian times. I can find no later traces of him and assume that he is one of those collaborators of Neugebauer who according to Vogel (private communication) fell in the war.

According to Neugebauer (MKT I, p. ix), Schott contributed intensely to MKT. He was also one of Neugebauer's intended collaborators in the publication of the corpus of Babylonian astronomical texts, planned around 1935 (see the description in (Neugebauer 1937)) – not realized immediately because of the war. Schott died at the end of the war in 1945 (Thompson 2010). He had also collaborated with the astronomer Paul Neugebauer on other aspects of Mesopotamian astronomy, and he translated the Gilgameš-epic in 1934 (eventually published with revisions by von Soden in 1958).

As we see, "mathematics proper" did not stay central to those three who had the possibility to make Assyriological work after 1936. Though also engagin in other matters, Thureau-Dangin was actually more tenacious as regards mathematics, as expressed in his (1940a, b).

[56] As Neugebauer tells with gratitude in (1927: 5), he has also been well counselled and trained by Anton Deimel during a fairly long stay at the Pontificium Institutum Biblicum in Rome, as his initial interest in Mesopotamian mathematics (as a parallel elucidating the foundations of Egyptian mathematics) had first been stimulated by works of Thureau-Dangin (1898, 1921) and Deimel (1922).

Thureau-Dangin's starting point had been the classical stance of Assyriologists: in order to understand Mesopotamian sources and civilization, it was mandatory to understand metrology and mathematics. Reversely, for Neugebauer, the Göttingen seminar and the *Quellen und Studien* programme, understanding Babylonian mathematics was necessary for understanding *mathematics* as the product of an ongoing historical process. However, it was essential for the outcome that both left aside these motivations (or at least behaved as if they had), and took up "Babylonian mathematics" as a research project that was of major interest in itself and needed no further excuse.

For the fruitful outcome of the race it was also essential that the two, in spite of the unmistakeable animosity which gradually developed between them,[57] in general remained respectful when citing each other and constructive in their mutual criticism, and even allowed each other access to whatever material was needed.[58] Great moral models for all scholars, and giants on whose shoulders it was always a pleasure to stand.

References

AHw: von Soden, Wolfram. 1965–81. *Akkadisches Handwörterbuch*. Wiesbaden: Otto Harrassowitz.

Allotte de la Fuÿe, F.-M. 1915. Mesures agraires et formules d'arpentage à l'époque présargonique. *Revue d'Assyriologie et d'Archéologie Orientale* 12: 117–146.

Baqir, T. 1950a. An important mathematical problem text from Tell Harmal. *Sumer* 6: 39–54.

Baqir, T. 1950b. Another important mathematical text from Tell Harmal. *Sumer* 6: 130–148.

Baqir, T. 1951. Some more mathematical texts from Tell Harmal. *Sumer* 7: 28–45.

Baqir, T. 1962. Tell Dhiba'i: New mathematical texts. *Sumer* 18: 11–14, pl. 1–3.

Bezold, C. 1926. *Babylonisch-Assyrisches Glossar*. Heidelberg: Carl Winter.

Brünnow, R.E. 1889. *A classified list of all simple and compound cuneiform ideographs occurring in the texts hitherto published, with their Assyro-Babylonian equivalents, phonetic values, etc.*. Leiden: Brill.

CAD: 1964–2010. *The Assyrian dictionary of the Oriental Institute of Chicago*, 21 vols. Chicago: The Oriental Institute.

Cory, I.P. 1832. *Ancient fragments of the Phoenician, Chaldæan, Egyptian, Tyrian, Carthaginian, Indian, Persian, and other writers*, 2nd ed. London: William Pickering.

CT IX: 1900. *Cuneiform texts from Babylonian tablets, &c., in the British Museum*, Part IX. London: British Museum.

Damerow, P., and R.K. Englund. 1987. Die Zahlzeichensysteme der Archaischen Texte aus Uruk. In *Zeichenliste der Archaischen Texte aus Uruk*, ed. M.W. Green and H.J. Nissen, 117–166, Band II, Kapitel 3 (ATU 2). Berlin: Gebr. Mann.

[57] "They hated each other", I was told by Olaf Schmidt, Neugebauer's assistant during his stay in Copenhagen. Schmidt, too gentle to hate anybody as far as I can judge, may have mistaken animosity for genuine hatred.

[58] Given the general unreliability of Evert Bruins, I permit myself to disregard what he claimed in a letter to me: that Thureau-Dangin took care that Neugebauer should not get access to the mathematical texts from Susa, which had been found in 1933.

de Genouillac, H. 1939. Allotte de la Fuÿe (1844–1939). *Revue d'Assyriologie et d'Archéologie Orientale* 36: 41–42.

Deimel, A. 1922. *Die Inschriften von Fara. I, Liste der archaischen Keilschriftzeichen*, Wissenschaftliche Veröffentlichungen der Deutschen Orient-Gesellschaft 40. Leipzig: J.C. Hinrichs'sche Buchhandlung.

Delaporte, L. 1911. Document mathématique de l'époque des rois d'Our. *Revue d'Assyriologie et d'Archéologie Orientale* 8: 131–133.

Delitzsch, F. 1914. *Sumerisches Glossar*. Leipzig: J. C. Hinrichs'sche Buchhandlung.

Englund, R.K. 1990. *Organisation und Verwaltung der Ur III-Fischerei*, Berliner Beiträge zum Vorderen Orient 10. Berlin: Dietrich Reimer.

Epping, J., and unter Mitwirkung von P.J.N. Strassmaier. 1889. *Astronomisches aus Babylon*. Freiburg im Breisgau: Herder.

Falkenstein, A. 1953. Die babylonische Schule. *Saeculum* 4: 125–137.

Fincke, J.C. 2003. The Babylonian texts of Nineveh: Report on the British Museum's *Ashurbanipal Library Project*. *Archiv für Orientforschung* 50: 111–149.

Fossey, C. 1907. *Manuel d'assyriologie*. Tome premier. *Explorations et fouilles, déchiffrement des cunéiformes, origine et histoire de l'écriture*. Paris: Leroux.

Frank, C. 1928. *Straßburger Keilschrifttexte in sumerischer und babylonischer Sprache*, Schriften der Straßburger Wissenschaftlichen Gesellschaft in Heidelberg, Neue Folge, Heft 9. Berlin/ Leipzig: Walter de Gruyter.

Friberg, J. 1978. The third millennium roots of Babylonian mathematics. I. A method for the decipherment, through mathematical and metrological analysis, of Proto-Sumerian and Proto-Elamite Semi-Pictographic inscriptions. *Department of Mathematics, Chalmers University of Technology and the University of Göteborg* No. 1978–9.

Friberg, J. 1979. The early roots of Babylonian mathematics. II: Metrological relations in a group of semi-pictographic tablets of the Jemdet Nasr Type, Probably from Uruk-Warka. *Department of Mathematics, Chalmers University of Technology and the University of Göteborg* No. 1979–15.

Friberg, J. 1982. A survey of publications on Sumero-Akkadian mathematics, metrology and related matters (1854–1982). *Department of Mathematics, Chalmers University of Technology and the University of Göteborg* No. 1982–17.

Friberg, J. 1993. On the structure of cuneiform metrological table texts from the -1st millennium. *Grazer Morgenländische Studien* 3: 383–405.

Gadd, C.J. 1922. Forms and colours. *Revue d'Assyriologie et d'Archéologie Orientale* 19: 149–159.

Goetsch, H. 1968. Die Algebra der Babylonier. *Archive for History of Exact Sciences* 5(1968–69): 79–153.

Goetze, A. 1945. The Akkadian dialects of the old Babylonian mathematical texts. In *Mathematical cuneiform texts*, American oriental series, vol. 29, ed. O. Neugebauer and A. Sachs, 146–151. New Haven: American Oriental Society.

Gundlach, K.-B., and W. von Soden. 1963. Einige altbabylonische Texte zur Lösung »quadratischer Gleichungen. *Abhandlungen aus dem mathematischen Seminar der Universität Hamburg* 26: 248–263.

Heuzey, L. 1906. À la mémoire de Jules Oppert. *Revue d'Assyriologie et d'Archéologie Orientale* 6(1904–07): 73–74.

Hilprecht, H.V. 1906. *Mathematical, metrological and chronological tablets from the temple library of Nippur*, The Babylonian expedition of the University of Pennsylvania. A: Cuneiform texts, XX, 1. Philadelphia: Department of Archaeology, University of Pennsylvania.

Hincks, E. 1854a. Cuneiform inscriptions in the British Museum. *Journal of Sacred Literature*, New Series 13(October 1854): 231–234, reprint after *The Literary Gazette* 38(1854): 707.

Hincks, E. 1854b. On the Asssyrian mythology. *Transactions of the Royal Irish Academy* 22(6): 405–422.

Hommel, F. 1885. *Geschichte Babyloniens und Assyriens*, Allgemeine Geschichte in Einzeldarstellungen 2. Berlin: Grote'sche Verlagsbuchhandlung.

Høyrup, J. 2002. *Lengths, widths, surfaces: A portrait of old Babylonian algebra and its kin*. Studies and sources in the history of mathematics and physical sciences. New York: Springer.

Kramer, S.N. 1963. *The Sumerians: Their history, culture, and character*. Chicago: Chicago University Press.

Kugler, F.X. 1900. *Die babylonische Mondrechnung. Zwei Systeme der Chaldäer über den Lauf des Mondes und der Sonne*. Freiburg im Breisgau: Herder.

Labat, R. 1963. *Manuel d'épigraphie akkadienne (signes, syllabaire, idéogrammes)*, Quatrième edition. Paris: Imprimerie Nationale.

MCT: Neugebauer, O., and A. Sachs. 1945. *Mathematical cuneiform texts*, American Oriental Series 29. New Haven: American Oriental Society.

Meißner, B. 1920. *Babylonien und Assyrien*, 2 vols. Heidelberg: Carl Winther, 1920, 1925.

MKT: O. Neugebauer. *Mathematische Keilschrift-Texte*, 3 vols. Quellen und Studien zur Geschichte der Mathematik, Astronomie und Physik. Abteilung A: Quellen. 3. Band, erster-dritter Teil. Berlin: Julius Springer, 1935, 1935, 1937.

Neugebauer, O. 1927. Zur Entstehung des Sexagesimalsystems. *Abhandlungen der Gesellschaft der Wissenschaften zu Göttingen. Mathematisch-Physikalische Klasse*, Neue Folge 13: 1.

Neugebauer, O. 1928. Zur Geschichte des pythagoräischen Lehrsatzes. *Nachrichten von der Gesellschaft der Wissenschaften zu Göttingen, Mathematisch-physikalische Klasse* 1928: 45–48.

Neugebauer, O. 1929. Zur Geschichte der babylonischen Mathematik. *Quellen und Studien zur Geschichte der Mathematik, Astronomie und Physik.* Abteilung B: *Studien* 1(1929–31): 67–80.

Neugebauer, O. 1930a. Beiträge zur Geschichte der Babylonischen Arithmetik. *Quellen und Studien zur Geschichte der Mathematik, Astronomie und Physik.* Abteilung B: *Studien* 1(1929–31): 120–130.

Neugebauer, O. 1930b. Sexagesimalsystem und babylonische Bruchrechnung, *Quellen und Studien zur Geschichte der Mathematik, Astronomie und Physik.* Abteilung B: *Studien* 1(1929–31).

Neugebauer, O. 1931a. Über die Approximation irrationaler Quadratwurzeln in der Babylonischen Mathematik. *Archiv für Orientforschung* 6(1931–32): 90–99.

Neugebauer, O. 1931b. Sexagesimalsystem und babylonische Bruchrechnung II, *Quellen und Studien zur Geschichte der Mathematik, Astronomie und Physik.* Abteilung B: *Studien* 1: 452–457.

Neugebauer, O. 1931c. Sexagesimalsystem und babylonische Bruchrechnung III, *Quellen und Studien zur Geschichte der Mathematik, Astronomie und Physik.* Abteilung B: *Studien* 1(1929–31): 458–463.

Neugebauer, O. 1932a. Zur Transkription mathematischer und astronomischer Keilschrifttexte. *Archiv für Orientforschung* 8(1932–33): 221–223.

Neugebauer, O. 1932b. Studien zur Geschichte der antiken Algebra I. *Quellen und Studien zur Geschichte der Mathematik, Astronomie und Physik.* Abteilung B: *Studien* 2(1932–33): 1–27.

Neugebauer, O. 1932c. Sexagesimalsystem und babylonische Bruchrechnung IV. *Quellen und Studien zur Geschichte der Mathematik, Astronomie und Physik.* Abteilung B: *Studien* 2(1932–33): 199–410.

Neugebauer, O. 1933a. Babylonische »Belagerungsrechnung«. *Quellen und Studien zur Geschichte der Mathematik, Astronomie und Physik. Abteilung B: Studien* 2(1932–33): 305–310.

Neugebauer, O. 1933b. Über die Lösung kubischer Gleichungen in Babylonien. *Nachrichten von der Gesellschaft der Wissenschaften zu Göttingen, Mathematisch-physikalische Klasse* 1933: 316–321.

Neugebauer, O. 1934a. *Vorlesungen über Geschichte der antiken mathematischen Wissenschaften. I: Vorgriechische Mathematik*, Die Grundlehren der mathematischen Wissenschaften in Einzeldarstellungen 43. Berlin: Julius Springer.

Neugebauer, O. 1934b. Serientexte in der babylonischen Mathematik. *Quellen und Studien zur Geschichte der Mathematik, Astronomie und Physik.* Abteilung B: *Studien* 3(1934–36): 106–114.

Neugebauer, O. 1936. Zur geometrischen Algebra (Studien zur Geschichte der antiken Algebra III). *Quellen und Studien zur Geschichte der Mathematik, Astronomie und Physik.* Abteilung B: *Studien* 3(1934–36): 245–259.

Neugebauer, O. 1937. Untersuchungen zur antiken Astronomie I. *Quellen und Studien zur Geschichte der Mathematik, Astronomie, und Physik.* Abteilung B: *Studien* 4(1937–38): 29–33.

Neugebauer, O., and W. Struve. 1929. Über die Geometrie des Kreises in Babylonien. *Quellen und Studien zur Geschichte der Mathematik, Astronomie und Physik.* Abteilung B: *Studien* 1(1929–31): 81–92.

Neugebauer, O., J. Stenzel, and O. Toeplitz. 1929. Geleitwort. *Quellen und Studien zur Geschichte der Mathematik, Astronomie und Physik.* Abteilung B: *Studien* 1(1929–31): 1–2.

Oppert, J. 1872. L'étalon des mesures assyriennes fixé par les textes cunéiformes. *Journal asiatique*, sixième série 20(1872): 157–177; septième série 4(1874): 417–486.

Oppert, J. 1885. Les mesures assyriennes de capacité et de superficie. *Revue d'Assyriologie et d'Archéologie Orientale* 1(1884–85): 124–147.

Oppert, J. 1894. Les mesures de Khorsabad. *Revue d'Assyriologie et d'Archéologie Orientale* 3(1993–95): 89–104.

Powell, M.A. 1976. The antecedents of old Babylonian place notation and the early history of Babylonian mathematics. *Historia Mathematica* 3: 417–439.

Powell, M.A. 1982. Metrological notes on the Esagila tablet and related matters. *Zeitschrift für Assyriologi* 72: 106–123.

Powell, M.A. 1990. Maße und Gewichte. *Reallexikon der Assyriologie und Vorderasiatischen Archäologie* VII, 457–516. Berlin/New York: de Gruyter

Rawlinson, H. 1855. Notes on the early history of Babylonia. *Journal of the Royal Asiatic Society of Great Britain and Ireland* 15: 215–259.

Rawlinson, G. 1871. *The five great monarchies of the ancient Eastern World*, 3 vols. London: John Murray.

Sayce, A.H. 1887. Miscellaneous notes. *Zeitschrift für Assyriologie und verwandte Gebiete* 2: 331–340.

Sayce, A.H. 1908. *The archaeology of cuneiform inscriptions*, 2nd ed, revised. London: Society for Promoting Christian Knowledge.

Scheil, V. 1915a. Les tables igi x gal-bi, etc. *Revue d'Assyriologie et d'Archéologie Orientale* 12: 195–198.

Scheil, V. 1915b. Le calcul des volumes dans un cas particulier à l'époque d'Ur. *Revue d'Assyriologie et d'Archéologie Orientale* 12: 161–172.

Scheil, V. 1916. Notules. XX.–Le texte mathématique 10201 du Musée de Philadelphie. *Revue d'Assyriologie et d'Archéologie Orientale* 13: 138–142.

Schuster, H.-S. 1930. Quadratische Gleichungen der Seleukidenzeit aus Uruk. *Quellen und Studien zur Geschichte der Mathematik, Astronomie und Physik.* Abteilung B: *Studien* 1(1929–31): 194–200.

Steinkeller, P. 1979. Alleged GUR.DA = ugula-géš-da and the Reading of the Sumerian Numeral 60. *Zeitschrift für Assyriologie und Vorderasiatische Archäologie* 69: 176–187.

Struve, V.V. 1930. Mathematischer Papyrus des Staatlichen Museums der Schönen Künste in Moskau. *Herausgegeben und Kommentiert. Quellen und Studien zur Geschichte der Mathematik.* Abteilung A: Quellen 1, Band. Berlin: Julius Springer.

Struve, V.V. 1934. The problem of the genesis, development and disintegration of the slave societies in the ancient orient. Abbreviated translation of an article from 1934. In 1969. *Ancient Mesopotamia, socio-economic history. A collection of studies by soviet scholars,* ed. I.M. Diakonoff, 17–67. Moskva: «Nauka» Publishing House, Central Department of Oriental Literature.

Thompson, G.D. 2010. The recovery of Babylonian astronomy. (9) The Pinches Era – Otto Neugebauer and Abraham Sachs (and Theophilus Pinches). http://members.westnet.com/Gary-David-Thompson/babylon9.html. Accessed 29 Oct 2010.

Thureau-Dangin, F. 1897. Un cadastre chaldéen. *Revue d'Assyriologie et d'Archéologie Orientale* 4(1897–98): 13–27.

Thureau-Dangin, F. 1898. *Recherches sur l'origine de l'écriture cunéiforme*. 1re partie + Supplément à la 1re partie. Paris: Leroux, 1898–99.

Thureau-Dangin, F. 1909. L'u, le qa et la mine. *Journal asiatique*, 13ième série 13: 79–111.

Thureau-Dangin, F. 1921. Numération et métrologie sumériennes. *Revue d'Assyriologie et d'Archéologie Orientale* 18: 123–142.

Thureau-Dangin, F. 1922. *Tablettes d'Uruk à l'usage des prêtres du Temple d'Anu au temps des Séleucides*, Musée de Louvre – Département des Antiquités Orientales. Textes cunéiformes 6. Paris: Paul Geuthner.

Thureau-Dangin, F. 1926. *Le Syllabaire Akkadien*. Paris: Geuthner.

Thureau-Dangin, F. 1931. Notes sur la terminologie des textes mathématiques. *Revue d'Assyriologie et d'Archéologie Orientale* 28: 195–198.

Thureau-Dangin, F. 1932a. *Esquisse d'une histoire du système sexagésimal*. Paris: Geuthner.

Thureau-Dangin, F. 1932b. Notes assyriologiques. LXIV. – Encore un mot sur la mesure du segment de cercle. LXV. – BAL = «raison (arithmétique ou géométrique)». LXVI. – *Warâdu* «abaisser un perpendiculaire»; *elû* «élever un perpendiculaire». LXVII. – La mesure du volume d'un tronc de pyramide. *Revue d'Assyriologie et d'Archéologie Orientale* 29: 77–88.

Thureau-Dangin, F. 1933. Notes Assyriologiques. LXIIV. – *Igû* et *igibû*. LXXV. – La tablette de Strasbourg n° 11. LXXVI. – Le nom du «cercle» en babylonien. *Revue d'Assyriologie et d'Archéologie Orientale* 30: 183–188.

Thureau-Dangin, F. 1934. La tablette de Strasbourg n° 11. *Revue d'Assyriologie et d'Archéologie Orientale* 31: 30.

Thureau-Dangin, F. 1939. Sketch of a history of the sexagesimal system. *Osiris* 7: 95–141.

Thureau-Dangin, F. 1940a. Notes sur la mathématique babylonienne. *Revue d'Assyriologie et d'Archéologie Orientale* 37: 1–10.

Thureau-Dangin, F. 1940b. L'Origine de l'algèbre. *Académie des Belles-Lettres. Comptes Rendus* 1940: 292–319.

TMB: Thureau-Dangin, F. 1938. *Textes mathématiques babyloniens*, Ex Oriente Lux, Deel 1. Leiden: Brill.

TMS: Bruins, E.M., and M. Rutten. 1961. *Textes mathématiques de Suse*, Mémoires de la Mission Archéologique en Iran, XXXIV. Paris: Paul Geuthner.

Ungnad, A. 1916. Zur babylonischen Mathematik. *Orientalistische Literaturzeitung* 19: 363–368.

Ungnad, A. 1917. Lexikalisches. *Zeitschrift für Assyriologie und Vorderasiatische Archäologie* 31(1917–18): 38–57.

Ungnad, A. 1918. Lexikalisches. *Zeitschrift für Assyriologie und verwandte Gebiete* 31(1917–18j): 248–276.

von Ranke, L. 1885. *Geschichten der romanischen und germanischen Völker von 1494 bis 1514*, Dritte Auflage. Leipzig: Duncker & Humblott.

von Soden, W. 1936. Leistung und Grenze sumerischer und babylonischer Wissenschaft. *Die Welt als Geschichte* 2(411–464): 507–557.

von Soden, W. 1937. Review of MKT. *Zeitschrift der Deutschen Morgenländischen Gesellschaft* 91: 185–203.

von Soden, Wolfram. 1939. Review of TMB. *Zeitschrift der Deutschen Morgenländischen Gesellschaft* 93: 143–152.

von Soden, W. 1952. Zu den mathematischen Aufgabentexten vom Tell Harmal. *Sumer* 8: 49–56.

von Soden, W. 1964. Review of TMS. *Bibliotheca Orientalis* 21: 44–50.

Waschow, H. 1932a. Verbesserungen zu den babylonischen Dreiecksaufgaben S.K.T. 8. *Quellen und Studien zur Geschichte der Mathematik, Astronomie und Physik*. Abteilung B: *Studien* 2(1931–32): 211–214.

Waschow, H. 1932b. Angewandte Mathematik im alten Babylonien (um 2000 v. Chr.). Studien zu den Texten CT IX, 8–15. *Archiv für Orientforschung* 8(1932–33): 127–131, 215–220.

Waschow, H. 1936. *Babylonische Briefe aus der Kassitenzeit*. Inaugural-dissertation. Gräfenhainichen: A. Heine.

Weidner, E.F. 1916. Die Berechnung rechtwinkliger Dreiecke bei den Akkadern um 2000 v. Chr. *Orientalistische Literaturzeitung* 19: 257–263.

Zimmern, H. 1916. Zu den altakkadischen geometrischen Berechnungsaufgaben. *Orientalistische Literaturzeitung* 19: 321–325.

François Thureau-Dangin and Cuneiform Mathematics

Béatrice André-Salvini

Abbreviations

ISA	Thureau-Dangin 1905.
MKT	Neugebauer 1935–1937.
RA	*Revue d'Assyriologie et d'Archéologie Orientale.*
SAK	Thureau-Dangin 1907.
TCL	*Textes cunéiformes du Louvre.*
TMB	Thureau-Dangin 1938

François Thureau-Dangin[1] played an exceptional role in the rediscovery of the history and civilization of ancient Mesopotamia. A philologist and historian, his research on how the cuneiform writing system worked, and on deciphering the Sumerian language, form the basis of our current knowledge in Assyriology. The interest he developed for cuneiform scientific texts may be explained by the course of his career and his research.

In 1895, he joined the Louvre, assisting Léon Heuzey, the curator of the Department of Oriental Antiquities, who entrusted him with the study of the large number of texts discovered by Ernest de Sarzec in Tello, Mesopotamia. He carried out several missions to Constantinople to classify the texts, so they could be divided between the Imperial Ottoman Museum and the Louvre. This abundance of unedited documents, mostly written in Sumerian, a difficult language whose decipherment was then still uncertain, gave Thureau-Dangin the opportunity to trace and

[1] Born in Paris, January 3rd 1872 – died in Paris, January 24th 1944. For a recent biography of F. Thureau-Dangin, cf. André-Salvini 2012, 2013.

B. André-Salvini (✉)
Musée du Louvre, Paris, France
e-mail: beatrice.andre-salvini@orange.fr

A. Jones et al. (eds.), *A Mathematician's Journeys*, Archimedes 45,
DOI 10.1007/978-3-319-25865-2_6

understand the origin and evolution of cuneiform signs through patient and painstaking work that prepared him for the later study of mathematics texts. In his *Recherche sur l'origine de l'écriture cunéiforme* (Research on the Origin of Cuneiform Writing) published in 1898–1899[2] he established a catalogue of 600 archaic characters identified for the first time in the inscriptions found at Tello, and he established their equivalence with more recent cuneiform signs. He developed a classification for the signs which permitted a single mode of transcription for Sumerian and Akkadian, by drawing up statistics on the use of each of the syllabic values, by period.[3] He differentiated the homophones using a rational notation system consisting in attributing a numerical index to the values according to their order of frequency, which enables the transcribed sign to be recognized immediately. This system is still in use today.

His book, *Les Inscriptions de Sumer et d'Akkad (*Inscriptions from Sumer and Akkad*)*, published in 1905,[4] in which he transcribed, translated and established the grammatical rules of all the known Mesopotamian royal inscriptions, from the archaic Sumerian period until the beginning of the second millennium BC, places him among the decipherers of Sumerian writing and the language itself, following in the footsteps of his professor Jules Oppert, whose classes he attended at the *Collège de France*. His work bears witness to a concern for completeness in the establishment of the catalogue of texts and, for the first time, a real understanding of the literary language of ancient Sumer. In this work, Thureau-Dangin demonstrates the complete command of a thorough, methodological and intuitive spirit, as analytic as it was synthetic. *ISA* became the tool of reference for orientalists the world over, putting an end once and for all to the arguments over the origin of cuneiform writing.[5] On this book, his friend Edouard Dhorme wrote: "The author's mathematical instinct is revealed by the desire to achieve a fastidious completeness in the understanding of the words and the phrases, as in determining the phonetic value of the ideograms".[6]

Alongside his fundamental research on understanding the cuneiform system, his very first articles showed his interest in numbers and calculation procedures which he came to from tablets dealing with administrative accounting and surface area calculations, found in Tello. His first article, published in 1895, is devoted to

---

[2] Thureau-Dangin 1898–1899.

[3] This research would lead to two important later publications: *Le SyllabaireAccadien*, in 1926 (Thureau-Dangin 1926) and *Les Homophones Sumériens* in 1929 (Thureau-Dangin 1929). He was preparing new editions of these works when he died suddenly, cf. also his posthumous article: "*Les graphies rompues en akkadien*" (Broken Spelling in Akkadian) (Thureau-Dangin 1946†).

[4] Thureau-Dangin 1905.

[5] Thureau-Dangin 1907: The work was immediately translated into German and published in Leipzig in 1907, under the title: *Die Sumerischen und Akkadischen Kônigsinschriften*. This research (abr. SAK), launched the series "*Vordersiatische Bibliothek*".

[6] "*L'instinct mathématique de l'auteur se révélait par le souci d'atteindre une scrupuleuse exactitude dans l'intelligence des mots et des phrases, comme dans la détermination de la valeur phonétique des idéogrammes*", Dhorme 1946, p. 8.

"*La comptabilité agricoleen Chaldée au IIIe millénaire*".[7] In 1898, a paper entitled "Fractional numbers in archaic Babylonian writing"[8] introduced a field of study that has an important place in his later work.

His taste for the sophisticated Babylonian number system grew as a result of his experiences in the Great War. Volunteering in November 1914, initially posted to a Territorial regiment guarding railway lines, and then to the Supply Corps, he asked to be transferred to the Salonika front. In 1917 Brigadier Thureau-Dangin participated in archeological research with the Eastern Army (*l'armée française d'Orient AFO*).[9] He returned to France in 1918, when he was recruited by the Ministry of Foreign Affairs to work in the Cryptography Section – the Cipher Department or "*Cabinet Noir*" – thus using his gift for deciphering and his analytical mind to decode encrypted documents during a period of time crucial for the cessation of the armed conflict. The art of encrypting and decrypting information was, at that time, experiencing a major advance, primarily through the implementation of "Kerckhoffs' principle", established in 1883 by philologist August Kerckhoffs, whose research provided the founding principles of scientific cryptology, due to its clarity of expression and the quality of its system of decryption, established from sources that were relevant and verified.[10] Until his demobilization in January 1919, Thureau-Dangin's philological skills were put to use and blossomed in the particularly delicate situation of the period which followed the Armistice. His exemplary conduct earned him the *Légion d'honneur*, the *Médaille militaire* and the *Croix de guerre 14–18*, awarded for "services rendered in the East".

After the interruption to his academic work during the First World War, the orientation of his research changed. While not abandoning Sumerian, the subject of his first studies, he devoted himself to more recent Akkadian texts. This interest was provoked by the renewed development of archeological excavations in Mesopotamia and Syria and the discovery of numerous tablets from the Old Babylonian and later periods which intrigued his pioneering spirit in the quest for unedited texts and encouraged his desire to gather as much documentation as possible in his analysis of the sources. In some ways, this mindset accords well with the principles of Mesopotamian science, based as it is on casuistic reasoning. His contribution extends archeology and all domains of Assyriology.[11] He developed, in brilliant syntheses, his known and particular fascination for religious literature, chronology,

---

[7] Thureau-Dangin 1895.

[8] Thureau-Dangin 1898.

[9] Cf. Mendel 1918.

[10] Kerckhoffs 1883.

[11] He became a member of the "Consultative committee for excavations and archeological research in East Asia" ("*Commission consultative pour les fouilles et recherches archéologiques dans l'Asie occidentale*"), created in 1920. A journey to Syria and Mesopotamia, in September–October in the company of E. Dhorme allowed him to locate the archeological sites and gave him the desire to participate in the discoveries. On resigning from the Louvre in August 1928 he became an archeologist and, until 1931, directed the excavations at Arslan Tash (ancient Hadatu) and at Tell Ahmar (ancient Til-Barsip), in Syria. He published the results of these excavations in two volumes: Arslan-Tash in 1931 and Til-Barsib in 1936 (Thureau-Dangin 1931, 1936a).

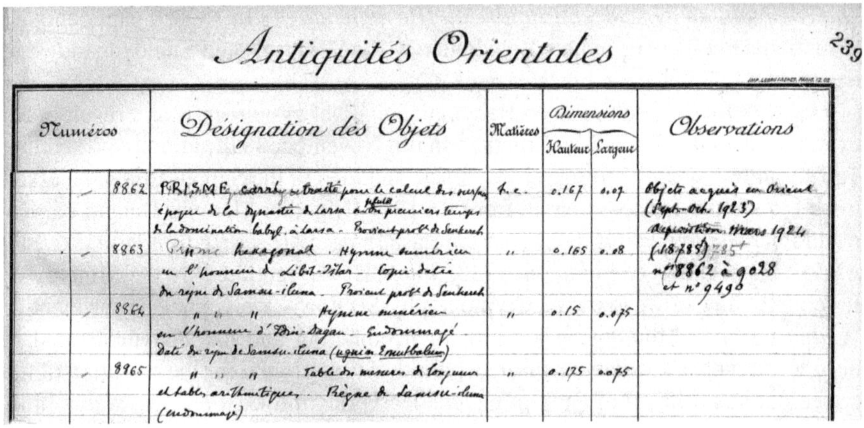
Antiquités Orientales 239

Numéros	Désignation des Objets	Matières	Dimensions Hauteur	Dimensions Largeur	Observations
8862	PRISME carré ... traité pour le calcul des surfaces, époque de la dynastie de Larsa ... premiers temps de la domination babyl. à Larsa. Provenant prob. de Senkereh	t. c.	0.167	0.07	Objets acquis en Orient (Sept.-Oct. 1923) Acquisition Mars 1924 (18385) nos 8862 à 9028 et nº 9490
8863	Prisme hexagonal. Hymne sumérien en l'honneur de Libit-Ištar. Copie datée du règne de Samsu-iluna. Provenant prob. de Senkereh	"	0.165	0.08	
8864	" " " Hymne sumérien en l'honneur d'Būr-Dagan. Endommagé. Daté du règne de Samsu-iluna	"	0.15	0.075	
8865	" " " Table des mesures de longueur et tables arithmétiques. Règne de Samsu-iluna (endommagé)	"	0.175	0.075	

Fig. 1 Extract of the Inventory of the Department of Oriental Antiquities AOII, 1924, hand-written by François Thureau-Dangin, recording the entry on mathematical prisms AO 8862 (Thureau 1932b) and AO 8865 (Thureau-Dangin 1930, Neugebauer *MKT* 1, p. 69sqq., Proust 2005)

and for scientific, mathematical and astrological texts, many of which came to the collection in the Louvre thanks to him. Throughout his career he was interested in chronology, the fundamental principle of history, which remained one of his favorite fields until the end of his days. He used the synchronisms provided by the astronomical texts known at the time to establish relative dating. His first important work on chronology appeared in 1918, after his return to Paris and normal life, it was also the subject of his final dissertation.[12] His early interest for metrology, numeration and Babylonian science progressed with the discoveries he made from the harvest of thousands of tablets he had access to. From 1909, but more so after 1918, he regularly published articles and numerous "Assyriological notes" on the sciences, and particularly in each issue of the *Revue d'assyriologie et d'archéologie orientale* of which he became co-director alongside Father Scheil, in 1910. In the years leading up to the Second World War especially, he devoted a vast amount of time to this field, which gave rise to many monographs.

His métier as curator (Fig. 1) and his long career at the Louvre,[13] during which texts of all genres and from all epochs were at the disposition of his insatiable curiosity, allowed him to acquire a taste for classification and discovery, as much as a profound intimacy with the tablets, and a keen eye for challenging readings. Thus, in 1909, on a mission to Constantinople in order to catalogue and study the Tello

[12] Cf. in particular: Thureau-Dangin 1918a, b, 1942. In this last article, in light of new data, he revised the chronology and lowered the date from what had hitherto been accepted.

[13] Joining the Louvre as a volunteer "free attaché" in 1895, he then became a paid attaché in 1902, and was appointed assistant curator of the Department of Oriental Antiquities in 1908. Thureau-Dangin became the curator in charge in 1925. On August 10th 1928 he handed in his resignation on health grounds, as he was suffering from progressive hearing loss. However, he continued to work on the museum's collections of cuneiform texts and contributed to the enlargement of the holding through the creation of the "Thureau-Dangin Fund" for acquisitions.

tablets, he wrote to Léon Heuzey: "I have been here in Constantinople for just over eight days. I have found the 1904 tablets still in their crates. My first concern was to unpack and classify them. I then searched the cellars and drawers and have been able, little by little, to find all the tablets from the earlier excavations... Today I opened exactly 170 drawers (you read correctly; one hundred and seventy) before finding what I was looking for...".[14] The quality, as well as the quantity of his sources brought considerable benefits to this tireless, tenacious worker who as an editor of texts was second to none. From his earliest publications, one can see a taste for clarity and simplification, coupled with a keen attention to detail. He was always seeking access to new documents, particularly from the constant enrichment of the Louvre collection as a result of the generosity of personal donations and the informed and thoughtful acquisition of coherent lots of tablets from many origins and diverse in nature – administrative, literary and scientific. In 1910 he created the "*Textes cunéiformes du Louvre*" series to assure the publication[15] of the collection. His powers of observation were remarkable and his autographs elegant, precise and accurate. His concern for reproducing exactly what he saw was such that even today it is rare for corrections to be made to copies of cuneiform texts he made from original documents. This exceptional gift was underscored by Henri Maspero, in the following terms: "an assurance of reading and an almost divinatory feeling for the engraved sign, even when it is all but invisible, which must make him one of the most remarkable epigraphists...".[16] Whenever he had the chance he travelled to copy tablets in foreign museums. The only approximate interpretations that can be attributed to him are those taken from poor quality photographs, which were sent to him as access to the original was difficult for him. Such was the case for the trans-

[14] "*Me voici depuis un peu plus de huit jours installé à Constantinople. J'ai trouvé les tablettes de 1904 encore dans les caisses. Mon premier soin a été de les déballer et de les classer. J'ai ensuite fouillé les caves et les tiroirs, et j'ai pu retrouver à peu près toutes les tablettes des fouilles antérieures... Aujourd'hui j'ai ouvert exactement 170 tiroirs (vous lisez bien cent soixante-dix) avant de retrouver ce que je cherchais...*". Letter of September 8th 1909, written at the Pera Palace, Constantinople to Léon Heuzey (Extract of the correspondence between *Mr Fr. Thureau-Dangin and Mr Heuzey*, Archives of the *Institut de France*, Paris).

[15] He was himself the editor of several of these volumes, published in Paris by Geuthner, whose titles show his command of all fields of Assyriology: *Lettres et contrats de l'époque de la première dynastie babylonienne, (Letters and Contracts of the first Babylonian Dynasty) TCL* I., 1910; *Une relation de la huitièmecampagne de Sargon (An Account of Sargon's Eighth Campaign) (714 av. J.-C.)* and *TCL* III, 1912; *Tablettes d'Uruk à l'usage des prêtres du temple d'Anu au temps des Séleucides*, (TabletsfromUruk Used by Priests in the Temple of Anu in the Time of the Seleucid) *TCL* VI, 1922; *Lettres de Hammurapi à Šamaš-Hâsir*, (Letters from Hammurapi to Šamaš-Hâsir) TCL VII, 1924; *Les Cylindres de Goudéa découverts par Ernest de Sarzec à Tello*, (The Gudea Cylinders Discovered by *Ernest de Sarzec*) *TCL* VIII, 1925; *Tablettes cappadociennes. Deuxième série*, (Cappadocian Tablets, Second series) *TCL* XIV, 1928.

[16] "*une sûreté de lecture et un sens presque divinatoire du signe gravé, même quand il est presque invisible, qui devaient faire de lui le plus remarquable des épigraphistes...*".*Funeral eulogy given* on February 4th 1944 at the *Académie des Inscriptions et Belles-Lettres*. Cf. Maspero 1944, p. 56.

literations of the Yale mathematical texts, which he carried out in 1938, following those done by Neugebauer in 1937, from the same photographs.[17]

He addressed mathematics as both a philologist and a historian in numerous preparatory articles published in two monographs in 1932 and 1938,[18] devoted to the two major categories of Babylonian mathematical texts.[19] In the first, which he entitled *Esquisse d'une histoire du système sexagésimal (An Outline of a History of the Sexagesimal System)*, developing his first works, he shows that the Babylonian system of numeration was a legacy from the Sumerians, as he had previously said and done in 1913, in his review of a book by J. Halévy: "The sexagesimal system is one of the most original features of the Assyrian-Babylonian civilization. Yet it is not, it cannot be a Semitic invention".[20] He concluded that the Sumerians created a numerical algebra thanks to which the Babylonians were able to solve second degree equations. That same year, 1932, he dedicated 13 articles or notes to these subjects.[21] He continued this work in the years that followed and in 1938 published under the title *Textes mathématiques babyloniens transcrits et traduits* (Babylonian Mathematical Texts Transcribed and Translated)[22] a collection bringing together an enormous documentation of 623 problems on first and second degree equations. Their translation and interpretation are introduced by an essay on the history and principles of Babylonian mathematics. He accompanied the text with philological commentaries and provided a list of Akkadian mathematical terms and numbers written phonetically or as ideograms with their Sumerian and Akkadian value. The mastery of the translations, the manner of the reasoning and the sober and simple style of this book attracted the interest of historians of science (Fig. 2). Abel Rey, a philosopher, and professor of the history of science at the Sorbonne, devoted an enthusiastic review to him in 1940, in the *Journal des Savants*. He noted: "Without doubt, Mr Thureau-Dangin has a knowledge of the language of Chaldean mathe-

---

[17] Cf. On this subject and à propos the relations between the two great scholars, the article by Ch. Proust in this book.

[18] Thureau-Dangin 1932a, 1938 (*TMB*).

[19] *TMB*, p. IX. "*Les textes mathématiques babyloniens qui nous sont parvenus sont ou des tables destinées à faciliter le calcul ou des exercices pratiques, des problèmes. Ce sont exclusivement les textes de cette seconde catégorie qui forment l'objet du présent travail. ...L'originalité de la mathématique babylonienne réside pour une bonne part dans son système de numération...*". "The Babylonian mathematical texts that have survived are either tables destined to facilitate computations or practical exercises, problems. The subject of this present work is exclusively the texts from this second category... The originality of Babylonian mathematics lies largely in its system of numeration".

[20] Halévy, J., "Précis d'allographie assyro-babylonienne" (Paris: Leroux, 1912); Thureau-Dangin 1913, p. 195: "... *Le système sexagésimal est l'un des traits les plus originaux de la civilisation assyro-babylonienne. Or il n'est pas, il ne peut être une invention sémitique*".

[21] All published in *RA* XXIX, 1932. During his career, he wrote around eighty contributions – monographs, articles and notes, – on cuneiform mathematics (Cf. F. Thureau-Dangin's bibliography *in* Dhorme 1946, p. 19–35).

[22] *TMB*. This volume launched the publications of the *Société orientale* « Ex Oriente Lux » by Leyden.

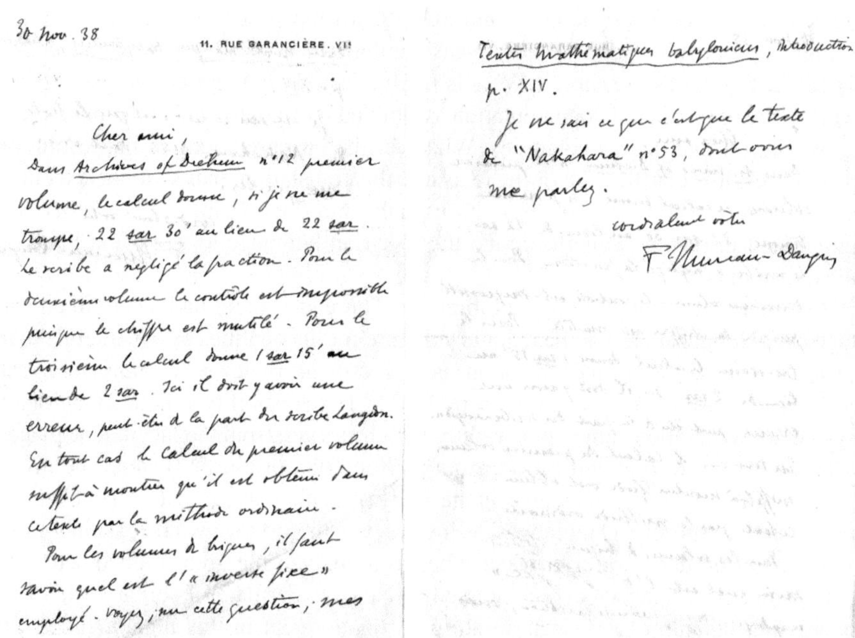

Fig. 2 Hand-written letter by François Thureau-Dangin (probably to Léon Legrain or Charles-François Jean). Dated November 30th 1938, replying to a request concerning a text edited by Stephen Langdon. *Tablets from the Archives of Drehem*, Paris, Geuthner, 1911 n°12. The tablet recounts volume calculations for the digging of a canal, but certain calculations contain errors. (Archive of the Department of Oriental Antiquities, Louvre Museum, Thureau-Dangin Fund)
Nov. 30 1938
Dear friend,
In the n°12, first volume of the Drehem Archives, the calculation gives, if I am not mistaken, 22 sar 30' instead of 22 sar. The scribe forgot the fraction. For the second volume the verification is impossible, as the figure is mutilated. For the third, the calculation gives 1 sar 15' instead of 2 sar. Here there must be an error, maybe by the scribe Langdon. In any case, the computation in the first volume is enough to show that it is obtained in this text by the ordinary method.
For the volumes of bricks, one should know what the "fixed inverse" used was. See my "Textesmathématiquesbabyloniens" on this question, introduction p. XIV.
I do not know the "Nakahara" text n°53 you mention.
Yours sincerely,
F. Thureau-Dangin
(Author's note: "Nakahara": Nakahara, Y., 1928, The Sumerian Tablets in the Imperial University of Kyoto. Tokyo).

30 nov. 38
Cher ami,
Dans Archives of Drehem, n°12 premier volume, le calcul donne, si je ne me trompe, 22 sar 30' au lieu de 22 sar. Le scribe a négligé la fraction. Pour le deuxième volume le contrôle est impossible puisque le chiffre est mutilé. Pour le troisième le calcul donne 1 sar 15' au lieu de 2 sar. Ici il doit y avoir une erreur, peut-être de la part du scribe Langdon. En tout cas le calcul du premier volume suffit à montrer qu'il est obtenu dans ce texte par la méthode ordinaire.
Pour les volumes de briques, il faut savoir quel est l'« inverse fixe » employé. Voyez; sur cette question; mes Textes mathématiques babyloniens, introduction p. XIV.
Je ne sais ce que c'est que le texte de « Nakahara » n°53, dont vous me parlez.
Cordialement vôtre,
F. Thureau-Dangin

matics that is second to none"[23] and also: "...to me, nearly all of Mr Thureau-Dangin's propositions appear more plausible than those of Messrs Neugebauer and Vogel, or at least as plausible...What is really masterly in the great Assyriologist's work is the unparalleled documentation assembled on Babylonian mathematics, and above all, on 2nd degree equations. Whatever the interpretations of the resolution procedures, this clarification of the texts and this translation, thanks to the excellent glossary and the philological observations in the introduction, allow those otherwise uninitiated in reading cuneiform and the Assyrian language to work first hand on the documents".[24]

On the point of the philological thinking, Thureau-Dangin had a certain advantage over Neugebauer, who he recognized more for his qualities as a mathematician than as a philologist and historian in his critical reviews of *Mathematische Keilschrift-Texte* (*MKTI*) from 1936 to 1937.[25] His sensibility for the Babylonian system of numeration came, quite rightly for a historian, from the fact that he placed it in the history of the civilization, from its Sumerian origins.[26] Beyond the divergence of their methods and views on the study of mathematical texts, Neugebauer's reproach that Thureau-Dangin was, as it were, wanting to review his texts in a spirit of competition does not hold water when you analyze the great French scholar's personality. The letter Goetze addressed to Neugebauer dated February 14, 1937, expressing his surprise at such an attitude[27] is eloquent in this regard. Indeed, it seemed that for Thureau-Dangin, the spirit of rivalry existed purely in terms of the demands of science. The editor of the *TMB* presented it thus in the book's preface: "... After presenting works for which Assyriology and in particular our knowledge of the Sumerian language owe so much, Mr. Thureau-Dangin now gives the result of his studies on Babylonian mathematics. With his unparalleled accuracy and precision, and his reassuring caution, he is one of the great scholars...who continue the glorious history of exactitude and good taste in French philology".[28]

---

[23] "*M. Thureau-Dangin possède, comme sans doute personne au monde, la langue mathématique chaldéenne*", Rey 1940, p. 16.

[24] "*... Les propositions de M. Thureau-Dangin me paraissent presque toutes ou plus plausibles que celles de MM. Neugebauer et Vogel, ou au moins aussi plausibles... Ce qui est vraiment magistral dans l'œuvre du grand assyriologue, c'est l'incomparable documentation rassemblée sur la mathématique babylonienne et surtout sur l'équation du 2me degré. Quelles que soient les interprétations des procédés de solution, cette mise au point des textes et cette traduction, grâce à l'excellent lexique et aux remarques philologiques de l'Introduction, permettent aux profanes en lecture cunéiforme et en langue assyrienne de travailler à même les documents de première main.*", Rey 1940 *Loc. cit.* p. 18 and 21.

[25] Neugebauer, 1936–1937 (c.r. of Thureau-Dangin 1936b et 1937). Cf. the article in this book by Christine Proust, note 16.

[26] Ch. Proust notes that sometimes Thureau-Dangin and Neugebauer provide philological insights for the mathematician and mathematical insights for the Assyriologist. Loc. cit.

[27] Ch. Proust, *Loc. cit.*

[28] "*... Après avoir mis au jour des ouvrages dans lesquels l'Assyriologie et en particulier la connaissance de la langue sumérienne lui doivent tant, M. Thureau-Dangin donne maintenant le résultat de ses études sur la mathématique babylonienne. Avec son acribie sans pareille et sa*

The esteem and the respect he inspired from all were masterfully précised by Henri Maspero, during the funeral eulogy that he gave as President of *l'Académie des inscriptions et belles-lettres* which had made Thureau-Dangin a member in 1917, while he was fighting in the East.[29] "Thureau-Dangin's career was an admirable career of a great scholar: honest, brilliant, irreproachable…a born enemy of all brilliant but baseless theories, his only wish was to write that which he was sure of and for which he could provide textual proof…he stood out by the strength of his work, by the rigor of his method, by the pertinence of his deductions and conclusions, by the precision with which he knew how to bring out from the texts what they contained, and nothing more. Never seeking personal advancement, but only the advancement of science, fame came to him without him looking for it… fame due to the profound esteem that his character and his science engendered from all his fellow scientists".

In his final years, until his sudden death in 1944 on his return from Fresnes prison where he had been held with other *Académiciens, résistants* like him, his hearing, damaged by a shell burst near his position as he was fighting on the Eastern front, deteriorated leaving him totally deaf. It is very likely that the heightened concentration from being immured in silence played a role in his scholarly reflection and the intensity of his scientific production. He worked to the end in the library of the *Cabinet d'Assyriologie* at the *Collège de France*, on a second, extended edition of the *Syllabaire accadien* in which he updated the values attributed to certain syllabic signs. Intuitive but rigorous and clear in his working methods, his contemporaries recognized in him a very humane personality and an exceptional scholar.

References

André-Salvini, B. 2012. François Thureau-Dangin. In *Dictionnaire critique des historiens de l'art actifs en France de la Révolution à la Première Guerre mondiale*, ed. Ph. Sénéchal, and C. Barbillon. Electronic publication: www.inha.fr/spip.php?article2552

André-Salvini, B. 2013. Thureau-Dangin, François. *Reallexikon der Assyriologie und Vorderasiatischen Archäologie* (*RLA*), Band 13/7–8: 640–641.

prudence rassurante il est un des savants … qui continuent la vieille gloire d'exactitude et de bon goût de la philologie française." (*TMB*, foreword).

[29] Maspero 1944, *Loc. cit.*: "*La carrière de Thureau-Dangin a été une admirable carrière de savant, droite, lumineuse, sans aucune brisure… Ennemi né de toutes les théories brillantes mais mal fondées, il n'a voulu écrire que des choses dont il fut sûr et auxquelles il put apporter des textes pour preuves … Il s'est imposé par la force de son travail, par la rigueur de sa méthode, par la justesse de ses déductions et de ses conclusions, par la précision avec laquelle il savait faire sortir des textes ce qu'ils contiennent, et rien de plus. Ne cherchant jamais un accroissement personnel, mais seulement l'accroissement de la science, la notoriété lui est venue sans qu'il la cherchât… la notoriété due à l'estime profonde que son caractère et sa science imposaient à tous ses confrères*".

Dhorme, E. 1946. *Hommage à la mémoire de l'éminent assyriologue François Thureau-Dangin (1872–1944)*, 7–14 et "Liste des publications de F. Thureau-Dangin", 19–35, Mémoires de la Société orientale "Ex Oriente Lux", vol. 8. Leiden: Brill.
Kerckoffs, A. 1883. La cryptographie militaire ou les chiffres usités en temps de guerre. *Journal des sciences militaires* IX (January 1883): 5–33 and (February 1883): 161–191.
Maspero, H. 1944. Éloge funèbre de M. François Thureau-Dangin. *Comptes rendus des séances de l'Académie des inscriptions et belles-lettres* 88(1): 55–63.
Mendel, G. 1918. Les travaux du service archéologique de l'armée française d'Orient. *Comptes rendus des séances de l'Académie des inscriptions et belles-lettres* 62(1): 9–17.
Neugebauer, O. 1935–1937. *Mathematische Keilschrift-Texte* I-III. Quellen und Studien zur Geschichte der Mathematik, Astronomie und Physik. Berlin: Springer.
Proust, C. 2005. A propos d'un prisme du Louvre: aspects de l'enseignement des mathématiques en Mésopotamie. *Sources and Commentaries in Exact Sciences (SCIAMVS)* 6: 3–32.
Rey, A. 1940. A propos des 'Mathématiques babyloniennes'. *Journal des savants,* janvier-mars 1940/1: 16–26. (C.r. de Fr. Thureau-Dangin, *Textes mathématiques babyloniens transcrits et traduits*. Un volume in-8°, XL-293 pages. Leyde: E.J. Brill, 1938.)
Thureau-Dangin, F. 1895. La comptabilité agricole en Chaldée au IIIe millénaire. *Revue d'Assyriologie et d'Archéologie Orientale* III: 118–146.
Thureau-Dangin, F. 1898. Les chiffres fractionnaires dans l'écriture babylonienne archaïque. *Beiträge zur Assyriologie* III: 588–589
Thureau-Dangin, F. 1898–1899. *Recherches sur l'origine de l'écriture cunéiforme, Ie partie: Les formes archaïques et leurs équivalents modernes.* Paris: Leroux, 1898; *Supplément à la Ire partie.* Paris: Leroux, 1899.
Thureau-Dangin, F. 1905. *Les Inscriptions de Sumer et d'Akkad.* Paris: Leroux.
Thureau-Dangin, F. 1907. *Die Sumerischen und Akkadischen Königsinschriften*, "Vorderasiatische Bibliothek" I. Leipzig: Hinrichs.
Thureau-Dangin, F. 1913. Compte rendu de « J. Halévy.- *Précis d'allographie assyro-babylonienne*, Paris, Leroux, 1912 », *Revue d'Assyriologie et d'Archéologie Orientale* X: 195–196.
Thureau-Dangin, F. 1918a. *La Chronologie des dynasties de Sumer et d'Accad.* Paris: Leroux.
Thureau-Dangin, F. 1918b. La chronologie de la dynastie de Larsa. *Revue d'Assyriologie et d'Archéologie Orientale* XV: 1–58.
Thureau-Dangin, F. 1926. *Le Syllabaire accadien.* Paris: Geuthner.
Thureau-Dangin, F. 1929. *Les Homophones sumériens.* Paris: Geuthner.
Thureau-Dangin, F. 1930. La graphie du système sexagésimal. *Revue d'Assyriologie et d'Archéologie Orientale* XXVII: 73–78.
Thureau-Dangin, F. 1931. *Arslan-Tash*, vol. I: texte, vol. II: plates. Collab. de Frédéric Barrois, Georges Dossin, Maurice Dunand. Paris: Geuthner.
Thureau-Dangin, F. 1932a. *Esquisse d'une histoire du système sexagésimal.* Paris: Geuthner.
Thureau-Dangin, F. 1932b. Le prisme mathématique AO 8862. *Revue d'Assyriologie et d'Archéologie Orientale* XXIX: 1–10.
Thureau-Dangin, F. 1936a. *Til-Barsip*, vol. I: texte; vol. II, plates. Collab. de Maurice Dunand. Paris: Geuthner.
Thureau-Dangin, F. 1936b. [Review of Neugebauer, *Mathematische Keilschrift-Texte* I-II]. *Revue d'Assyriologie et d'Archéologie Orientale* XXXIII: 55–62.
Thureau-Dangin, F. 1937. [Review of Neugebauer, *Mathematische Keilschrift-Texte* III]. *Revue d'assyriologie et d'archéologie orientale* XXXIV: 87–92.
Thureau-Dangin, F. 1938. *Textes mathématiques babyloniens transcrits et traduits*, Mémoires de la Société orientale Ex Oriente Lux 1. Leyden: Brill.
Thureau-Dangin, F. 1942. La Chronologie de la Première Dynastie Babylonienne. Examen critique des solutions récemment proposées. *Mémoires de l'Académie des Inscriptions et Belles-Lettres* 43.
Thureau-Dangin, F. 1946†. Les graphies rompues en accadien. *Mémoires de la Société orientale Ex Oriente Lux* 8: 17–18

Mathematical and Philological Insights on Cuneiform Texts. Neugebauer's Correspondence with Fellow Assyriologists

Christine Proust

Abbreviations

MCT	Neugebauer and Sachs 1945
MKT	Neugebauer 1935–1937
TMB	Thureau-Dangin 1938
YUL	Yale University Library
YBC	Yale Babylonian Collection

One of the most remarkable features of mathematics from ancient Mesopotamia and Egypt is that it has reached us from archaeological sources. We appear to have before us texts written and used by the ancient scholars themselves, and not, as is most often the case with texts transmitted through a long written tradition, which may have profoundly transformed the original work. Does this mean that historians of the ancient Near East have direct, unbiased access to the original ancient texts? It would be an illusion to believe this on at least two counts. First, archaeological sources themselves, including the oldest, are for the most part the result of textual transformation such as compilations, copies, translations, dictations, and many other processes, the history of which is difficult to reconstruct. Second, for the historian, access to his sources can only be gained through the mediation of the work of the scholars who made the texts available through process such as cataloging, copying, and editing. This chapter scrutinizes the effects of such intermediaries.

When reading an ancient text, we are deeply indebted to readings made previously by scholars who edited the texts and made them available. The process of

C. Proust (✉)
Laboratoire SPHERE, CNRS & Université Paris Diderot,
Case 7093, 5 rue Thomas Mann, 75205 Paris Cedex 13, France
e-mail: christine.proust@univ-paris-diderot.fr

A. Jones et al. (eds.), *A Mathematician's Journeys*, Archimedes 45,
DOI 10.1007/978-3-319-25865-2_7

decipherment, transliteration, translation, and commentary does not provide us with the raw material, but rather with interpretations of these texts. These reading keys are not always made explicit in the publications, but they appear more clearly in the editors' working documents, such as drafts, letters, annotations or notebooks.

In recent years, working documents produced and used by Otto Neugebauer when he was preparing the earliest editions of mathematical cuneiform texts became accessible to the researchers. These archives show how Neugebauer worked day by day and reveal the issues he faced while discovering cuneiform mathematics. Among Neugebauer's papers, his correspondence with other Assyriologists is of particular interest as it indicates how Neugebauer, a mathematician, benefited from the expertise of philologists either through deep friendship, such as with Albrecht Goetze and Abraham Sachs, or through competition as with François Thureau-Dangin. Thus, this documentation allows us to follow and to understand the project that led to the publication of *Mathematische Keilschrifttexte* (Neugebauer 1935–1937 – *MKT* hereafter), and later *Mathematical Cuneiform Texts* with Sachs (Neugebauer and Sachs 1945, *MCT* hereafter).

The aim of this chapter is to show some aspects of the impact of Neugebauer's work on our current understanding of cuneiform mathematics. The first part provides a brief overview of Neugebauer's papers relating to mathematical cuneiform texts. The second part presents some examples showing how the works of Neugebauer and his close colleagues (mainly Goetze, Sachs and Thureau-Dangin) shaped the editions of cuneiform mathematics.

Overview of Neugebauer's Archives Used in this Chapter

Neugebauer's papers are divided between several archives kept at Yale University, Institute for Advanced Study in Princeton, New York University, Brown University, Michigan University and other places in the United States and Europe. More details on the archives containing papers related to cuneiform mathematics are provided in the following.

*Yale Babylonian Collection (*YBC *Archives)*

The Yale Babylonian Collection includes archives, especially the letters exchanged by the Curators with scholars who worked on the tablets, as well as with dealers from whom tablets were purchased. These archives contain, among others items, two folders of interest for us: the correspondence with Neugebauer and with Sachs.

The "Neugebauer 1931-1957" folder contains 124 letters between the Curators and Neugebauer. They deal mainly with issues ranging from authorization for publication, to photographs of mathematical and astronomical tablets kept at

Yale. Some of the letters also contain discussions on scientific issues, for example about how Babylonian mathematicians solved quadratic equations.

The "Sachs 1942-1955" folder contains 49 letters exchanged by the curator, Ferris J. Stephens, and Abraham Sachs, concerning the inventory, indexing, and photographs of the tablets kept at Yale.

*Yale University Library, Manuscripts & Archives (*YUL *Archives)*

274 letters exchanged by Neugebauer (and sometimes Sachs) and Goetze are kept among the Manuscripts & Archives of Yale University Library (Albrecht Goetze papers, MS # 648, box # 15 – *YUL* thereafter). They deal with political, scientific as well as more personal topics. The political issues are related to the situation in Europe and the United States before and during World War II, the assistance offered to Jewish scholars dismissed from their positions in Germany,[1] the condition of immigrants from Europe in the United States, anti-Semitism in American Universities and other matters. The scientific discussions focus on the mathematical tablets in the Yale collection, that is, the search for new texts, as well as on collations and philological assistance provided by Goetze to Neugebauer.

The Shelby White and Leon Levy Archives Center, Institute for Advanced Study (IAS Archives)

The Institute for Advanced Study in Princeton hosts the bulk of the personal library and papers left by Neugebauer. The collection was subsequently enriched by documents donated by Edward S. Kennedy, and later by John P. Britton on behalf of Asger Aaboe.[2] The Shelby White and Leon Levy Archives Center, created at the IAS in 2010 (Di Bella 2010), is now in charge of the preservation of the papers.[3]

---

[1] Leo Oppenheim's escape, in particular, was a major concern for Neugebauer, and he worked hard to help Oppenheim emigrate to the USA.

[2] "Neugebauer left instructions that the bulk of his library be left to the Institute for Advanced Study for disposition after his death. The notebooks were found among the items in his library. He had previously donated the majority of the volumes of publications (library acquisition 84-B1019), with one small volume added by the Historical Studies-Social Science Library in 1993. The files related to Astronomical Cuneiform Texts and his Copernicus notes were donated by John P. Britton on behalf of the family of Asger Aaboe in 2007. (Neugebauer had given the materials to Aaboe for disposition after his death.) The diary and correspondence with Edward S. Kennedy were donated by Kennedy in 1997." (IAS website http://library.ias.edu/finding-aids/neugebauer)

[3] Neugebauer's papers are divided into 14 boxes. The archives include notebooks, manuscripts and working papers related to his major publications, published articles, correspondence with Kennedy (1950–1990), as well as a diary that Neugebauer kept while in the Austrian army during World War

Among the working papers, let us also mention the manuscript of *Über vorgriechische Mathematik*, written in 1934, and *Babylonische Rechentabellen*, undated (both available on the IAS website). *Babylonische Rechentabellen* is a 174 page typewritten document. It contains tables of calculation in sexagesimal place-value notation, necessary for the reading of cuneiform mathematical and astronomical texts. More than half of this document is devoted to reciprocal tables for 1, 2, 3, 4, 5 and 6 sexagesimal place-value numbers. The reciprocal tables are noted without using trailing zeroes or signs separating the fractional part. This important detail shows how Neugebauer used floating-point notation for recording data and, presumably, calculating, in a manner comparable to that of ancient scribes.

Aaboe-Britton Archives

Neugebauer entrusted some personal notes to Asger Aaboe, who himself later bequeathed them to John Britton, who in turn made them available to ISAW researchers and visitors during the 2009–2010 academic year. These archives consist of several hundred folders. Each folder is devoted to a single mathematical or astronomical cuneiform tablet and contains transliterations, translations, collations, photos, and sometimes relevant letters. There are 206 folders for mathematical tablets, covering almost all the tablets published in *MKT* and *MCT*. Besides the items described above, these folders contain numerous collations provided by Goetze, as well as several originals of Goetze's letters, the carbon copies of which belong to *YUL* archives.

The main protagonists of the letters are Ferris J. Stephens, Albrecht Goetze, Abraham Sachs, and Otto Neugebauer. The discussions between them concern mainly photographs of the tablets, collations of the parts of the texts unclear in the photographs, for example of the signs written on the edges, and the problem of regional dialectal variations ("dialect business", to quote Goetze[4]). These discussions shed light on the major role played by Goetze in the edition of cuneiform mathematics. The scientific issues involved by the "dialect business" are of great importance for the history of science since they address the uniformity of mathematics in Mesopotamia. Before entering the very content of the letters, a short presentation of the main actors involved may be useful.

I on the Italian front. The notebooks contain extremely thorough lecture notes taken by Neugebauer during mathematics seminars held in Göttingen, Graz and Munich between 1919 and 1926 (see D. Rowe's chapter in this volume).

[4] YUL, Goetze to Neugebauer, 1942/08/17, 1942/11/14 (see Appendix section Dialect Business). Note that the references to the letters are given in this chapter according to the following form: ARCHIVE, sender to recipient yyyy/mm/dd.

The Actors

Raymond P. Dougherty was Curator of the Yale Babylonian Collection from 1926 to his death in 1933, an event referred to in several letters.[5] He succeeded Albert T. Clay, the founder of Yale Babylonian Collection (Curator 1910–1925). Ferris J. Stephens, Curator from 1933 to 1962, was Neugebauer's main interlocutor for everything concerning access to the tablets.

Albrecht Goetze (1897–1971) was Professor of Semitic Languages and Comparative Linguistics in Heidelberg from 1927 to 1930, Professor of Semitic Languages and Ancient Oriental History in Marburg from 1930 to 1933. In 1933, he was dismissed by the Nazi government as "politically unreliable." After a brief stay in Copenhagen and Oslo, he came to Yale University in 1934 as invited professor of Assyriology and was to stay with the Yale Babylonian Collection until he retired in 1965. We can follow through his letters to Neugebauer his adaptation to the American way of life and his long path from the precarious status of a recent immigrant to the highest academic positions in Yale University: William Laffan Professorship in 1936, Sterling Professorship in 1956. The correspondence between Goetze and Neugebauer is written in German until June 1940. From this date, only 1 year after Neugebauer's arrival in the U. S., the two German scholars communicated only in English. As noted above, the letters are of interest from many points of view: political, sociological and of course scientific. In the limited scope of this paper, only some of the scientific aspects are addressed. As we shall see later, Goetze played an essential role in the discovery and decipherment of mathematical texts in the Yale Collection. This role is little known since Goetze did not publish much on mathematical subjects. Besides his seminal contribution on dialects in *MCT*,[6] the papers he wrote on subjects linked to mathematical texts are: "Numbers idioms in Old Babylonian" (1946), "A mathematical compendium from Tell Harmal" (1951); and, of course, a "Review of Mathematical Cuneiform Texts" (1948). Essentially, Goetze's contribution to cuneiform mathematics took the form of his close collaboration with Neugebauer. In fact, this collaboration had been imposed to Neugebauer by Ferris J. Stephens, who was then Curator, as a condition for permission to publish new texts (*YBC* Stephens/Neugebauer 1934/05/29, 1934/06/12, 1934/07/09 – see Appendix section "Photo Business").

This condition was warmly received by Neugebauer and Goetze who had known each other since 1929 at least (*YUL* Goetze 1929/05/18) and had followed parallel paths from their flight from Nazi Germany to their arrival at U.S. Universities, via Copenhagen. Goetze identified most of the mathematical texts in the Yale Babylonian

[5] Stephens to Neugebauer 1933/06/12, 1934/01/04; Neugebauer to Stephens 1934/01/31 (the latter letter is provided in Appendix section Photo Business).

[6] Goetze 1945. In his review of *MCT*, Jacobsen (1946: 18) stresses the importance of this work: "Before concluding this review we would once more call the attention of Assyriologists specifically to Goetze's important contribution, chapter IV. The criteria for provenance of the tablets which he there establishes have a bearing far beyond the mathematical texts." We will come back on this essential point below in the section on "Dialect Business".

Collection that Neugebauer later published (see for example *YUL* Goetze to Neugebauer 1942/02/04 in Appendix section "Dialect Business"). In addition, Goetze helped Neugebauer to address the philological difficulties the mathematician faced everyday, a role increasingly assumed by Sachs from 1941. The numerous collations made by Goetze on behalf of Neugebauer testify his tireless assistance. But Goetze's role was important on another level. As a specialist in Semitic linguistics,[7] he became expert in Old Babylonian dialects and applied his knowledge of Akkadian orthography and phonology to mathematical texts.

Abraham Sachs (1915–1983) met Neugebauer for the first time in 1941 while working at the Oriental Institute of Chicago on the Assyrian Dictionary. A few months later, the young Assyriologist became Neugebauer's assistant.[8] Although Sachs' work is well known through his own letters,[9] we don't have any letters between Sachs and Neugebauer. Of course, the daily contacts between the two friends explain this absence. But many letters found in the Goetze folder at Yale University Library are signed "MCT Inc.", meaning Neugebauer and Sachs. Moreover, "the Owl", nickname[10] given to Sachs by Neugebauer, is omnipresent in Neugebauer's correspondence. Despite the prominent role of Sachs in the edition of mathematical texts, he remained in the shadow of Neugebauer, and rare are the scholars, such as Oppenheim (1947, pp. 126, 128), who do justice to his contribution.

Photo Business

Neugebauer's letters to Dougherty and Stephens include endless requests for photographs. His primary concern was the quality of the pictures. Neugebauer occasionally provided detailed technical advice on this matter (*YBC*, Neugebauer to Stephens 1934/01/31 – see Appendix section "Photo Business" and Fig. 1).

---

[7] Finkelstein 1972, p. 199.

[8] Swerlow 1993, p. 152: "Sachs was interested in Neugebauer's work, about which he already knew something, and he could read any text no matter how obscure or damaged. Neugebauer decided immediately that this was the person to continue the great project of publishing all the astronomical texts, and on the way back to Providence, he stopped in New York to discuss the matter with the Rockefeller Foundation. In the fall Sachs came to Brown as a Rockefeller Foundation Fellow with his wife Janet, who worked both at the university and at MR [*Mathematical Reviews*]. Then in 1943 Neugebauer received a 10-year grant from the Rockefeller Foundation, mostly to pay for a research associate, and Sachs became the Research Associate. And when the Department of the History of Mathematics was formed in 1947, Sachs joined the faculty, becoming an associate professor in 1949 and a professor in 1953. For more than 40 years Sachs was Neugebauer's closest colleague and closest friend. While they collaborated on a number of publications, this in itself gives no idea of the depth of their working relation."

[9] Lieberman 1991.

[10] From 1946, Neugebauer and Goetze began to use nicknames: "Nujipuri", and "Elephant" for Neugebauer; "Hippopotamus", and sometimes "Rhinoceros" for Goetze (see for example YUL, Goetze to Neugebauer, 1946/06/22; Neugebauer to Goetze 1946/05/31, 1946/07/17). Neugebauer sometimes signed his letters with the picture of an elephant.

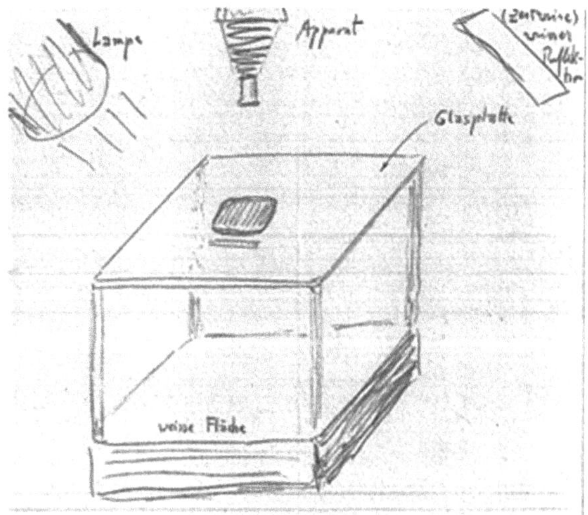

Fig. 1 *YBC*, Neugebauer to Stephens 1934/01/31

His second concern resulted from the fact that clay tablets are tri-dimensional objects, unlike the flat surface of a book's pages. Thus, signs written on the edges of the tablets do not appear on the photos of the obverse or the reverse of the tablets. Consequently, Neugebauer often required additional images of the edges and suggested methods for improving the depth-of-field of the photos.

Through these exchanges, we also follow in detail the stages of tablet restoration: cleaning, baking, and rinsing (*YUL* 1942/09/02 and 1942/09/11). This preliminary stage of editorial work is generally completely absent from publications.

The letters show how meticulous and precise Neugebauer was in his reading of texts. Any slightly obscure cuneiform sign was subject of long discussion by correspondence with his colleagues at Yale, and he did not hesitate to inundate them with relentless requests for collations. For example, the reading of YBC 7164 occupies most of the correspondence between Neugebauer and Stephens during the fall 1942 (letters dated 1942/09/22, 1942/10/13, 1942/10/27, 1942/11/03). Goetze was also involved: "*Both Goetze and I examined the signs circled in red on your manuscript. We made our drawings independently and compared results only after both had completed the drawings. I am glad to see that we agree as to what is to be seen on the tablet.*" (*YBC*, Stephens to Neugebauer 1942/11/03). See Fig. 2.

One can imagine the time spent by all of them in the collective decipherment of tablets by correspondence. Thus, it is surprising that Neugebauer only rarely examined the tablets himself. While travel was impossible at the beginning of his collaboration with Stephens and Goetze when he was living in Europe, it was easy from the time he moved to Providence, which is only 1 h by train from New Haven.

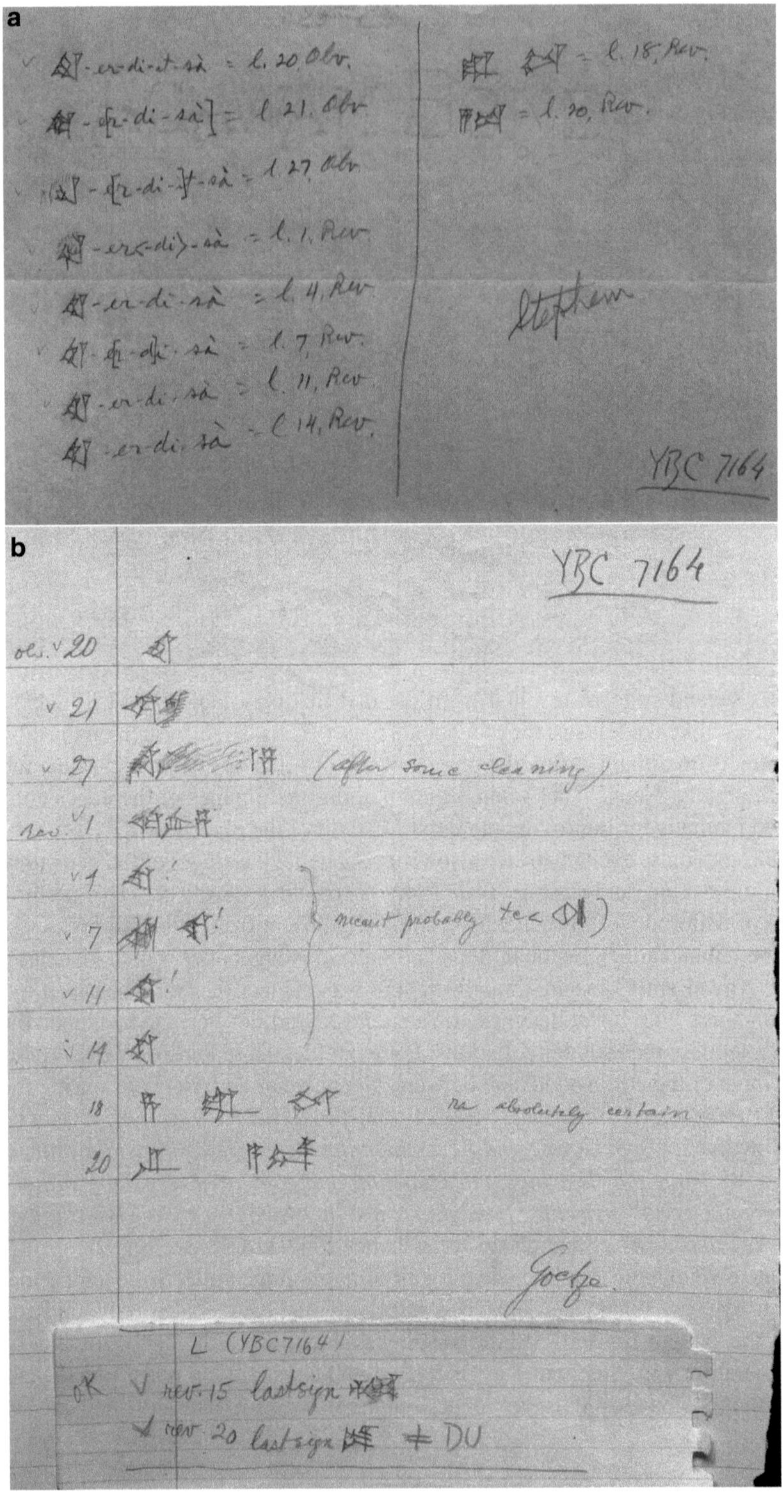

Fig. 2 Collations of YBC 7164 made independently by Stephens and by Goetze, probably in October 1942 (Aaboe-Britton Archives, folder YBC 7164)

He very rarely made this short journey. Neugebauer seems to have preferred to work on photos and leave the collations in the good care of his competent colleagues, perhaps, paradoxically, due to his concern for detail and accuracy. However, the habit of working only on photographs had some drawbacks, as Goetze sometimes kindly remarked to his friend: *Obviously the photograph did fool you.* (*YUL*, Goetze to Neugebauer 1943/01/21 – See complete letters in Appendix section "Dialect Business"). Indeed, it could happen that the photograph "fooled" Neugebauer for the reading not only of individual signs, but also for entire sections. In a few cases, bracketed text in transliteration (that is, text considered as damaged by Neugebauer) is in fact perfectly preserved. Examining the photos used by Neugebauer, which are mostly kept in the Aaboe-Britton folders, one realizes that a "damaged text" for Neugebauer may often simply be a blurred photo (see YBC 4710 #4 in Appendix section "Damaged Tablet or Damaged Photo? YBC4710). In other cases, the blurring of the photo combined with some dirt pasted on the clay surface make the text almost unreadable, although the tablet is currently in perfect condition (see YBC 4668, #33 in Appendix section "Dirty Tablet and Blurred Photo: YBC 4668"). It is interesting to note that the unnecessary brackets are found at the same place in Thureau-Dangin's transcriptions. This detail proves that Thureau-Dangin used the very same photographs as Neugebauer (see Appendix B.2).

Comparing the text of "damaged" sections restored between brackets by Neugebauer and the actual text on the original tablet shows how deep his understanding of the texts was. Indeed, in most of the cases, the restoration by Neugebauer is correct. However, in some exceptional cases, such as YBC 4710 #4, a new reading can restore crucial information and improve the interpretation notably (see Appendix section "Damaged Tablet or Damaged Photo? YBC 4710").[11]

Beside the daily work on the tablets, the correspondence sheds light on scientific discussions between Neugebauer and the Assyriologists which are of great interest in the history of ancient mathematics. A good example is a long explanation by Neugebauer in reply to a question by Stephens: "did the ancient scribes really solve the equations, or did they simply guess the solutions by trial and error?". Another example, on which the following section focuses, is a discussion on the dialectal variations in mathematical texts. This discussion should become an additional chapter of *MCT* written by Goetze.

Dialect Business

Beyond philological assistance to Neugebauer, Goetze's contribution to the history of mathematics is crucial for he demonstrated the existence of dialectal differences within the Old Babylonian mathematical corpus.

[11] This is also the case for tablet YBC 4696: Neugebauer used a poor photograph, and actually, the reading of the text can be improved by examining the original tablet.

In 1942, Goetze writes to Neugebauer in a post scriptum:

> Did I ever tell you that the mathematical tablets in Old Babylonian can be divided in a northern and a southern group on linguistic grounds? The evidence in most cases confirms the information as to provenance given by the dealers from whom the tablets were purchased. (YUL, Goetze to Neugebauer 1942/02/04 – see complete letter in Appendix section "Dialect Business")

Neugebauer immediately shows his interest:

> You never mentioned anything about the dialect in the mathematical texts. This interests me of course very much, and I hope to hear more when I am down at Yale. (YUL, Neugebauer to Goetze 1942/02/10)

However, Neugebauer and Sachs underlined the difficulty of the project as far as the Yale tablets were concerned, since most of them are catalogues and series texts written with Sumerograms and use very few syllabic notations of Akkadian words.

> … we are herewith taking advantage of your kind affirmative answer to our request to look at the dialect(s) of our new Akkadian mathematical texts. Enclosed are the transcriptions of YBC 4608, 4662, 4663, and 4675, the only texts written in Akkadian except for the Plimpton tablet, on which the only two Akkadian words we can read with certainty are ṣi-li-ip-tim and [in]-na-as-sà-hu-ú-[ma]. (YUL, MCT Inc to Goetze1942/08/01 – See complete letter in Appendix section "Dialect Business")

Some 9 months after announcing his idea to Neugebauer, Goetze details the method he intends to use:

> The right procedure, I feel, would be this: numerate the characteristics which allow the classification (giving a number to every item). Then, listing the signature of the texts and adding the number of the characteristic with the necessary references. (YUL, Goetze to Neugebauer 1942/11/14)

He completed the work in a short time since, as early as January 21, 1943, he announces to Neugebauer and Sachs:

> I have just typed a 14 page statement concerning the "dialect" of the mathematical tablets. It needs going over and checking and will then be mailed together with your manuscript. (YUL, Goetze to Neugebauer and Sachs 1943/01/21)

A couple of days later, Goetze sent the manuscript to Neugebauer, but not without expressing some reservations about the results:

> Here enclosed you will find the statement on the "dialects" of the mathematical tablets which I promised you some time ago. At the same time I am returning the pages of your manuscript which you so kindly placed at my disposal.
>
> Not that I am entirely satisfied with the result. You will see that I felt compelled to attempt some grouping of the texts. On this point I expect your criticisms. It is my feeling that I rather encroached on your domain. I would feel much better, if you could be persuaded to handle this subject in a special chapter which should precede mine. As it is, I could give only some hints in footnotes. You have expressed yourself the intention of doing some grouping at the head of your glossary. And I think the subject calls for some fuller treatment. (YUL, Goetze to Neugebauer 1943/01/25 – See Appendix section "Dialect Business")

Neugebauer and Sachs' reaction is extremely interesting in that it reveals an appreciation of the diversity of cuneiform mathematics quite different from Goetze's (the following is an extract, but the whole letter deserves reading – see YUL, MCT Inc. to Goetze 1943/02/17; note that the letter is erroneously dated 1942 by the author).

> As you know, we are unable to establish any local distinction in our material –[...]. It is therefore of great interest to learn that a clear "southern" group can be isolated. Unfortunately, we cannot contribute anything to this view of yours from the point of view of content. Our arrangement (A, B, C, D, etc.) is purely arbitrary according to content e.g., geometrical problems, irrigation problems, etc. The more material we get, the more we begin to realize to how great an extent we are at the mercy of the accidental character of the excavation and preservation of our texts. Grouping which seemed to be quite reasonable in MKT (e. g. "series texts") disappear more and more. All we can say at present is that the content of the Old-Babylonian mathematical texts is so homogeneous and uniform that from this point of view one cannot make any classification with regard to origin or time (of course, the clear distinction from Seleucid material remains). (See complete letter in Appendix section "Dialect Business")

'Goetze's analysis of dialectal variations opened perspectives that Neugebauer and Sachs had not considered, convinced as they were that cuneiform mathematics was highly homogeneous and probably came from very few different centers producing cuneiform mathematics. The geographical and chronological variations were not taken in account in *MCT*. Indeed, the organization of *MCT* does not reflect Goetze's groups in any way. Neugebauer and Sachs decided to publish Goetze's contribution in a separate chapter that would not affect the organization and content of the rest of the book. It is surprising that Neugebauer and Sachs did not say anything in *MCT* on their opinion of Goetze's point of view, despite an express request from the latter (*I would feel much better, if you could be persuaded to handle this subject in a special chapter which should precede mine* – Goetze in the letter of 1943/01/25 cited above). This silence by Neugebauer and Sachs might reveal a lack of confidence concerning the difficult problem of the uniformity of cuneiform mathematics, on which they seem to have more convictions than strong arguments, or simply a reluctance to disagree publically with their friend Goetze.

Today, this decision might seem amazing. Indeed, Goetze's chapter on dialects turned out to be a seminal work that has not been surpassed, and still remains a landmark for Assyriologists. After World War II, the discovery and publication of mathematical texts from Ešnunna, a city located in the northern part of Mesopotamia, and from Susa, in western Iran, provided new evidence showing the diversity of Old Babylonian mathematics. In modern historiographical trends, a greater sensitivity to the diversity of mathematical cultures is observed. Goetze's groups provided a solid basis for the studies on cuneiform mathematics, particularly the work on mathematical cultures in Mesopotamia by Høyrup and Friberg.[12]

[12] Høyrup 2000, 2002: ch. 9 and Friberg 2000.

Several reasons explain Neugebauer and Sachs's decision not to use Goetze's groups to reorganize *MCT*. The first and most evident is that *MCT* was almost finished when Goetze sent his chapter to "MCT Inc." The end of 1942 was devoted to corrections of *MCT*, and most of the letters between Yale and Brown focused on corrections of the final manuscript. The second reason is, as stated by Goetze himself, that the dialectal groups could not include all of the tablets since the texts written with sumerograms, namely most of the Yale sources, could indeed not be analyzed on the basis of Akkadian dialects. The third and more fundamental reason is Neugebauer and Sachs's conviction that Old Babylonian mathematics was strongly homogeneous. Indeed, when Goetze proposed his project to them, they did not believe that the results would change their perception of things. Their favorable welcome to the "dialect business" seems to have been more polite than enthusiastic. The correspondence between 1942/02/04 and 1943/01/25, focused on correcting *MCT*, does not show excessive concern from Neugebauer and Sachs about the potential results of Goetze's investigations on dialects.

In order to understand Neugebauer and Sachs' view, we have to keep in mind that a striking feature of Old Babylonian mathematics is the standardization of the metrological systems. This feature gives the mathematical texts an apparent unity. But the uniformity of notation is the result of the relative uniformity of education in the scribal schools, which, during the Old Babylonian period, formed a dense network, spread out over a large part of the Ancient Near East.[13] Onto this common elementary knowledge, shared by the erudite scribes in Mesopotamia and beyond, were grafted specific scholarly traditions which could differ deeply one from the other. In a way, with their opposite views, Neugebauer and Goetze were describing different aspects of the same complex *realia*.

Neugebauer and Thureau-Dangin: A Mathematical Approach versus a Philological Approach?

In the correspondence between Neugebauer and Goetze, we see with great precision how the mathematical and philological skills of two scholars with completely different backgrounds were able to complement and inspire each other and to produce editions of texts of outstanding quality that has not been surpassed since.

The cooperation between Neugebauer and Thureau-Dangin was of a quite different nature. The Yale correspondence echoes the tensions between the two prominent specialists of cuneiform mathematics:

[13] Old Babylonian scribal schools have been the subject of many studies; see for example Veldhuis 1997 with its bibliography.

> With surprise and honest regret I have come to know that a certain animosity has developed between Thureau-Dangin and yourself. Personally, I feel it is most unfortunate since I have always treasured Thureau-Dangin and never had the slightest reason to doubt his excellent character. Isn't this a rare case among Assyriologists. It almost seems as if Assyriology ruins one's character! (YUL, Goetze to Neugebauer 1937/02/14 – See complete letters in Appendix section "Competition with Thureau-Dangin")

This animosity probably goes back to some months (see Appendix section "Competition with Thureau-Dangin"). During the spring of 1936, Stephens asked Neugebauer for permission to send the photos of the tablets to Thureau-Dangin, who was willing to "check" his transliterations and translations (*YBC*, Stephens to Neugebauer 1936/04/03, 1936/06/16 – See complete letters in Appendix section "Competition with Thureau-Dangin"), probably for his review of *MKT I* (Thureau-Dangin 1936). It is easy to understand that Neugebauer was not very happy with this request. In his reply, 3 months later, the competition between the two scholars became even more apparent (*YBC*, Neugebauer to Stephens 1936/06/27 – See complete letters in Appendix section "Competition with Thureau-Dangin").

As a matter of fact, Thureau-Dangin subsequently extended his project beyond his initial plan to "check" Neugebauer's work. He published *Textes Mathématiques Babyloniens* (hereafter *TMB*) in 1938, which is not far from being a re-edition of the texts published in *MKT*, where the French Assyriologist corrected and substantially improved the readings by Neugebauer.[14] These improvements had been listed in detail in his quite critical reviews of *MKT* (Thureau-Dangin 1936, 1937). But it is clear that Thureau-Dangin was fully confident in Neugebauer's understanding of the mathematics: *L'interprétation mathématique est, dans l'ouvrage de N.* [*MKT I*], *beaucoup meilleure que l'interprétation philologique. L'auteur est là sur son terrain.* (Thureau-Dangin 1936: 59).[15] The philologist did not change the meaning of the texts, except for two texts (Str. 362 and VAT 8528), in which he corrected the mathematician on his own ground (Thureau-Dangin 1936: 59–60). An important consequence of this story is that *MKT* and *TMB* have become inseparable. Today scholars cannot use *MKT* without, at the same time, also referring to *TMB*, and vice-versa.

Thureau-Dangin read the tablets from the photos sent to him by Stephens, who is warmly thanked in many footnotes in *TMB*, and he used the same material as Neugebauer. Thus, the uncertain readings due to a blurred picture in *MKT* remain uncertain readings between brackets in *TMB*, as already noted above and in Appendix B.

The difficult but fruitful cooperation between Neugebauer and Thureau-Dangin upsets simplistic ideas whereby mathematicians are only capable of mathematical understanding, and philologists of philological insight. Neugebauer and Thureau-

---

[14] It seems that the "animosity" between the two scholars was to spread later. In his *Hommage à la mémoire de l'éminent assyriologue François Thureau-Dangin*, Edouard Dhorme presents *TMB* as an *ouvrage unique au monde* without any word on the work of Neugebauer (Dhorme 1946).

[15] "The mathematical interpretation in N's work [MKT 1] is much better than philological interpretation. The author is more at home".

Dangin's case shows an interesting reversal of roles. As early as 1932, Neugebauer criticized the transcription system used by Thureau-Dangin:

> "While the transcription system of numerals is merely a matter of convention, the question concerning the transcription of ideograms of mathematical expressions belongs to the field of philology, where a solid tradition has been established. I am aware, that with the following I am going against the *opinio communis*, but nonetheless would like to open a discussion, which to me seems factually important.
>
> An example. Thureau-Dangin transcribes (a) *GAM* (b) *DU-ma* with (a) *adi* (b) *tubalma*. One is to note here: why the second person and not the imperative? Both are used in mathematical texts, and the selection concerning *DU-ma* is merely arbitrary (same goes for the verb forms, I or II and the like). Why *DU=abâlu* and not for example=*šakânu* (the latter being frequently used in mathematical texts)? And similar questions occur in other cases.
>
> The matter touched upon here is not negligible but a totally essential point. First, one can no longer conclude which symbol was written in the text from the transcription (a fact which could be significant in the case of questions concerning how to put together certain text groups[16]). Furthermore, how should one to proceed if entire Sumerian phrases are contained in Akkadian texts? Why should one Akkadize them? By the same rule, one would eradicate the Latin and French in essentially German writings of scholars of the 17th century. And finally, with the process of Akkadization one destroys the fundamental role of the ideograms: namely that they function entirely like mathematical symbols and that the Akkadization essentially does the same as the substitution of e- $^{x\,2}$ with "e-superscript-minus-*x*-squared" or (because one generally does not know how to pronounce such a symbol) the fabrication of a "real" grammatical sentence. The engagement with this text genre made it more and more apparent to me that the ideograms play a crucial role for the remarkably strong algebraic character of Babylonian mathematics (if originally intended or not is not of importance in this instance)." (Neugebauer 1932–1933: 222- Translated in English by Sandra Hoehn)

In fact, the Akkadization of the transcription by Thureau-Dangin (see examples in Appendix B) did not generally hide the original ideograms, since he used different transcriptions for Akkadian words noted phonetically in the tablets, and Akkadian words corresponding to ideograms. The original ideogram could be identified thanks to his "Lexique" (*TMB*, pp. 215–243). However, as noted by Neugebauer, this Akkadization could sometimes be ambiguous and, more seriously, could hide specific mathematical meaning conveyed by ideographic writing. This issue is particularly marked for texts that use Sumerograms almost exclusively, such as series texts. In this matter, Neugebauer's philological sensitivity is obvious, and subsequent philological tradition was to prove him right.[17]

For his part, Thureau-Dangin showed a remarkable acuteness in understanding the nature of the numbers used in mathematical cuneiform texts. As early as 1930, he offered a penetrating analysis of cuneiform sexagesimal place value notation.[18]

---

[16] I guess that by "groups of texts" (Textgruppen) Neugebauer refers to the series texts that he labeled "Serientexte" in further publications.

[17] As an evidence of the great influence of the French assyriologist, we must observe that for mathematical texts, Goetze used Thureau-Dangin's method of transcription, and not Neugebauer's (Goetze 1951).

[18] Thureau-Dangin 1930.

This work was followed by his seminal *Esquisse d'une histoire du système sexagésimal* (1932),[19] where interesting parallels are drawn between Babylonian abstract numbers and new concepts of numbers introduced by mathematicians Simon Stevin (1548–1620) and John Wallis (1616–1703). He suggests that the sexagesimal place value notation, which includes both integers and fractions in a unified system, and was used in astronomy in early modern Europe, may have inspired Simon Stevin's generalization of decimal place value system. Referring to "Cantor, *Vorlesungen über Gesch. D. Math. 1ᵉ éd. II, p. 563 ss.*", he states:

> The Hindu system concerned itself only with the expression of the integers. The Babylonian system, which assimilated the integers and the fractions, emanated from an extremely wide and comprehensive conception of the number. The idea of applying to the fractions the same progression as to the integers, but in decreasing order, has not been realized in our system of numeration prior to the dawn of the modern times. Simon Stevin was the first to give a clear exposition of it in a treatise published by Plantin, at Leyden, in 1585, under the title: "La Disme, enseignant facilement expedier par nombres entiers sans rompuz, tous comptes se rencontrans aux affaires des Hommes."[20]

Thureau-Dangin's deep understanding of numbers is probably related to the fact that he gave great importance to metrological texts, which were discarded by Neugebauer as not "really mathematical". The study of Mesopotamian metrology is one of the most important of Thureau-Dangin's works. His earliest papers deal with these topics (see Thureau-Dangin 1893, 1896), as do most of his articles published between 1928 and 1934. Conversely, Neugebauer constantly insisted on the fact that he was not interested in metrological lists and tables, which nevertheless were to appear later as an essential component of mathematical education in Old Babylonian scribal schools.[21]

As we see, in some cases, deep mathematical sensitivity came from Thureau-Dangin, the philologist, and rigorous philological methodology from Neugebauer, the mathematician.

---

[19] See Høyrup's chapter, section on sexagesimal place value notation. In 1939, Thureau-Dangin published *Sketch of a History of the Ssexagesimal System*, an English version of *Esquisse* (Thureau-Dangin 1932). Analyzing both versions shows that the publication of MKT in 1935 slightly modified his approach to metrological tables.

[20] Thureau-Dangin 1939, 140–141. The original French version (Thureau-Dangin 1932, 80) is: *Le système Indou, tel du moins que nous l'avons emprunté, ne concernait que l'expression des entiers. Le système babylonien, qui assimilait entiers et fractions, procédait d'une conception du nombre autrement large et compréhensive. L'idée d'appliquer aux fractions la même échelle qu'aux nombres entiers n'a été réalisée dans notre système de numération qu'à l'aube des temps modernes. Simon Stevin l'a, le premier, clairement exposée dans un traité imprimé à Leyde, chez Plantin, en 1585, sous le titre : "La Disme, enseignant facilement comment expedier par nombres entiers sans rompuz, tous comptes se rencontrans aux affaires des Hommes."*.

[21] See for example letter YBC, Neugebauer to Stephens 1934/01/31 in Appendix section "Photo Business" where he describes the texts he is interested in, excluding metrological lists and tables. Similar claims can be found in many other letters (among them: YBC Neugebauer to Stephens 1932/08/12, 1934/07/27, 1935/05/31, 1935/12/28). Neugebauer tried to entrust metrological matters to Sachs (YUL Goetze to Neugebauer, 1945/01/22, Neugebauer to Goetze 1945/04/02). For more developments on the opposite approach of metrological texts by Thureau-Dangin and Neugebauer, see Proust 2010.

Toward Other Horizons

This selective review of Neugebauer's correspondence underlines the collaborative aspects of the work which was to lead to the publication of *MKT* and *MCT*. Thanks to Neugebauer's scientific rigor and his ability to work collectively, a set of highly reliable sources are available for research. However, scrutinizing the way these sources were shaped allows us to use them in a critical way. For example, as already said, we cannot work with *MKT* without using also *TMB*. The fact that Neugebauer (and, to some extent, Thureau-Dangin) had access to the tablets almost exclusively through photographs should encourage the modern researchers to examine, as far as possible, the original tablets.

Neugebauer was interested essentially in the publication of primary sources. As his energy was mobilized mainly by the search for new texts, he never returned to them once the mathematical cuneiform texts had been published. He opened extremely fruitful avenues of research, such as exploring and using ancient systems of classification of mathematical material (YUL, Neugebauer to Goetze 1935/03/26). However, he did not follow them through, and his ambitious program still awaits completion, as if he had left the work of deeper interpretation to future generations. He somewhat abandoned the history of mathematics in 1945, after the publication of *MCT*, to devote himself to astronomy.

Acknowledgements I would like to thank my friend and colleague Françoise Rougemont for the time she spent with me studying the contents of the Neugebauer correspondence as well as his publications in German, and for her careful reading of the drafts of this article. Without this collaboration, the present chapter would not have been possible. The material used in this study was collected thanks to a grant offered by the Institute of Advanced Study (Otto Neugebauer Fund, fall 2009) and the Institute for the Study of the Ancient World, NYU (spring semester 2010). My deep gratitude goes to these two institutions, and thanks are especially due to IAS librarians, Christine Di Bella and Erica Mosner. I would like above all to thank warmly Ulla Kasten and Benjamin Foster who pointed out the interest of the correspondence between the Curators of the Yale Babylonian Collection and Neugebauer. This work benefited greatly from the enthusiastic support of the late John Britton, to whom I am immensely grateful.

Appendices

Appendix A: Letters

German letters were translated in English by Sandra Hoehn.

Photo Business

YBC, Neugebauer to Stephens 1934/01/31 (in German)
Copenhagen, January 31, 1934

Dear Professor Stephens,

I would like to thank you for your letter of the 4 January and much to my honest regret, I was sorry to hear of the passing of Professor Dougherty.

The matter of the photographs of your collection was kindly arranged by Professor Flexner, New York, and led Professor Dougherty to provide me in 1932 with photographs of mathematical texts of his collection. These photographs are important for my entire enterprise on mathematical cuneiform texts. Thus far, he has forwarded me photographs of the following: YBC 4692, 4709, 4710, 4713 and promised me to send me further texts, which unfortunately did not happen due to his illness. At the time, he sent me three types of texts as photographs in order for me to ascertain which ones are important to me. They are the following: (1) metrological texts (several lists of measures such as YBC 4701), (2) multiplication and division tables such as YBC 4692, (3) purely mathematical texts such as YBC 4709/10/13.

Of these types of texts only the last two are of interest to me, while there are also texts which could be grouped in the first as well as the second type; these are needless to say also important to me.

Considering the still relatively small number of mathematical texts compared to the rest of the cuneiform literature, it is of the utmost interest to me to become familiar with even the smallest and poorly preserved fragments. Based on purely mathematical reasons, which I cannot explain in few words, I find even the seemingly straightforward tables interesting (if you are interested in peculiarities, maybe I can direct you to my research on the "Sexagesimal and Babylonian Fractions 1 to 4" in "Quellen und Studien zur Geschichte der Mathematik B 1 und 2").

I would be very grateful if you could ensure that the photographs also contain the text margins and, if possible, include a scale. Furthermore, I would like to ask you not to mount the photographs. It goes without saying that I am prepared to pay the costs for the photographs and I would like you to inform me about this.

I hope I am not causing you too much inconvenience with my requests and would like to take the liberty to add another. Could you please get the photographs prepared soon, since I am quite advanced in my work and would shortly like to reach a relatively final overview of the material. It may be interesting for you to know that the Yale texts YBC 4709/10/13 especially represent a new and very interesting type of text (systems of quartic equations) and it is thusly of the utmost historic interest to find out how the other texts from your collections are to be classified in respect to this group.

I am looking forward to hearing from you again and would like to thank you once again for your kind efforts.

My warmest regards,

Yours respectfully

P.S. Certainly Professor Flexner will be willing to offer more information. Hence I am taking the liberty of sending him a copy of this letter.

YBC, Stephens to Neugebauer 1934/05/29

New Haven, May 29, 1934

Dear Professor Neugebauer:

After considering from every angle the question of your publishing our mathematical texts, I may now give you the following decision. You have my permission to make full use of the photographs which Professor Dougherty sent to you. They may be published in transcription, translation, autographed copied and photographic reproduction, if you desire. After examining the rest of the mathematical texts in our Collection it is very obvious that Professor Dougherty selected the best one to be photographed. Seeing that you have had some difficulty in reading even these photographs, I am sure that the majority of the texts could not be satisfactorily read from photographs without the help of someone who could have direct access to the tablets themselves. I do not feel like assuming full responsibility for doing the work that would be necessary in collating and copying many of these tablets. If you could find it possible to come in person to New Haven to study our texts at first hand I should be very glad to place them at your disposal. If this is not possible I have one other suggestion. You probably know that Professor Albrecht Goetze is to be Visiting Professor of Assyriology at Yale University next year. It has occurred to me that you may be able to secure his consent to give you the necessary assistance in handling these texts. I suggest that you get in touch with him at the following address: Esperance Alle 18, Kobenhavn-Charlottenlund. If he is willing to give you the assistance you will require I shall give my consent to the full use of all of the texts. Otherwise it seems to me that the balance of them should be reserved for publication by someone who have direct access to the tablets themselves.

I am returning to you herewith the copy of your manuscript which you kindly sent to me some time ago.

Yours very sincerely,

Ferris J. Stephens, Acting Curator

YBC, Neugebauer to Stephens 1934/06/12 (in German)

Copenhaguen, June 12, 1934

Dear Professor Stephens,

Thank you very much for your letter of the 29 May and the return of my manuscript. Moreover, I would like to thank you for putting the photographs, which are already in my possession, at my disposal and for allowing me to work on further mathematical texts. Personally, I know Professor Götze very well. By this time he is writing to you of his commitment to support me in the collation of problematic parts of the text as well as in any other way possible.

From a purely technical point of view, I would like to make the proposition that at this stage you already procure photographs of the texts in questions. It would very much facilitate and accelerate my work if I could already begin the preparatory work. In addition, it would allow me to discuss fundamental questions with Professor Götze as long as he is still here. Needless to say, I am intending to wait with any kind of publication until Professor Götze is in New Haven. I am convinced that this proposition raises no concerns, the more so because photographs of these texts are without doubt required for a general examination of the condition of the texts.

My warmest gratitude for your endeavours,

Yours respectfully

YBC, Stephens to Neugebauer 1934/07/09

New Haven, July 9, 1934

Dear Professor Neugebauer,

Since I have received your letter of the 12th of June and that of Prof. Götze of the 16th of June, in which he promises to help you with the necessary collations of our mathematical texts, I have had photographs of six additional tablets made. There are still several others which have not been photographed. I shall have them done and sent to you later. I have not sent them all at once for two reasons. First, I wish to learn if you have any suggestions concerning the technique of photographing. Second, some of the remaining tablets are in need of cleaning and perhaps baking. This will require some time. You perhaps realize that in the process of baking and cleaning tablets some damage may result to the tablet in spite of all the care we can exercise. In cases where I think it is necessary to bake a tablet do you wish to go to the additional expenses of having photograph made both before and after the baking process?

There will be no expense to you for the baking and cleaning of tablets, of course. We shall charge you only for the photographs. You need not pay for these until you have received all of them. The photographs which I am sending you today have been made at a cost of $10.00. I have a record that Prof. Dougherty once sent you photographs costing $5.00. This makes $15.00 in all.

Trusting that you will find these new photographs satisfactory, I remain,

Yours very sincerely

Ferris J. Stephens

Acting Curator

Dialect Business

YUL, Goetze to Neugebauer 1942/02/04
February 4, 1942

Dear Neugebauer:

The mathematical tablets newly found among the unbelievable richness of our collection number eight (tables excluded). It may well be that I have still overlooked one or the other piece. It is not easy task to go through unclassified material and picks out what you are looking for. So this will be for the present the best I can do for you. The tablets are in relatively good condition. Photographs have been already been taken. Stephens will take care of cleaning and baking as soon as possible. After that new photographs may be taken, or you may come down personally to make a collation.

The tables are less exciting. I think they can wait until you come down for your lecture.

I am sorry that the oriental Club cannot take advantage of your coming. But it was only a very slight chance anyway. Have provision be made for living quarters? If not, don't forget that my house is always at your disposal. This is valid also for Sachs, if he should like to accompany you.

Cordially yours

Did I ever tell you that the mathematical tablets in Old Babylonian can be divided in a northern and a southern group on linguistic grounds? The evidence in most cases confirms the information as to provenance given by the dealers from whom the tablets were purchased.

YUL, Neugebauer and Sachs to Goetze 1942/08/01
Providence, August 1, 1942

Dear Goetze:

We hope this finds you recuperated from the Summer Session of the Linguistic Society because we are herewith taking advantage of your kind affirmative answer to our request to look at the dialect(s) of our new Akkadian mathematical texts. Enclosed are the transcriptions of YBC 4608, 4662, 4663, and 4675, the only texts written in Akkadian except for the Plimpton tablet, on which the only two Akkadian words we can read with certainty are *ṣi-li-ip-tim* and [*in*]-*na-as-sà-hu-ú*-[*ma*]. In the enclosed transcription please disregard all underlining, since we used some of the carbon copies to make a vocabulary and underlined words to make sure that we didn't miss any.

We also hope you won't mind settling a friendly dispute which has arisen between us. Enclosed on a separate sheet is a translation of the first five lines of YBC 5037. One of us wishes to delete all the words which are in red, the other to keep them. We have decided to abide by your decision as to which of the two translations you would like to read.

We are waiting for several months now with bated breath for the new photographs of the Yale texts while the draft creeps up relentlessly toward one of us and while the other one is being loaded with more and more University duties. Stephens wrote us some weeks ago that the photographs had long been taken but the photographer did not have the time to make the prints. If you, with your usual discretion, could hurry this up, we would be very grateful. Cordially

MCT Inc.

For your convenience, the following numbers are the pages on which the beginnings of the transcriptions of the Old-Babylonian Akkadian texts will be found:

MKT I: 108, 124, 126, 137, 143, 194, 219, 239, 244, 248, 257, 259, 267, 269, 270, 274, 278, 287, 289, 294, 303, 311, 314, 317, 319, 335, 341, 346, 351, 353, 368, 373, 516.
MKT II: 37, 43, 60.
MKT III: 1, 22, 29.

YUL, Goetze to Neugebauer 1942/11/14
November 14, 1942

Dear Neugebauer,

I had your letter concerning my contribution to your and Sachs' Mathematical texts.

I am quite willing to make good promise, but I am afraid I can hardly do before the Xmas recess. There are too many various things whirling around us to make possible the necessary concentration on that task. Furthermore, if I do it, would not it be advisable to include all the texts you are going to publish and not merely the four texts the notes on which you are returning to me? This would mean that I need your transliterations of all those texts (at any event the transliteration of YBC 4675, 4608, 4662, and 4683).

The right procedure, I feel, would be this: numerate the characteristics which allow the classification (giving a number to every item). Then, listing the signature of the texts and adding the number of the characteristic with the necessary references.

Stephens has to go to Indiana suddenly in family affairs (nothing tragic), and may thereby have been prevented from mailing the photos as he intended to. He will be back next Tuesday, and I shall talk to him as soon as I see him. I have seen the copies and they were very good indeed.

Let me know your decision on the dialect business.

Cordially yours

p. 2

YBC 4663

Southern Old Babylonian (mimation inconsistent):>

PI = pi / e:	né-pe-šu (obv. 6, 13, 19, rev. 25); ḫé-pe (rev. 7, 19)
DU = ṭù:	pu-ṭù-ur (obv. 9, 22, 24, 30, rev. 3).
ZU = sú:	ú-sú-uh (rev. 9).
For AZ = úṣ see remark to YBC 4662.	
p. 3	

YBC 4675

Southern Old Babylonian:

PI = pe:	te-ḫe-pe-e-ma (obv. 8, rev. 9).
DA = ṭa:	ta-pa-ṭa-ar (obv. 9, rev. 10).
ZA = sà:	sà-ni-iq (obv. 7, rev. 6, 16).

Complement in a.šà-lam (obv. 3).

Repeated vowel for simple length : zu-ú-uz (obv. 3), ga-me-ru-ú-tim (obv. 7), ki-la-a-al-le-e-en (obv. 8). Also ša-ni-i-tim (obv. 5), [a-ra-ka-re-e-em (rev. 2, 12).]

Nazalization in i-na-an-di-kum (obv. 11, rev. 1).

YBC 4608

Southern Old Babylonian (mimation inconsistent):

DA = ṭa: i-pa-aṭ-ṭa-ar (obv. 16), ip-pa-aṭ-ṭa-ar (rev. 18).

Repeated vowel in a-ma-ri-ka (obv. 22, 28).

Construction aššum…amāri-ka (obv. 21 f., 28).

For mali cf. TCL XVII 58 37; XVIII 117, 8; UMBS VIII 2 125 11); TS 71 2 (all southern); but also CT VIII 38a etc.; 50d 10; YBT II 42 22 (northern).

YBC 4662

Southern Old Babylonian (mimation inconsistent):

PI = pí/e: hé-pe (obv. 17, 30).

Peculiar the employment of AZ in ḫu-ru-úṣ (obv. 22, 34) which recurs in YBC 4663 […] and in AO 8862 [MKT II Taf. 35 ff.], both southern texts (Fig. 3).

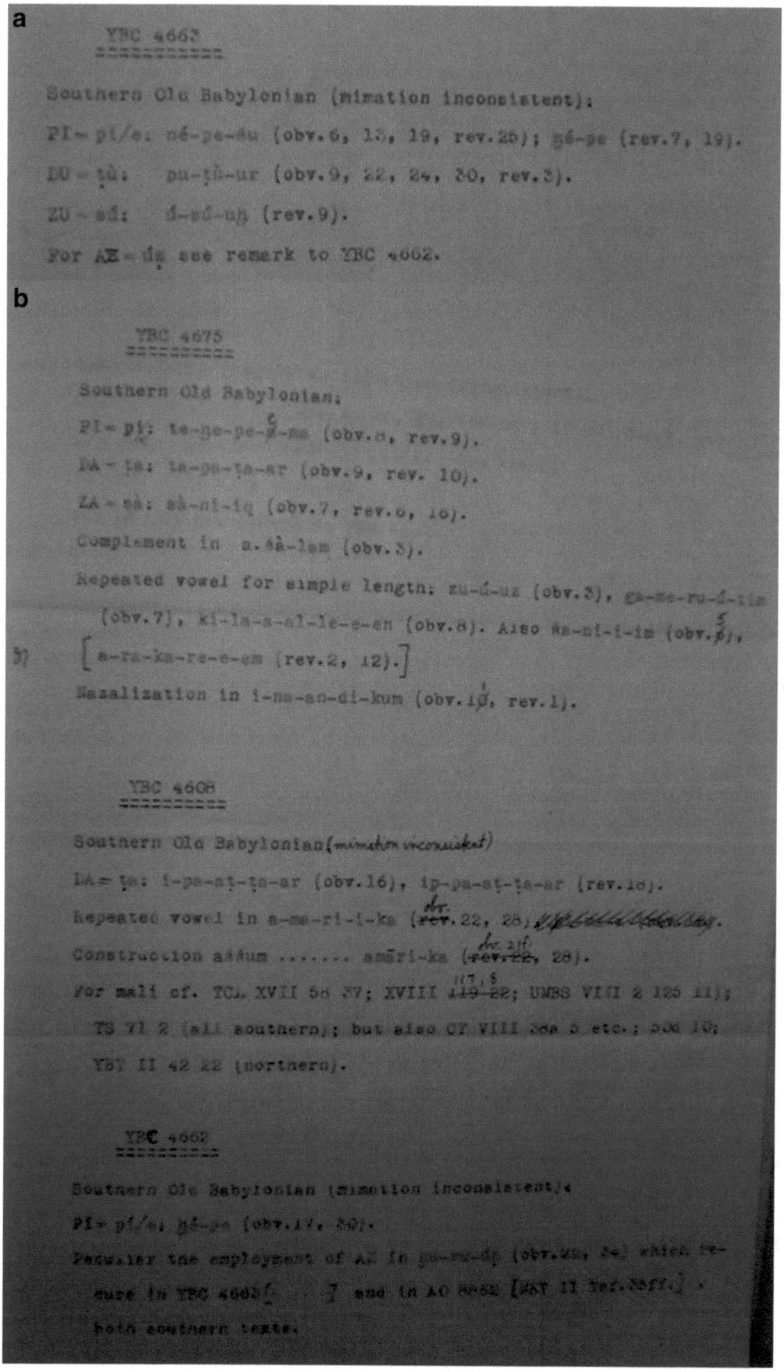

a

YBC 4663

Southern Old Babylonian (mimation inconsistent):

PI = pí/e: né-pe-šu (obv.6, 13, 19, rev.26); pé-pe (rev.7, 19).

DU = ṭù: pu-ṭù-ur (obv.9, 22, 24, 30, rev.3).

ZU = sú: ú-sú-uḫ (rev.9).

For AZ = ṣá see remark to YBC 4662.

b

YBC 4675

Southern Old Babylonian:

PI = pí: te-pe-pe-e-ma (obv.8, rev.9).

DA = ṭa: ta-pa-ṭa-ar (obv.9, rev. 10).

ZA = sà: sà-ni-iq (obv.7, rev.6, 10).

Complement in a.šà-lam (obv.3).

Repeated vowel for simple length: zu-ú-uz (obv.3), ga-me-ru-ú-tim (obv.7), ki-la-a-al-le-e-en (obv.8). Also ša-ni-i-im (obv.5),

[a-ra-ka-re-e-em (rev.2, 12).]

Nasalization in i-na-an-di-kum (obv.11, rev.1).

YBC 4608

Southern Old Babylonian (mimation inconsistent)

DA = ṭa: i-pa-aṭ-ṭa-ar (obv.16), ip-pa-aṭ-ṭa-ar (rev.10).

Repeated vowel in a-ma-ri-i-ka (obv.22, 28).

Construction aššum .... amāri-ka (obv.21f, 28).

For mali cf. TCL XVII 56 37; XVIII 117, 5; UMBS VIII 2 125 11; TS 71 2 (all southern); but also CT VIII 38a 5 etc.; 50d 10; YBT II 42 22 (northern).

YBC 4662

Southern Old Babylonian (mimation inconsistent).

PI = pí/e: pé-pe (obv.17, 20).

Peculiar the employment of AZ in pa-ṣa-dp (obv.22, 24) which recurs in YBC 4663[ ] and in AO 8862 [MKT II ...] both southern texts.

Fig. 3 YUL, Goetze to Neugebauer and Sachs 1942/11/14

January 21, 1943

Dear Neugebauer and Sachs,

I have just typed a 14 page statement concerning the "dialect" of the mathematical tablets. It needs going over and checking and will then be mailed together with your manuscript.

You did not answer my last question: namely whether in Stbg. 366 obv. A reading sa-am-da-ka-am "triangle" makes sense.

I have include a short note on ki-i a-a in my manuscript. I certainly shall take no offense, if you are not convinced.

E [YBC 8633]. Rev. 9: I am reading now i-ta-di-... (perhaps –kum after all). You remember that the whole thing is corrected and partially squeezed in. The [...] form agrees with the same form in VAT 8512 which belongs in the same group because of concidences in terminology etc. The same form should also be restored in rev. 7 (and obv. 13?).

L [YBC 7164]: I took another look at the alleged te-er-di-it-sà and also had Stephens look at it. We both agree that the correct reading is te-er-di-iz-za. The iz has the same dimensions as in is-sú-uh (l. 18) and is markedly different from the it in ú-sa-mi-it (ll. 22, 24). [...] iz signs are quite normal in this period. When you look up Fossey you will sign that originally the sign was rather wide and became gradually narrower. Moreover te-er-di-it-sà is (as far as I can judge) an "Unform" in OB.

M: eh-re-e is necessary just as ep-te-e and el-qé-e among other. In Babylonian the Umlaut (caused be "sharp laryngeals") affects every preceding or following a and causes shift into e. This is one of the differences between Babylonian and Assyrian.

P: At the ends of ll. 3, 4 and 6a are erasures. In l. 3 the ma seems to be erased. In l. 6a I read an-nu-um-ma as-su with an erasure following. The an is certain. Your translation "now (?)" is hardly correct. The particle should be (and is almost everywhere else) a-nu-um-ma. The spelling with nu points to a form of the demonstrative pronoun.

The chief passage for tallum in my omen texts is YBC 4629 II 48 ff. There, one finds omina which begin šumma ta-al-lu. If one does not know what it is, one can hardly learn it from there.

Since I have not yet made out slips of all texts, I can quote at present only one passage for tarahhum (perhaps there is only this one). Iit is YBC 4678 IV 51 ff.: šum-ma mar-tum (52) li-pi-a-am (53) ta-ra-ha-a-sa u-ka-al-la "supposing the gall-bladder, its two tarahhu hold "fat".

[?]: In l. 13 of the obverse can be no doubt as to the reading ne-pu-šu. The pu is perfect, i. e. contains four Winkelnaken. Obviously the photograph did fool you.

Cordially yours

YUL, Goetze to Neugebauer and Sachs 1943/01/25

January 25, 1943

Dear Neugebauer:

Here enclosed you will find the statement on the "dialects" of the mathematical tablets which I promised you some time ago. At the same time I am returning the pages of your manuscript which you so kindly placed at my disposal.

Not that I am entirely satisfied with the result. You will see that I felt completely to attempt some grouping of the texts. On this point I expect your criticisms. It is my feeling that I rather encroached on your domain. I would feel much better, if you could be persuaded to handle this subject in a special chapter which should precede mine. As it is, I could give only some hints in footnotes. You have expressed yourself the intention of doing some grouping at the head of your glossary. And I think the subject calls for some fuller treatment.

Otherwise, I think, it turns out neatly enough.

Expecting your reaction,

cordially yours

YUL, Neugebauer and Sachs to Goetze 1943/02/17 (the letter is erroneously dated 1942)
Providence, February 2, 1942[sic]

Dear Goetze,

We still owe you a detailed reaction to your contribution to MCT in addition to the short words of thanks we have already sent you. We should have done this long ago, but we were both pretty tired out, partly from work on MCT (preparing the final manuscript and copying the texts) and stupid teaching in ever increasing amount. Fortunately enough, Brown University ran out of oil, and this gave us a chance to recover and to write to you under less unreasonable conditions.

First of all, we must emphasize that we are very glad that we asked you for this contribution, which brings a very interesting new element into our own discussion. As you know, we are unable to establish any local distinction in our material – an attempt (MKT 387 f.) to localize the "Series Texts" at Kish must now be abandoned because new evidence gained from analysis of an MCT text shows that sig_4 never means "volume", but always "brick(s)". It is therefore of great interest to learn that a clear "southern" group can be isolated. Unfortunately, we cannot contribute anything to this view of yours from the point of view of content. Our arrangement (A, B, C, D, etc.) is purely arbitrary according to content e.g., geometrical problems, irrigation problems, etc. The more material we get, the more we begin to realize to how great an extent we are at the mercy of the accidental character of the excavation and preservation of our texts. Grouping which seemed to be quite reasonable in MKT (e. g. "series texts") disappear more and more. All we can say at present is that the content of the Old-Babylonian mathematical texts is so homogeneous and uniform that from this point of view one cannot make any classification with regard to origin or time (of course, the clear distinction from Seleucid material remains). We therefore intend to incorporate your contribution with no essential alterations as a separate chapter at the end of the large section dealing with problem texts. As the title of this chapter we suggest: "The Akkadian Dialects of the Old-Babylonian Mathematical Texts. By A. Goetze". On the title page of MCT we would like to mention your name in the form "MCT by O.N. and A.S. with a chapter by A.G." Please let us know if you are in agreement.

We might suggest a few minor alterations to conform with our own manuscript. We will send you the final copy of your chapter with our suggestions incorporated when it is typed; at that time, you will have the opportunity of approving all details or making any alteration that you may wish.

We are slowly approaching the end of our work – provided that you don't discover new material. In the meantime, we wish to repeat our warmest thanks for your manuscript, which will contribute considerably to the rounding out of MKT and MCT.

Cordially yours
MCT, Inc.

Competition with Thureau-Dangin

YBC, Stephens to Neugebauer 1936/04/03
New Haven, April 4, 1936

Dear Prof. Neugebauer:

Thureau-Dangin has requested me to send him photographs of eight of the mathematical tablets which you published in transcription, but without the autographed copy, in MKT. These are YBC 4668, 4669, 4673, 4695, 4696, 4697, 4698, and 4711. These are eight of the nine texts concerning which you wrote me under date of 28.12.35 that you intend to publish the autographed copies, and that you then had your manuscript practically ready for the press. In my letter of December 6, 1935 I stated to you that, "our policy is to allow no one to study a tablet which has been assigned for publication, without the consent of the one to whom it had been assigned, until after it has been published." Th-D writes, "je n'ai nullement l'intention de publier ces photos. Je désire seulement être en mesure de contrôler les copies ou transcriptions de Neugebauer". Nevertheless, it becomes my duty to refuse his request, unless you give your consent to his having the photos of the above mentioned tablets. The decision is in your hands; please let me know your pleasure as soon as possible.

Yours sincerely,
Ferris J. Stephens
Acting Curator.

YBC, Stephens to Neugebauer 1936/06/16
New Haven, June 16, 1936

Dear Prof. Neugebauer:

Under date of April 3, 1935 I wrote you to decide whether or not Thureau-Dangin should be given photographs of 8 of the mathematical tablets which you intent to publish in autograph in your Nachtragsheft to MKT. He has stated that he has no intention to publish the photographs, but only wishes to be able to control the transcriptions as already given by you in MKT. I think he feels that he should be permitted to have the photographs. Nevertheless I am bound not to furnish them to him without your consent, because of our policy not to allow any one to study a tablet which has been assigned for publication without the consent of the one to whom has been assigned.

May I hear from you at an early date concerning this matter, and also concerning the progress of your Nachtragsheft?

Yours very sincerely,
Ferris J. Stephens
Acting Curator.

YBC, Neugebauer to Stephens 1936/06/27 (in German)
Copenhagen 27.6.36

Dear Professor Stephens,

Thank you very much for your letter of the 16 June. Within the next weeks I am going to send off my supplement on the MKT for printing. Unfortunately, I was incapable to work due to illness for an extended period of time and I have only now finished the autographs. Hence, I would be indebted to you if you could refrain from releasing photographs of these texts to Mr. Thureau-Dangin until the publication of my supplement (which is likely to happen this autumn). The texts in question are YBC 4668, 4669, 4673, 4695, 4696, 4697, 4698 and 6504.

You may be wondering why I am asking you to refrain from releasing the texts at the moment, but I have unfortunately had some very strange experiences with Mr. Thureau Dangin and would prefer to complete my work without his interference. He himself has made some texts from the British Museum, which are a direct extension of my texts, unobtainable to me. Moreover, he is so inclined to beat me at every corner that in return I see no reason to facilitate his run in this race, which he started despite me repeatedly communicating my urge for friendly but factual cooperation.

I hope that I do not cause you any discomfort. Needless to say, I would not like you to understand my message in any other way than as an expression of my personal wish, which is in no way binding to you. Please act solely at your own discretion.

With kind regards and warm gratitude for your courtesy,
Yours respectfully

YBC, Stephens to Neugebauer 1936/07/17
New Haven, July 7, 1936

Dear Prof. Neugebauer:

I have your two letters of the 18th and 27th of June. I am sorry to learn that you have been hindered by illness, and hope that you have now fully recovered.

Be assured that your wishes will be respected concerning the giving of photographs to Thureau-Dangin. I have allowed him to have photographs of the tablets whose texts you have already published in autographed copy. The rest has been with-held.

In regard to your earlier question concerning YBC 4697 I beg to report that, while it has not been baked in the furnace, it is as clean as it can be made. The surface is badly preserved on both sides, and this accounts for the numerous spots on the photographs. Ne cleaning would do any good at these places for the writing is obliterated. I do not think anything at all can be done to improve the legibility of this tablet.

With kindest greetings,
Yours sincerely,
Ferris J. Stephens, Curator.

YUL, Goetze to Neugebauer 1937/02/14 (in German)
February 14, 1937

Dear Mr. Neugebauer,

My deepest gratitude to you for sending me the third volume of your mathematical cuneiform texts. It is a worthy extension of your earlier work. Many thanks for this valuable gift.

With surprise and honest regret I have come to know that a certain animosity has developed between Thureau-Dangin and yourself. Personally, I feel it is most unfortunate since I have always treasured Thureau-Dangin and never had the slightest reason to doubt his excellent character. Isn't this a rare case among Assyriologists. It almost seems as if Assyriology ruins one's character!

It is astonishing how deeply you have acquainted yourself with the Assyriology (of your texts). All reasonable persons will surely forgive minor philological oversights and simply be pleased there are not more of them. Philologists have always been incapable of tackling these texts and thus they must be delighted with what you have done with them.

P. Schaumberger dropped by. He is trying to find astronomical texts in America. I don't know, if he will be successful. The American collections are to a large extent assembled by purchase. However, mathematical and astronomical tablets are extremely rare. I think Chicago has some; maybe one can also expect some in Philadelphia. But unfortunately, a vast quantity of tablets are still stowed in boxes and the chances are that, no one truly knows, what they contain. The museum in Philadelphia would need an Assyriologist solely for the inspection and publication of the tablets.

With kind regards,
Yours

YUL, Neugebauer to Goetze 1938/03/21 (in German)
Copenhagen, March 21, 1938

Dear Mr. Götze,

I have just been made aware of the fact that Thureau-Dangin is looking for someone to edit the Mari texts. Could you be so kind as to write to Thureau-Dangin and recommend Oppenheim? After all, he is the ideal man for this job. I would rather avoid doing so and needless to say, I do not wish to be mentioned at all. I suppose Thureau-Dangin is not too fond of me and therefore I dare say that my involvement in the matter would have the opposite effect.

In haste, my best regards,
Your

Appendix B: Photos and Transliterations

Damaged Tablet or Damaged Photo? YBC4710

The following pictures show problem 4 on YBC 4710 (obv. col. i, li. 14–24) in the photo used by Neugebauer, the copy in MKT, and the current photo (see Fig. 4).

The following transliteration of YBC 4710 #4 shows that Neugebauer and Thureau-Dangin used the same blurred photo. The text is, in fact, perfectly preserved, and doesn't require the use of brackets.

MKT I, p. 402	**TMB, p. 149**	**After collation (Proust 2009)**
14. a-ša$_3$ 1(eše$_3$) GAN$_2$ uš *ša*	14. *eqlum ebel ikîm šiddim*	14. a-ša$_3$ 1(eše$_3$) GAN$_2$ uš-ta
15. *a-na* ba-zi nu-zu	15. *mala assuhu ul(a) îde*	15. a-na ba-zi nu-zu
16. *šá* u[š-ta b]a-zi	16. *šá ina šiddim uštakil*	16. nig$_2$ uš-ta ba-zi
17. KI íb-t[ag$_4$ u]š ì-kú	17. *itti šapitti šiddim assuhu*	17. ki ib$_2$-taka$_4$ uš i$_3$-gu$_7$
18. sag [íb-si$_8$ *ù* 1 (eše)] gán (?)	18. *a[na pû]tim [aš]ši*	18. sag-še$_3$ bi$_2$-il$_2$
19. a-š[à (?) K]I (?) *šá* uš ba-zi	19. *eq[lam] ù šá* <*ina*> *šiddim assuhu*	19. a-ša$_3$ u$_3$ nig$_2$ uš ba-zi
20. ì-kú-ma (?)	20. *uštakil-ma*	20. i$_3$-gu$_7$-ma
21. ì-kú u-gù ì-k[ú]	21. ì-kú *eli* ì-kú	21. i$_3$-gu$_7$ ugu i$_3$-gu$_7$
22. 1,48 d[irig]	22. 1.48 *î*[*ter*]	22. 1.48 diri
23. [*š*]*á* uš-ta ba-[zi]	23. *šá i[na] šiddim as[suhu]*	23. nig$_2$ uš-ta ba-⌈zi⌉
24. [u]-gù íb-t[ag$_4$ u]š 6 dir[ig]	24. [*e*]*li ša[pilti šid]dim* 6 *îter*	24. ⌈ugu⌉ ib$_2$-taka$_4$ ⌈uš 6 diri⌉

In the Goetze's collations (notes in Aaboe-Britton Archives, folder YBC 4710 dated February 3, 1935), it is clear that the text was better preserved than thought by Neugebauer (Fig. 5):

Dirty Tablet and Blurred Photo: YBC 4668

The following pictures show problem 33 on YBC 4668 (obv. col. ii, li. 59–61) in the photo used by Neugebauer, the copy in MKT, and the current photo (Fig. 6):

Neugebauer's transliteration (MKT I, p. 425)

59. a-[ša$_3$ 1(eše$_3$) GAN$_2$ igi-3-g]al$_2$ uš
60. [igi-4-gal$_2$ sag *a-na* uš ugu sag diri]
61. [a-ra$_2$] 2 e-tab-ma uš sag

Thureau-Dangin's transliteration (TMB, p. 169)

59. *eq*[*lum ebet ikîm šaluš*]*ti šiddim*

Aaboe-Britton Archives, folder YBC 4710

Neugebauer Handcopy, MKT II, pl. 58

C. Proust Photo, 2009

Fig. 4 Three pictures of problem 4 on YBC 4710 (obv. col. i, li. 14–24)

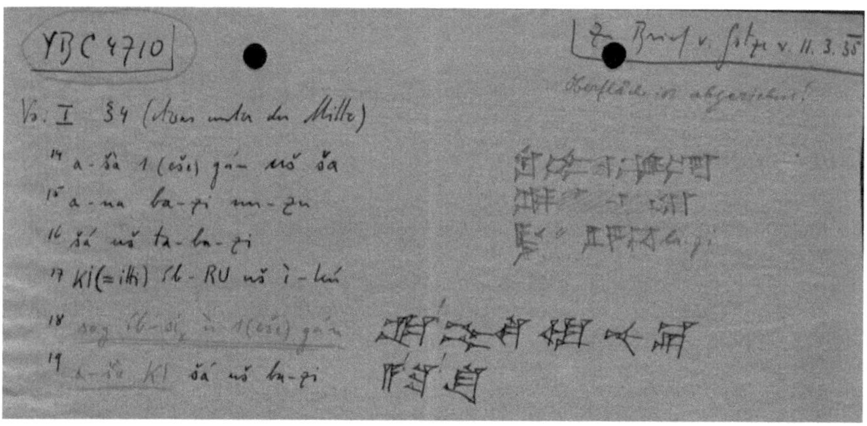

Fig. 5 Goetze's collations of YBC 4710, Aaboe-Britton Archives

60. [*rabiat pûtim mala šiddum eli pûtim îteru*]
61. [a-ra$_2$] *2 ^{e}êṣip-ma šiddum pûtum*

Translation after recent collation (made by C. Proust in 2009)

59. a-ša$_3$ 1(eše$_3$) GAN$_2$ igi-3-gal$_2$ uš
60. igi-4-gal$_2$ sag a-na uš<ugu sag diri>
61. a-ra$_2$ 2-e tab-ma uš sag

Aaboe-Britton Archives, folder YBC 4668 — Neugebauer Handcopy, MKT III, pl. 2 — C. Proust Photo, 2009

Fig. 6 Three pictures of problem 33 on YBC 4668 (obv. col. ii, li. 59–61)

References

Archives

Aaboe-Britton Archives: currently at the Institute for the Study of the Ancient World, New York University, New York, USA.

IAS Archives: The Shelby White and Leon Levy Archives Center, Institute for Advanced Study, Princeton, USA.

YBC Archives: Yale Babylonian Collection, Yale University, New Haven, USA. Two folders are used here: Neugebauer 1931–1957, and Sachs 1942–1955.

YUL Archives: Yale University Library, Manuscripts & Archives Yale University, New Haven, USA. The folder used here is Albrecht Goetze papers, MS # 648, box # 15.

Bibliography

Di Bella, C. 2010. From the Shelby White and Leon Levy Archives Center. *The Institute Letter* 3(Spring): 8.

Dhorme, E. 1946. *Hommage à la mémoire de l'éminent assyriologue François Thureau-Dangin (1872–1944)*, Mémoires de la Société Orientale "Ex Oriente Lux", vol. 8. Leiden: Brill.

Friberg, J. 2000. Mathematics at Ur in the Old Babylonian period. *Revue d'Assyriologie* 94: 98–188.

Goetze, A. 1945. The Akkadian dialects of the Old-Babylonian mathematical texts, chapter 4. In *Mathematical cuneiform texts*, ed. O. Neugebauer and A.J. Sachs. New Haven: American Oriental Series & American Schools of Oriental Research.

Goetze, A. 1946. Numbers idioms in Old Babylonian. *Journal of Near Eastern Studies* 5: 185–202.

Goetze, A. 1948. Review of *Mathematical cuneiform texts*, by Neugebauer and Sachs. *Journal of Cuneiform Studies* 2: 33–37.

Goetze, A. 1951. A mathematical compendium from Tell Harmal. *Sumer* 7: 126–155.

Høyrup, J. 2000. The finer structure of the Old Babylonian mathematical corpus. Elements of classification, with some results. In *Festschrift für Joachim Oelsner anläßlich seines 65. Geburtstages am 18. Februar 1997*, ed. J. Marzahn and H. Neumann. Münster: Ugarit Verlag.

Høyrup, J. 2002. *Lengths, widths, surfaces. A portrait of Old Babylonian algebra and its kin.* Berlin/London: Springer.

Jacobsen, T. 1946. Review of *Mathematical cuneiform texts*, by Neugebauer and Sachs. *Bulletin of the American Schools of Oriental Research* 102: 17–18.
Neugebauer, O. 1932–1933. Zur transcription mathematischer und astronomischer Keilschrifttexte. *Archiv fur Orientforschung* 8: 221–223.
Neugebauer, O. 1935–1937. *Mathematische Keilschrifttexte I-III*, Quellen und Studien zur Geschichte der Mathematik, Astronomie und Physik. Berlin: Springer.
Neugebauer, O., and A.J. Sachs. 1945. *Mathematical cuneiform texts*. New Haven: American Oriental Series & American Schools of Oriental Research Original Edition.
Oppenheim, A.L. 1947. Review of *Mathematical cuneiform texts*, by Neugebauer and Sachs. *Journal of Near Eastern Studies* 6: 126–128.
Proust, C. 2010. Mesopotamian metrological lists and tables: Forgotten sources. In *Looking at it from Asia: The processes that shaped the sources of history of science*, ed. F. Bretelle-Establet. New York: Springer.
Swerdlow, N.M. 1993. Otto E. Neugebauer (26 May 1899–19 February 1990). *Proceedings of the American Philosophical Society* 137: 139–165.
Thureau-Dangin, F. 1893. La comptabilité agricole en Chaldée. *Revue d'Assyriologie* 3: 118–146.
Thureau-Dangin, F. 1896. Quelques mots de métrologie. *Zeitschrift für Assyriologie und Vorderasiatische Archäologie* 11: 428–432.
Thureau-Dangin, F. 1930. Nombres concrets et nombres abstraits dans la numération babylonienne. *Revue d'Assyriologie* 27: 116–119.
Thureau-Dangin, F. 1932. *Esquisse d'une histoire du système sexagésimal*. Paris: Geuthner.
Thureau-Dangin, F. 1936. Review of *Mathematische Keilschrifttexte I*, by Neugebauer. *Revue d'Assyriologie* 33: 55–62.
Thureau-Dangin, F. 1937. Review of *Mathematische Keilschrifttexte III*. *Revue d'Assyriologie* 34: 87–92.
Thureau-Dangin, F. 1938. *Textes Mathématiques Babyloniens*. Leiden: Ex Oriente Lux.
Thureau-Dangin, F.' 1939. Sketch of a history of the sexagesimal system. *Osiris* 7: 95–101.
Veldhuis N. 1997. *Elementary education at Nippur, the lists of trees and wooden objects*. Ph. D. dissertation, University of Groningen.

After Neugebauer: Recent Developments in Mesopotamian Mathematics

Duncan J. Melville

When Otto Neugebauer began writing on Old Babylonian mathematics in the late 1920s, despite a certain amount of pre-history and heroic efforts by early pioneers, it was still a little-studied and poorly understood area. Once he engaged with the subject, a torrent of papers followed, leading up to the publication of the monumental *Mathematische Keilschrift-Texte* (MKT) in three volumes in 1935 and 1937. The appearance in 1945 of *Mathematical Cuneiform Texts* (MCT), mostly concerned with publishing tablets from Yale that had not been available to him earlier in Europe, as well as the infamous Plimpton 322, essentially completed his project. Neugebauer had read, translated, understood and described in precise mathematical detail the known corpus of Old Babylonian problem texts, as well as giving a categorization of the various types of table texts. Neugebauer himself moved on and, while his work on astronomy continued for the rest of his life, he rarely published on mathematics again. What was there left to do?

A field develops when there are new discoveries, but also in response to new questions. A part of the work of each generation of scholars is finding the hidden assumptions of their predecessors and questioning them. Both new discoveries and new questions have profoundly changed the way we see Mesopotamian mathematics since 1945. In 1996, Jens Høyrup published a masterly paper on the historiographical developments in Mesopotamian mathematics from the 1930s (which he termed the 'Heroic Era') up to roughly contemporary events (Høyrup 1996). Some of the "changing trends" Høyrup identified have continued over the last 20 years, and they have been joined by others; together, they add up to a re-visioning of Mesopotamian mathematics.

While Høyrup provided an insider's view, the outsider's view is perhaps best accessed through the tertiary literature, the history of mathematics textbooks where the Babylonians get their half-hour in the sun. From there we typically learn that the

D.J. Melville (✉)
Department of Mathematics, St. Lawrence University, Canton, NY, USA
e-mail: dmelville@stlawu.edu

A. Jones et al. (eds.), *A Mathematician's Journeys*, Archimedes 45,
DOI 10.1007/978-3-319-25865-2_8

Babylonians left us three things: the sexagesimal system; multiplication tables, and word problems. These things were known, of course, to Neugebauer, and in rather more detail than appears in modern textbooks. But we know them differently now, and I would like to take each of those topics in turn, as well as some others, and explore the differences. This paper is not intended to be a comprehensive survey of half a century of scholarship in Mesopotamian mathematics, but rather to delineate some of the more salient features that have changed our understanding of the ancient world.

Number System: Origins, Structure and Use

The first aspect of Mesopotamian mathematics to be deciphered was the number system, or, more precisely, the sexagesimal place-value system widely used for computations from the Old Babylonian period (ca 2000–1600 BC) onwards. This decipherment was aided by the pedagogical practice of Mesopotamian scribes and students of generating lists of organized numbers such as multiplication tables (some hundreds of these are now known) and by virtue of the immense simplicity of the system: a symbol for 'one', 𒁹, a single vertical wedge, was bundled up to nine, then a new symbol, a corner wedge 𒌋, was used to the left of the ones to represent 'ten'. Bundling the two symbols recorded numbers up to 59. 'Sixty' was recorded as a 'one' in the next place, allowing numbers up to 599 to be written and then 'Six hundred' was a 'ten' symbol in the next place left, and so on. This notational system was immediately recognizable and familiar to Western eyes, and it was taken for granted that 'their' number system was like 'ours', only with different symbols. Thus, the most abstract and artificial construction in the history of Mesopotamian mathematics was taken as its most basic feature. With an understanding of the number system in place, it was possible to recover arithmetic and so gradually discover that the sexagesimal place-value system did not just record positive integers, but was a floating-point system capable of recording 'sexagesimal fractions' and what had been seen as division tables should be read as reciprocal or inverse tables. A consequence of this insight was that there was no mystical or philosophical interest by Mesopotamian scribes in $12{,}960{,}000 = 60^4$, the so-called 'Number of Plato', nor an inexplicable desire to construct multiplication tables for 160,000 (Hilprecht 1906). The technical features of the sexagesimal system were well-understood by Neugebauer's time, but not its pre-history.

The search for origins is an inevitable part of intellectual history (and not always a benign one), and the rediscovery of the Old Babylonian sexagesimal system naturally sparked concern for its origins. Indeed, Neugebauer's first published paper on Mesopotamian mathematics 'Zur Entstehung des Sexagesimal systems' was on this topic in (Neugebauer 1927). François Thureau-Dangin also sought out origins, most notably in his 'Esquisse d'une histoire du system sexagésimal' of 1932 (see Thureau-Dangin 1928, 1929, 1932), translated and expanded as a 'Sketch of a history of the sexagesimal system' in (Thureau-Dangin 1939). These early papers were necessarily speculative as scholars had little early material to draw on. One of the tremendous developments of the post-Neugebauer era has been the strides made in understanding the early development of mathematics, and this includes many of the questions on the origin and development of the sexagesimal system. Although much more remains to be learned about the details of third millennium numeration, the outlines of development are clear. Of course, the solution to one origin question tends to precipitate other, earlier, origin questions.

Both Neugebauer and Thureau-Dangin correctly identified the origins of the sexagesimal system in the preceding Sumerian systems, Neugebauer locating it in metrology, and Thureau-Dangin cheerfully asserting that 'the sexagesimal system [was] the common and exclusive mode of numeration with the Sumerians'. The full story turned out to be rather more complicated. While some numerical and metrological notation from the earliest tablets from Uruk was known to the pioneers, the complete unraveling of early metrology was far in the future.

In the 1950s, the broad parameters of Old Babylonian mathematics were thought to be understood. As new texts appeared, they provided fascinating new details, but did not change the overall picture. However, despite the publication of some few texts from earlier periods, Old Babylonian mathematics appeared to have sprung from the ground fully formed. In particular, the preceding Ur III period (ca. 2100–2000) was essentially a mathematical blank, despite the tens of thousands of economic and administrative documents that had been published. Beginning in the 1970s with the publication of a small collection of Sargonic (ca. 2350–2200) mathematical texts (Limet 1973) and the work of Marvin Powell broadening the discipline to include metrology and etymology, the hunt for origins was on (starting with (Powell 1971, 1972a, 1972b), continued in numerous publications and summarized in (Powell 1990)). As research into the third millennium sources continued, Old Babylonian mathematics came to be seen as having developed slowly over time from these precursors.

Starting in the late 1970s and still continuing, the work of the Berlin group (primarily Peter Damerow, Robert Englund, Jöran Friberg, and Hans Nissen) has completely revolutionized our understanding of archaic mathematics for the earliest periods (ca. 3100–2900 BC). Together with the steady infilling of texts spanning the rest of the third millennium, their work means we are beginning to comprehend the gradual development of mathematics and the slow pace of abstraction. Earlier theories of the development of number systems and the abstraction of number itself have needed to be completely revised.

According to Christopher Woods, writing has been invented only four times in history 'from scratch'. The study of the proto-cuneiform texts from late-fourth millennium Uruk demonstrated conclusively the administrative origins of writing in Mesopotamia, and the first visible traces of calculation that came with the management of goods. As Woods notes, 'literature play[ed] no role in the origins of writing in Mesopotamia' (2010, 34).

Early writing in Mesopotamia used the cheap, widely available medium of clay and a complex iconography to create incised records of economic transactions and the names or titles of those responsible for them. The current corpus of the earliest tablets presents a collection of some 5000 sources, the bulk of which are accounting documents. The tablets are mostly broken, often into numerous small pieces. Unsurprisingly, most surviving sources stem from the largest, most complicated economic units, the temples. However, the tablets have largely been recovered from secondary contexts, discarded and used as fill for building sites after their information was no longer needed, and because of the poor archeological context, cannot tell us much about their precise functions.

Control of the flow of goods in a written record required quantitative notation and the decipherment of proto-cuneiform showed that the metrological systems were tied closely to the underlying goods. Some 1200 different signs are recorded, of which about half have been deciphered. Among the information that needed to be recorded was quantities of goods. Presumably, before writing emerged around 3100 BC, basic goods and units had well-developed metrologies. These must have passed into the new notational system, because quantity information was closely allied to context. However, it is incorrect to interpret these symbols as 'numbers'. There are no abstract numbers in the proto-cuneiform sources, and the development of abstraction in the conception of number and the operations of arithmetic is a major part of the history of third-millennium mathematics.

The first breakthrough in the decipherment of proto-cuneiform metrology came from Friberg. His careful analysis of the early proto-cuneiform tablets, and in particular, the totals recorded on them showed that there were several different systems of notation in use and that the same symbols could take on different meanings depending on context. The precise categories understood by the ancient scribes are not always clear to us, but the forerunner of the sexagesimal system was used to count certain types of discrete items. Other systems were used for capacity, length and area measurement, for example. Altogether some dozen systems utilized around 60 signs with the simpler signs often appearing in different contexts with different meanings and relationships. Friberg's original publication was in a University of Göteborg preprint in 1978, and the early papers on the project were dense, technical, and not widely circulated. The most accessible exposition of the results is the book of Nissen, Damerow and Englund, *Archaic Bookkeeping* (1993). One of the field texts discussed there provides a nice example of the kinds of computations that had to be mastered (MSVO 1,2; see Nissen et al. 1993, Figures 47 and 48). The lengths, widths and areas of five (rectangular) fields are given, together with qualifying information designating appropriate officials.

Lengths, at least on the scale of fields, were measured in units of *nindan* (about 6 m.); multiples of *nindan* were recorded using the discrete notational system, the forerunner of the later sexagesimal system. For field areas the basic unit was the *iku*, equivalent to 100 *sar*, where a *sar* is a square *nindan*. Above the *iku* were the *eše* of 6 *iku*, the *bur* of 3 *eše*, the *bur'u* of 10 *bur*, and the *šar* of 6 *bur'u*. While the sign for *eše* did not appear in the length system, the archaic symbols for the basic units of *nindan* and *iku* were identical, as were the signs for units of 10-*nindan* for length and for the *bur* of 18 *iku*. Early quantity notation re-used a limited number of signs and the relationships in terms of multiples were context-dependent.

On MSVO 1,2, the area computation for the first field is essentially correct, although an additional 2 *iku* has been recorded as 2 *eše*. The calculation for the second field is either correct or nearly so, the result runs over the edge of the tablet and the surface is damaged there. The result for the third field is slightly rounded. The area of the fourth field would be correct if the small left-over area were interpreted as subtracted from the total rather than added, and the fifth, which ought to be the simplest computation of them all, has an area that is quite inexplicable. On the reverse of the tablet, the total area is given in two components, neither of which bears a clear relation to the field areas. The main area is then doubled and the grand total found of the doubled area, the original main area and the additional area. This final total is correct.

The heterogeneity of the computations, with some results rounded, others not, confusion of units and confusion of addition and subtraction suggests that area computations represented a significant mathematical challenge. We do not know how such calculations were performed, although it was almost certainly not the way we would proceed by multiplying the lengths and widths to get a result in square *nindan* and then converting to higher units, as neither the notation nor the evidence suggests such a procedure. We are equally ignorant of how the mathematics of the time was learned. The opacity of the totaling computations implies that we do not fully grasp the conceptual categories the scribes were employing.

Proto-cuneiform provided Old Babylonian mathematics with a background more than a millennium old. Joining the dots, tracing the development of mathematical thought and practice over the course of the third millennium, generated a much richer understanding of the later field. Research has been hampered by a lack of sources for certain periods and the inevitable difficulty of building a picture spanning a thousand years.

Some 500 years after the proto-cuneiform documents from Uruk and Jemdet-Nasr were written, the flourishing city of Šuruppak (also known by its modern name, Fara), seems to have been largely destroyed shortly after making preparations for war. Large numbers of tablets, most of which appear to cover a very short span of time, have been recovered.

Fara has been excavated twice, each time quite briefly. In 1902–03, the Deutsche-Orient-Gesellschaft (D.O.G.) organized a 7½ month expedition that swiftly generated large trenches and a haul of (among other things) some thousand third-millennium tablets and fragments, but little in the way of detailed provenance. In 1931, the University of Pennsylvania sponsored a more careful 3-month expedition that recovered several dozen additional Fara-period tablets, from a much smaller excavated area. One important distinction between the two collections of finds is that the later University of Pennsylvania expedition found a much greater percentage of small tablets, suggesting that the rapidity of the excavation of the earlier campaign may have resulted in smaller items being missed.

The finds from the D.O.G. expeditions were split between Berlin and Istanbul. The Berlin tablets were published by Deimel (1923, 1924); those in Istanbul by Jestin (1937, 1957). Tablets from the University of Pennsylvania expedition were only published by Martin et al. (2001). Following Martin's analysis of the archaeology of Fara (1988), Pomponio and Visicato have been re-editing the texts and attempting to understand the social and economic structure of the city (Pomponio and Visicato 1994; Visicato 1995, 2000). While they have been able to deduce some outlines, the fragmentary and partial nature of the evidence means that much still remains to be understood. By the Fara period, there was a term for 'scribe' (indeed, Visicato lists a hundred such individuals in the city (2000)) and since writing was what scribes did, practicing writing was how they learned their function and differentiated themselves from the rest of society.

Administrative texts from Fara continue to show the types of calculations needed in the archaic texts. In particular, there are numerous examples of texts recording allocations of goods to long lists of individuals, in some cases 200 or more, together with an accompanying total. The totals are usually correct, or very nearly so, and testify that addition, even of very large numbers of entries, was a skill well-mastered by the scribes, however it was performed.

There are only a few mathematical texts that are clearly not economic documents, and there are some additional cases whose status is uncertain. Mathematical lists are represented by a table giving columns of lengths, (equal) widths and resulting areas. The entries are arranged in descending order, from 9 (60-*nindan*) x 9 (60-*nindan*) = 2 (*šar*) 4 (*bur'u*) 2 (*bur*) on down to 5 (*nindan*) x 5(*nindan*) = ¼ (*iku*), where the tablet becomes too broken to restore any more lines. The plethora of archaic metrological systems of the turn of the third millennium had been reduced considerably 500 years later, but lengths and areas were still recorded in much the same way as before, showing the conservatism of metrological usage. The text thus records both an abiding need to determine areas of fields in an agrarian society, and the associated requirement for scribes to practice the linkages between the two systems of lengths and areas.

The Fara excavations have also supplied the world's oldest known mathematical word problems. Distributing rations to people and determining the total distributed was a standard administrative task. However, the inverse problem of determining how many people can be served from a given resource is much more artificial and this is a problem found, remarkably, on two tablets from Fara published by Jestin as TSŠ 50 and TSŠ 671 (1937). The implications of this evidence as support for various suggestions of arithmetical practice and pedagogy have been much-discussed in the literature, most recently by Friberg (2005). While TSŠ 50 is written neatly with the question stated and solution given, the other text is poorly written with many mistakes and an incorrect solution. The failure of the (presumed) student to determine the correct answer opens up the possibility for historians to understand the particular difficulty the student had and so perhaps understand the structure of the underlying processes. A correct solution to a mathematical problem gives no hint as to how the solution was arrived at.

A rough translation of the exercise is, 'A granary of barley. Each man received 7 *sila* of grain. Its men: 4(*šaru*) 5(*šar*) 4(*gešu*) 2(*geš*) 5(*u*) 1; 3 *sila* of barley remains.' This is a classic mathematical word problem of a type largely unchanged in the intervening 4500 years. The problem is grounded in the everyday world of the student, giving it 'relevance', while at the same time being a highly artificial construction. The problem is metrological, training the student in computation in the different metrological systems. It displays the Mesopotamian characteristic of mixing very large units (the granary, a capacity unit inferred from the problem but otherwise unattested at this period) with small, the *sila* (the smallest capacity unit, of approximately one liter or less). It uses awkward numbers (the ration unit of 7 *sila* is chosen for the arithmetic difficulty it introduces; it is not a ration quantity used in any administrative texts), and it has a complicated result, even involving a remainder. The pedagogical goals are clear from this problem and seem characteristic of the period, so far as we can tell from the relatively sparse number of sources available.

The erroneous solution underscores their appropriateness: the student has not only made arithmetical blunders, but has also confused metrological notation, misusing a large unit from area notation. Although precise provenance information for the tablets is not known, neither the archaeology nor the texts support an institution so formalized as a school at this time.

Much of the commentary on these two texts has focused on reconstructing the underlying arithmetical procedures used for the solution of the problem, and provides an index of the development of the historiography of the field. The earliest interpretations saw the problem as an exercise in long-division exactly analogous to the modern approach (Guitel 1963). Later, Powell, searching for precursors and origins of the sexagesimal system, was the first to identify the second, error-prone text and saw evidence of early sexagesimalization and use of place value (Powell 1976). Høyrup (1982) saw the problem as a 'formal division exercise' and proposed a procedure that accommodated the student's error, while avoiding the need for place value notation.

A later analysis, based in part on Høyrup's intervening work on categories of thought in Old Babylonian mathematics, suggested that 'multiplication' and 'division' in the Fara period were approached as problems of repeated addition, except in area problems (Melville 2002a; Friberg 2005). Improved understanding of both earlier and later sources requires continual re-evaluation of the assumptions underlying interpretations of material from intervening periods.

In the last third of the third millennium the rise to prominence of Sargon of Akkad radically altered the political structure of much of Mesopotamia and called forth new administrative and intellectual responses. Until Sargon, Mesopotamia had been characterized by city-states, with independent cities rising and falling in power and engaging in extended trade and military conflict. Although Sargon styled himself in this way, and based his rule in the city Akkad (probably located somewhere in Northern Babylonia; the site has not been identified), his extensive conquests involved a large increase in centralization of the administration and a system of governors for the regions dominated by each large city.

The central administration required a unified and standardized system of accounting for economic activity. Metrological systems were modified for scribal convenience, including the introduction of an Akkadian *gur* of 300 *sila*. The calendar was reformed to include a system of year-names that lasted for some 800 years. The old notation for metrological units had largely fallen away and quantities were now written in cuneiform, but in the Sargonic period the shape and size of central administrative tablets was standardized. In addition to these changes wrought by the increase in central bureaucracy, there was the problem of language. Cuneiform writing had been developed over the preceding centuries to record Sumerian, and in southern provincial areas many texts were still written in Sumerian, but the central administration of the Sargonic empire proceeded in Akkadian, a Semitic language that had no connection with Sumerian. Scribes were thus involved in a great deal of innovation and change. (See Van De Mieroop 2004 for more background information.)

Although we have more sources than in earlier periods, there is a dearth of Sargonic mathematical tablets. Until recently, only a dozen or so were known and they had not been intensively studied. No lists or tables are known, there is one tablet containing a geometrical diagram, and the rest state and/or solve a problem. Most have no detailed archaeological provenance but, except for one tablet from Nippur, are presumed to come from the region of Girsu (see the summary of arguments for provenance in Foster-Robson 2004). The most striking feature of these mathematical texts is that they are all concerned with field computations, finding the areas of squares, rectangles, or irregular quadrilaterals, or the inverse problem of determining one side of a rectangular field given the area and the other side. One sees once more the centrality of field computations in the scribal mathematical curriculum. It is also worth noting that the mathematical fields do not have realistic sizes compared to those recorded on economic documents.

Two recent publications, (Foster-Robson 2004) and (Friberg 2005), have increased the published corpus by three texts and include re-evaluations of the previously published texts by the respective authors. Their conclusions radically differ and illustrate some of the problems involved in trying to understand ancient mathematical concepts and practice. While the statement and correct solution to a mathematical problem give away nothing about the procedure, sometimes what students wrote is completely baffling. Some of the Sargonic mathematical texts have so far resisted satisfactory interpretation. The Foster-Robson text is particularly puzzling.

Area and length computations provide exercises in two of the key systems of metrology, and the Sargonic period was particularly rich in units, especially small length units. The question for the modern historian is how computations involving these units were conceived and carried out. As Foster and Robson note, 'division [is] an arithmetical technique whose manner of execution has not yet been satisfactorily explained … prior to the introduction of the reciprocal table in Ur III or later.' (2004, 5).

The simplest type of exercise is finding the area of a square. One example will illustrate the interests of the scribes when having students practice such problems. The text is in a terse format, '11 *nindan*, 1 *kuš-numun*, 1 *giš-bad*, 1 *zipah* [is the length]'. The relevant relationships are 2 *zipah* = 1 *giš-bad*, 2 *giš-bad* = 1 *kuš-numun*, 6 *kuš-numun* = 1 *nindan*. Clearly the length is chosen with an eye to cleverness and complication rather than realistic depiction of actual fields. The tablet is ruled across after the statement of the length. Below, the answer is given, 'its area: 1 *iku*, ¼ (*iku*)

2 ½ *šar* 6 *gin* 15 *gin-tur*. It was found.' The solution is either correct, or nearly correct, depending on how one interprets the *gin-tur*.

The Foster-Robson text belongs to the group calculating a side from the area. Alone among the entire corpus of Sargonic mathematics the text contains two lines that, while not indicating the division procedure that passes from area to length, does suggest a conversion from 'sexagesimal fractions' to the standard metrological units, and so hints that the computations were carried out in some form of sexagesimal place-value system. Unfortunately, the text is riddled with inexplicable errors that considerably complicate the interpretation. Regardless of the technical difficulties, the important point is that Foster and Robson see evidence for an arithmetical procedure, calculation with numbers. In this they join most previous commentators.

Friberg, on the other hand, rejects the Foster-Robson interpretation and, indeed, all the claimed evidence for sexagesimalization in the Sargonic period, proposing instead a form of 'metric division' based on a geometric realization of the problem. For the side-to-area problem described above, he suggests starting with a square and subdividing it to attain a series of simplified computations based around the metrological units and fractions of them. For the inverse problem, as in the Foster-Robson text, his proposal involves starting with a square of the given area and adjusting it until one achieves a rectangle with the specified side, at which point the remaining side can be read off from following the opposite actions that lead to the given side as the area remains constant. Friberg's suggestion has affinities with Høyrup's geometrical interpretation of Old Babylonian mathematics described below. However, Friberg's reading of the Foster-Robson text also requires several amendments that are difficult to justify.

The small number of texts available prevents any systematic description of Sargonic mathematics and recent intense analysis of the available sources has not managed to resolve some key problems; they must await further study. The sources do allow us to note an abiding interest in metrological problems, especially those to do with moving between area and length notation, and a certain interest in technical virtuosity, as evidenced by the difficulty of the computations involved.

The Sargonic dynasty lasted for a century and a half, but the rulers of the erstwhile city-states chafed under centralized rule and there were frequent rebellions. Eventually the state disintegrated and there was a resurgence of independent cities. This period did not last long before there were attempts at consolidation, culminating in the rise of Ur-Nammu, who unified all of Babylonia under his rule, initiating the Third Dynasty of Ur, based at Ur in the far south. The Ur III state reached its zenith under Ur-Nammu's successor, Šulgi, who ruled for almost 50 years.

The rulers of Ur III created a vast bureaucracy that captured in great detail the administrative flow of the empire. Over a hundred thousand tablets have been recovered from this brief century of rule, most of them from a period of 50 years; of these less than half have been published and fewer than a tenth subjected to detailed study. Supporting this administrative machinery required a large cadre of scribes trained in standardized method of organizing and reporting administrative data. As in the Sargonic period, the demands of bureaucracy called forth sweeping changes in administrative, political and economic spheres including the introduction of a new calendar, alterations in the writing system and metrological changes that improved efficiency of calculations within and between the metrological systems (Steinkeller 1987).

Frustratingly, among the wealth of economic and administrative documents, there are very few mathematical texts. The situation is particularly unfortunate, because it is during the Ur III period that the sexagesimal place-value system was developed along with its attendant apparatus of multiplication and reciprocal tables. In the sexagesimal (base 60) system, numbers of quantities are recorded using only two symbols, those for 'one' and 'ten'. Larger quantities are represented by the same symbols, instead of the archaic systems of a collection of unit notations. The notation therefore no longer carries signification of absolute size. The base unit, while often standardized in many situations, must be inferred from the source of the computation, and multiples and fractions of the base unit are given by their place-value. The sexagesimal system is thus a floating-point system, an original and unique contribution to abstraction of calculation.

The origins and development of this system are still poorly understood, hampered as we are by the absence of evidence, but it seems to have been derived from the confluence of a number of stimuli. First is the evolution of writing over the preceding thousand years. The archaic curviform representations of quantities had gradually been simplified and replaced with cuneiform equivalents. The physical distinction in notation between 'large' units and 'small' units had slowly been lost, implying the beginnings of a place value system. On the other hand, increasing control and theoretical computations of small quantities, especially of valuable commodities, seems to have led to a 'generalized fraction', a use of *gin* to stand for one sixtieth of a base unit. The repetition of this rule (as in the *gin-tur* example from the Sargonic period discussed above) allowed arbitrary precision of fractional quantities in a base-60 setting. Thus a system was arising in which both large and small quantities could be represented using the same notation.

Aligning metrological systems around multiples of 60 facilitated conversion between systems, both drawing from and providing impetus to, adoption of sexagesimal computations by scribes. The most powerful feature of the new system was that it provided a unified notational framework for an abstract form of multiplication that subsumed both repeated addition, as in ration computations, and length to area conversions without regard to the physical status of the base unit. The abstraction of fractions provided a solution to the problem of division, via the introduction of reciprocals, originally conceived as factors of 60, rather than 1. The advantages of the new system must have been felt as profound, because a floating-point system is uniquely ill-suited to performing addition and subtraction, since it does not contain any information about the relative sizes of the base units. The introduction of the sexagesimal place-value system may be compared to the later introduction of logarithms, an innovation that swept Europe in a few decades, but only affected those performing intensive multiplicative calculation, passing everyone else by. It needs to be stressed that the sexagesimal system was an abstract artifact, intended to facilitate (multiplicative) computation and was otherwise largely invisible, as final results were always stated in the standard relevant metrological units.

Powell (1976) was the first to notice sexagesimal notation in an Ur III (non-mathematical) text, in the context of conversions from calculations in a base-unit into standard metrological units. The numbers occur on a 'scratch pad', the surface of which would be scraped off and re-used, and so the intermediate results would largely disappear from the archaeological record. Powell's text has a colophon containing a year-formula and so can be dated to the fifth year of Šulgi's successor Amar-Suen, or 2043 BC.

Training in the new system would require learning multiplicative relationships between numbers. No multiplication tables of the type ubiquitous in the Old Babylonian record can be securely dated to Ur III, although it is notoriously difficult to date texts containing just numbers on paleographic grounds. A few reciprocal tables, somewhat different from the Old Babylonian exemplars are known and several have been published. These tables make it clear that they list factorization of 60 into *n*th parts with a formulation '*igi n gal*'. The Ur III texts typically list increasing numbers and their accompanying factor, as '*igi* 2 *gal* 30', '*igi* 3 *gal* 20', '*igi* 4 *gal* 15', or abbreviate as '*igi* 5 12', '*igi* 6 10' using columns to keep the factors clear. What distinguishes the early forms of these tables is the recording '*igi* 7 *nu*' to show that 7 does not have a companion factor for 60. That is, one cannot write one-seventh of 60 as a finite sexagesimal expression. It is not known how these negative results were obtained. Later Old Babylonian tables dispensed with the entries for which there was no reciprocal, contenting themselves with recording reciprocal pairs. They also seem to have more fully assimilated the reciprocal character and floating-point nature of the sexagesimal system, no longer making a distinction between 1 and 60. Beyond these few tables, there is little direct evidence for the range and scope of Ur III mathematics.

The happy choice of a largely imperishable recording medium in Mesopotamia provides us with a unique opportunity to trace the gradual development of abstraction in a culture. Proto-cuneiform computational practice supplies no evidence for an abstract concept of number. Quantities were very physical and quantity notation was limited. Conversions between systems, such as lengths to areas, and connections between systems, such a connecting the number of men, measured in one system and grain ration quantities, measured in another system, were major preoccupations. Gradually, over time, some of the more specialized systems dropped out of use and the reach of the sexagesimal system extended. The increasing use of valuable metals and a bureaucratic passion for exactness brought smaller units, including the use of the sub-unit *gin* as a sexagesimal fraction of a larger, older, unit. A series of metrological reforms that appear designed to aid sexagesimal calculations led, at some point during the Ur III period, to a stunning conceptual advance.

Research by historians and anthropologists into the world's numeral systems has created a reasonably comprehensive database that has begun attracting the attention of philosophers of mathematics, such as Chrisomalis (2010) and Schlimm and Widom (2012), seeking to construct typological frameworks for numeral systems and allowing us a good overview. Against this background, we can make some observations. Over a hundred different numeral systems are known and out of all of them, the use of a base 60, or rather a base of 60 with a subbase of 10, is unique to Mesopotamia. Further, while the original proto-cuneiform discrete sexagesimal system was indeed a counting system, or at least a way of recording quantities of discrete objects, the abstract sexagesimal system of the late third millennium was for calculation, not counting. Its 'floating point' nature is adapted beautifully to calculation, and in particular multiplication (in modern terminology it is better seen as a multiplicative monoid rather than a ring). The truly exceptional nature of this new calculational tool was noted long before its origins were understood by Thureau-Dangin in his *Sketch*, where he commented how, "It was a delicate instrument to handle, but, in return, it was of incomparable suppleness in the hands of an expert" (1939, 141).

Arithmetic and Table Texts

The new abstract sexagesimal system required new tools for training scribes, and ones that could be integrated into the existing curricular strategies. The result was the table texts. Scribal learning had long been based on extensive lists, and these included metrological lists giving the notation for quantities of various types from the smallest units to the largest. In addition to these lists, there were now developed metrological tables that gave conversions from the standard metrological units into abstract sexagesimal numbers. The glory of the new system was multiplication, and division via the multiplicative inverse, and for training in this, student scribes were subjected to wholesale memorization of inverse tables (reciprocal tables) and a set of some 40 multiplication tables for different head numbers.

The Ur III king, Šulgi, under whose reign the important administrative reforms took place, was keen to be seen as educated and cultured. In one of his praise hymns, Šulgi reports, "When I was small, I was at the academy, where I learned the scribal art from the tablets of Sumer and Akkad. None of the nobles could write on clay as I could. … I qualified fully in subtraction, addition, reckoning and accounting. …. I am an experienced scribe who does not neglect a thing" (Šulgi B, 13–20, translation from *ETCSL*). George quotes a later section of the same hymn,

> Downstream, at Ur, in the Pure Place (my song) is sung, the House of Wisdom of Starry Nissaba is (the place) of my song. Upstream, at Nippur, in the Great Place (my song) is established … The scribe shall come, his hand shall capture (the song in writing)… For all eternity the Edubba is never to change… (George 2005, 133)

However, the *edubba*, the school or scribal academy, did change. The academies in Ur and Nippur during the middle of the Ur III period seem likely to have been substantial state-run organizations training cadres of scribes in preparation for administrative and religious positions across Babylonia and Assyria. The unified demands of the central administration created what appear to be the only 'schools' in Mesopotamian history, in the sense of large organizations.

There is an Old Babylonian body of *edubba* literature such as Kramer's 'Schooldays' (1949), but it has been known for a long time that such works cannot be taken at face value. In a penetrating article, Andrew George (2005) argued compellingly that the *edubba* featuring in Old Babylonian scribal exercises refer back to the glory days of the academies in Ur and Nippur of the twenty-first century. He suggests that while such institutions may have continued, at least at Isin, for a century and a half after the fall of Ur, they do not appear to have survived into the Old Babylonian period. Instead, he argued that most Old Babylonian education took place on a smaller scale. The archaeology of educational settings also supports only a few students – one to five perhaps – at locations in Nippur and Ur. Old Babylonian scribal education seems to have taken place on an individual or family basis, with no more than a few students at any one time learning from a practicing scribe.

In the early stages of the recovery of Mesopotamian literature and mathematics scholars focused on reconstruction of the major texts and delimiting the boundaries of scribal knowledge. More recently, attention has turned to the scribal experience and understanding of curriculum and pedagogy. In this, the work of Steve Tinney, Niek Veldhuis, Eleanor Robson and, most recently, Christine Proust, has been enormously fruitful. The humblest tablets have often had the most to say.

The Old Babylonian period is conventionally dated from around 2000 to 1600, with the sack of Babylon in ca. 1595 marking the end of the period. Few mathematical texts are dated, those that are stem mostly from the period 1800 to 1600. In contrast to all other periods of Mesopotamian history, we have an abundance of mathematical texts from this era. Some thousand have been published, including many table texts and rough workings, but also including around 200 problem texts.

The corpus is not without its problems. Most of the tablets are undated; they can be assigned to the Old Babylonian period on paleographic grounds, but not more precisely than within a couple of centuries. Thus, there is little possibility of a diachronic analysis of Old Babylonian mathematics and so it tends to be treated as a single chronological layer. More importantly, the great majority of Mesopotamian mathematical tablets, especially the extensive problem texts, were bought on the antiquities market in the late nineteenth and early twentieth centuries and have no reliable archaeological provenance.

In contrast to the centralized bureaucracies of the Sargonic and Ur III periods and the dominance of the state in the sources available to us, many Old Babylonian tablets appear to come from private, individual or family archives. The collapse of the large state institutions and the devolving of scribal education into private hands imply the probability of regional variation, but the absence of archaeological context makes such analyses extremely difficult. Albrecht Goetze in *MCT* made the first concerted attempt to organize the unprovenanced mathematical texts on a geographic basis, principally through orthographic analysis. More recently, Høyrup (1998) and Friberg (2000) have revisited the issue with the inclusion of tablets published since 1945 and suggested some refinements to Goetze's overall schema but without serious departures from his original proposals. The surprising consequence of these studies is that we cannot confidently point to major regional differences in either subject matter or procedures. There does seem to be a fair degree of unification in overall approach, after linguistic differences have been taken into account. Isma'el and Robson have made a regional study of the tablets found at assorted sites in the Diyala region (and thus quite removed from the centers of Nippur and Ur), concluding that, while there are noticeable idiosyncrasies and regional differences, 'it becomes increasingly likely that all genres were known and used in Eshnuna' (2010, 161).

Only in the last 10 years have the outlines of the Old Babylonian scribal curriculum become clear, and there are still many unresolved problems. Furthermore, when one does get down to the fine detail, there do emerge considerable differences in education in different cities, and possibly even in different locations within a city, so it is unclear how far one can generalize from very specific studies.

Scribes wrote, and scribal training began with learning how to write, from making simple marks on clay, through basic signs and sign groups, to syllables and words, and then practicing writing long lists of nouns. The bulk of student's instruction was in Sumerian, by the Old Babylonian period no longer a living tongue, but one still very much alive as a scribal medium. Grammar was learned through ringing changes in simple sentences, repetition of short Sumerian proverbs and extracts of a canonical body of Sumerian literature. Most of this literature was presumably originally composed during the Ur III or Isin-Larsa periods and so in many cases three to five centuries old when being copied by a trainee scribe (Tinney 1998).

The place of mathematics in this curriculum was fixed by an observation of Veldhuis (1997) who noticed mathematical tables appearing on the reverse of tablets whose obverse contained Sumerian proverbs, indicating that students were reviewing their Sumerian and performing mathematics during the same day. Basic metrology including writing of the standard units and their multiples appears as part of the earlier sections of the syllabus, but it seems that students were not expected to learn calculations until quite late in their education. Christine Proust (2007) has made the most detailed and up-to-date study of the curriculum in her publication of the texts from Nippur that had languished unread in Istanbul for a century after excavation by Hilprecht. The summary below is based upon her work. If the ratio of surviving school texts is representative of the curriculum, then mathematics comprised perhaps 5–10 % of a student's work, and most of that was mastering the metrological systems.

First came learning basic cuneiform signs. The next stage was learning long thematic lists of words. At this point a student began working with metrological lists, mostly lists of capacity measurement. At around the time the Sumerian education shifted to acrographic lists (list ordered by similarities in sign-shape), a student would begin with the metrological lists for weights, then areas, and finally length units. Simultaneously, came the introduction of metrological tables of capacity, converting units into sexagesimal multiples. As students learned the sexagesimal notation from the metrological tables, they then began learning the sexagesimal tables, beginning with tables of inverses and then working through the series of multiplication tables. Robson (2004), on the basis of a pair of dated multiplication exercises written by one Suen-apil-Urim, has made the reasonable suggestion that working through the set of multiplication tables took students about a year.

As students completed their study of metrological lists, they began to work on the metrological tables for weights, areas, and lengths (the same order in which they had encountered the lists), and a new set of tables for heights. At the conclusion of the multiplication phase, some practiced square and square roots tables. At this point, the Sumerian portion of education moved into model contracts and proverbs, and the mathematical side began simple calculation exercises. The Nippur sources fade out as the elementary level of education was completed.

The core of Old Babylonian scribal arithmetic was thus fluency in multiplication and division via multiplicative inverses. That is, division in Old Babylonian mathematics is usually carried out as 'multiplication by the reciprocal'. In order to gain this fluency, students (those who advanced this far) copied out standard multiplication tables of which several hundred have been published. The tables have a principal number p and multiples are written out on separate lines as 'p times 1 p', 'times 2 $2p$', 'times 3 $3p$' and so on through 'times 20 $20p$'. Next are lines for 30 times, 40 times and 50 times, and sometimes a concluding line giving the square of p. The word for 'times' is the Sumerian *a.ra*, literally meaning 'steps of' and pointing to the repeated addition origin of this format. Repeated addition makes it very easy to fill out the table. These tables were all constructed of abstract sexagesimal numbers. There are no connections to physical quantities.

There were about 40 standard principal numbers, closely allied with the entries of the standard reciprocal table. Along with tablets containing single multiplication tables, there are ones that contain several together, and up to more or less complete sets. These combined tables are usually written more tersely and Robson (2002) has argued that these were compiled as a review after students had worked through the set of individual tables.

The central table for organizing Old Babylonian arithmetic was the inverse or reciprocal table. The standard inverse table listed all finite inverse pairs, that is pairs of numbers whose product is 1 (considered as a floating point unit), from the pair 2 and 30 to the pair 1,21 and 44,26,40. The latter pair perhaps being chosen by the fact that 1,21 can be read as the square of 9. Tables sometimes began with the phrasing, '1 its two-thirds 40', 'its half 30' and continued through the other pairs 'its nth part $\bar{n}$'. As in the case of multiplication tables, matters were often abbreviated to just the lists of number-pairs. Again, numbers were treated as abstract entities, divorced from metrological reality. The large tablets containing sequences of multiplication tables were often headed by a reciprocal table, emphasizing the unity of the approach to arithmetic.

A student who had mastered reciprocals and the multiplication tables would have all the training in abstract arithmetic needed to perform standard scribal mathematics; this manipulation of abstract numbers was built upon earlier training with tables giving conversions from metrological quantities to sexagesimal numbers. A scribe could then take any daily problem involving any types of quantity information, convert into sexagesimal, calculate in the abstract system and convert the result back into the appropriate units. The key technology, the use of reciprocals devised in the Ur III period, was largely hidden from outside view.

The development of the abstract sexagesimal system for multiplicative computations was a singular event and should not be interpreted to mean that conceptual distinctions between the different kinds of 'multiplication', principally repeated addition and constructing areas from lengths, were lost. In most mathematical texts, given the inputs and output of an arithmetical operation, one can deduce the operation and supply the appropriate modern terminology with little need for regard of the nuances of Akkadian and Sumerian. Such an approach, while unraveling the steps of a mathematical procedure and making it clear to the modern reader what has been done, obscures how the ancient scribe thought about the steps in the procedure. Over the course of some 15 years, Høyrup subjected the technical vocabulary of Old Babylonian mathematics to a detailed analysis culminating in (2002) and drawing some surprising and unassailable conclusions.

The first of his key findings lies in the conceptualization of arithmetic. We take the operations of addition, subtraction, multiplication and division applied to abstract numbers so completely for granted that it is hard to imagine other ways of thinking about arithmetic. However, Høyrup's analysis makes it clear that our categories of thought differ significantly from those of Old Babylonian scribes for whom arithmetical operations were conceived of in a much more concrete fashion. Høyrup has shown that, within the procedure texts, terminology is consistently used to distinguish two operations which we would call addition, a similar pair of subtractive operations and no less than four types of multiplication.

Høyrup's arguments are detailed and technical and only the outline of his results can be given here. In the case of addition, there was a concrete, physical addition of two entities sharing a common attribute, for example lengths or areas, which were being 'added'. Here the Akkadian verb *wasābum* was used, translated by Høyrup as 'to append'. The other type of addition Høyrup translates as 'to accumulate'; it used the verb *kamārum* for situations in which no physical object with the accumulated quantity resulted. For example, one 'accumulated' the sides of a rectangle to find the perimeter.

In the case of multiplication, Høyrup traces four different concepts. There was the repeated addition 'steps of' terminology used in the multiplication tables as well as in problem texts; there was a special term for 'doubling' with a generalization to represent repetition; there were a cluster of terms to do with constructing a concrete object via multiplication, for example a physical volume from the base and height. Finally, there was separate terminology for the very important construction of rectangles from their sides. Here, the key term was one Høyrup translates as 'to make hold', emphasizing the physical nature of the operation. Thus we see that the terminology of mathematics, and its conceptualization by Old Babylonian scribes, retained the distinction between repeated addition and measurement of areas, although the development of the abstract sexagesimal system had produced a technical tool suitable for providing calculations of either type of problem. Høyrup's investigations into the technical vocabulary of arithmetic revealed the physical nature of much of Old Babylonian mathematics that had been misunderstood by modern researchers.

Word Problems

The third pillar of Old Babylonian mathematics was computation carried out via word problems. Simple computations appear at the later stages of the elementary phase of education; more complicated problems belong to the advanced level, and raise questions about usage, which will be discussed below. First, a simple example from Proust (2007), to illustrate the changes wrought by the new, esoteric, sexagesimal system and the infrastructure of tables its invention had required. The problem is written on a small square-shaped tablet (Ist Ni 018) of the kind regularly used by students for computations and rough work. The question concerns calculating the area of a (square) shape, given the length of its sides. Of course, Mesopotamia was a predominantly agrarian society, and this problem had deep roots, stretching back over a thousand years to some of the early proto-cuneiform tablets. One of the earliest known tables, from mid-third millennium Suruppak converts lengths into square areas. Almost all known Sargonic mathematical problems concern computation of areas of quadrilateral fields, or the inverse problem of computing sides given the area. One could not have a better example to illustrate the differences in mathematical practice occasioned by the new technology.

The problem is stated in the lower-right corner of the tablet: 1/3 kuš 3 su-si, its side. What is its area?' The metrological units are converted to sexagesimal: 2,10 and the multiplication is carried out in the top left corner (where the scribe gets the wrong answer); and the abstract area is translated back into correct metrological units. 'Its area is 13 še and ¼ še'. Note that to anyone except a scribe, the abstract sexagesimal system is invisible. Training in the metrological tables for length allowed the scribe to make the first conversion; practice with the multiplication tables facilitated the sexagesimal multiplication, and metrological tables for area were needed for the conversion back into physical units.

Beyond simple examples such as this one, the available sources are some 200 published problem texts. Problem tablets come in multiple guises. They range from those including just one problem to long abbreviated lists that might contain hundreds of problems. In some cases only the statement of the problem is recorded, in others there is a statement and the solution, and in some there is a description of the procedure to be followed in solving the problem. These latter are typically written in the form, 'I (the teacher) did something, you (the student) follow these prescribed steps to find the solution'. The first statement gives the initial data and sets up the problem. The second section gives a step-by-step guide to finding the solution of the problem.

The extent to which the different kinds of content of problem texts, and their physical characteristics, reflect different usages has been a matter of some debate and is an area of current research. One tool for gaining a better understanding of the question is a finer-grained typology of problem texts. The first general analysis of shape and form of tablets used for elementary exercises (not specifically mathematical) was given by Miguel Civil (1979). Veldhuis (1997) clarified the relationships between shape and usage of tablets at the elementary level, and Tinney (1999) introduced the distinction between single and multiple-column tablets. Høyrup (2002) emphasized theme-texts with series of problems on the same topic, while Friberg (1996) coined the term 'recombination text' for the large multi-column tablets including problems on a wide variety of topics, which he saw as being drawn from collections of theme texts. Robson (2008, 99) and Proust (2012), have both introduced new classifications to help shape understanding of the usage of elementary and advanced texts.

These problem texts have been interpreted in a number of different ways over time. Since the problems always involve computing the value of some unknown variable, a natural approach for a modern mathematician is to write the unknown as x, and construct an equation from the stated procedure that solves for x. This method has the advantages that it casts the problems in familiar light and makes it easy to see what kinds of problems Old Babylonian mathematics was interested in, in terms of modern categories. This was essentially the approach favored by Neugebauer in his mathematical commentaries.

However, the static form of modern equations does not well represent the procedural nature of the text. The algorithmic features of Mesopotamian mathematics were most famously brought into view by Knuth (1972), with a deeper analysis in the work of Ritter (1995a, b, 2004). Another variation on exposing the formal procedures of some Old Babylonian problems is Melville (2005).

The most well-known and most dramatic revolution in our understanding of Mesopotamian mathematics is Jens Høyrup's geometric interpretation of word problems that arose out of his detailed analysis of mathematical terminology. The classic example to illustrate his approach is the first problem of the Old Babylonian problem text BM 13901. This tablet contains a compilation of 24 quadratic problems and has been much studied and analyzed since it was first translated by Thureau-Dangin in 1936. A translation of the first problem is:

I added the area and side of my square: 0,45.
   You, put down 1, the projection.
   Break 1 in half.
   Multiply 30 and 30.
   Add 15 to 45: 1
   1 is the square root (of 1).
   Subtract the 30 which you multiplied from 1: 30.
   The side (is) 30.

The initial interpretation by Thureau-Dangin was very much algebraic – and the problem can be easily translated into modern algebraic notation and the steps of the problem conform to algebraic manipulation. This approach led to the idea that although many of these problems were cast in geometric language, they were in fact manipulations of numbers designed to find an unknown. Høyrup showed that the terminology was consistent with a geometrical, physical, manipulation of the subject.

Høyrup's insight was to treat the square as physically present (the 'addition' in the first line is an 'accumulation') and so the projection is a 'broad line', an extension of the square of unit length. The projection was torn in half, and one half moved

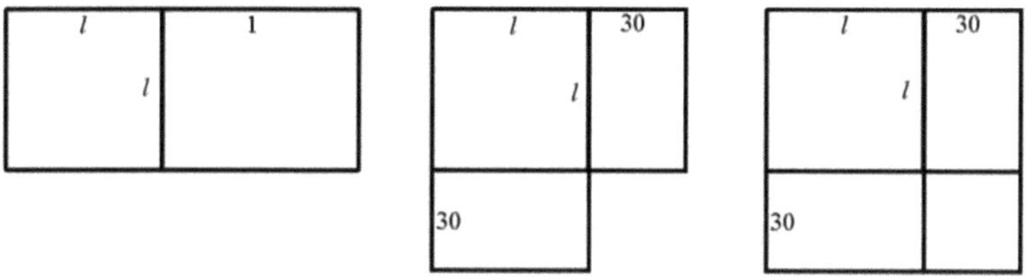

Fig.1 Completing the square

to the other side of the square to make a gnomon. Half of 1 is 30, the step multiplying 0;30 and 0;30 is thus completing the square, whose resulting area is the original 0;45 (since the tearing and moving did not affect the area), and the 0;15 from the small square in the corner. The square root of the area gives the side of the new square, from which the 0;30 from the half of the projection must be subtracted to find the side of the original square (Fig. 1).

Høyrup developed his arguments in a long series of articles culminating in (2002). Høyrup (and others) have shown that this geometrical methodology extends through a broad array of mathematical problems. Hence, we should consider Old Babylonian mathematics as dealing with very tangible objects, the subjects of the problems should be conceived of as really there, even when quite unrealistic parameters are involved.

The abstract nature of the sexagesimal system seems to have generated some interest in numbers, manifested in an assortment of tables giving sequences of squares, square roots, powers, etc, of numbers. The central importance of reciprocals led to the development of an algorithm for finding reciprocals of numbers not in the standard table, and the collection of sometimes large pairs of such numbers. For example, MLC 651 has the pair 1,20,54,31,6,40 and 44,29,37,50,15,20 (the second number should be 44,29,40,39,50,37,30). Regardless of these explorations of abstract numbers (and Robson (2002) has given a cut-and-paste procedure for determining reciprocal pairs), the bulk of Old Babylonian mathematics is grounded in manipulation of physical (even if idealized) objects.

While Høyrup was concentrating on the mathematics of the quadratic problem texts, others have been questioning the usage of the actual tablets themselves. The question of precise usage of mathematical tablets is complicated by the lack of archaeological provenance of many tablets. In particular, the larger and more impressive a tablet, the more likely it is that it was illicitly excavated and sold on the antiquities market in the late nineteenth and early twentieth century. Hence, this is a problem that particularly affects the more advanced material, which tended to be recorded on larger tablets.

For the elementary texts, we do have some good assemblages that were excavated in relatively careful fashion and which have been used to clarify the limits of the first phases of education. It can be very difficult to interpret a tablet with a few scratched computations, but some studies have been attempted. The most detailed studies have been made on collections of student tablets from Ur and Nippur.

Robson (2000, 2001, 2002) made a detailed and penetrating study on the tablets from House F in Nippur from around 1740, and this has now been supplemented by Proust's inclusion of the Nippur tablets held at Istanbul (2007) and Jena (2008). Robson located the mathematical and metrological texts within the overall curricular framework and showed how students were exposed to reciprocal and multiplication tables towards the end of the elementary phase of education. In contrast, she stated, "calculation – active mathematics – belonged to the advanced curriculum along with Sumerian literature". The active mathematics she identified was confined to calculation of squares (such as the example given above) and reciprocals using the standard procedure. As Robson stated in her conclusion, "House F provides no evidence, direct or indirect, for the use of mathematical problem texts" (2002, 361). While the picture from Nippur seems clear, one must be careful of hasty generalization. The Ur texts, dated some 50 years earlier than the Nippur sources, do seem to show some more sustained operations with multiplication, but nothing that can be identified with the advanced problem texts (Robson 1999; Friberg 2000). The range of formats and contents in the problem texts argues for a nuanced response to their probable usages in antiquity.

Those problem texts that provide lists of exercises on related topics, especially those with graded sequences of problems such as YBC 4652 are the ones that most clearly seem intended for issuing to students (Melville 2002b). Proust (2012), following earlier terminology of Friberg and Høyrup, has termed these 'catalogue' texts and made a recent study of their contents. Problems with long histories appear: the first problem of YBC 4612 (published by Neugebauer and Sachs in *MCT*) reads, '3,45 *nindan* is the length, 1,20 *nindan* is the width. What is its area? Its area is 1 *bur'u*'. Despite its terse style, computing the area of a rectangle or field has a long lineage. In contrast to earlier area problem texts, YBC 4612 proceeds with the inverse problems of given the area and either the length or the width find the remaining variable and then into more complicated problems such as finding the length and width having been given the area and the sum or difference of the two sides. Proust has suggested that in contrast to procedure texts, where the whole step-by-step solution algorithm is written out, the catalogue texts represent an organizing principle, perhaps by the same scribes who first created the problems, but aimed at a different purpose. Proust argues, 'the catalogues seem to have been elaborated in order to classify and archive mathematical material used in advanced education' (2012).

Even terser and more difficult to unravel are the series and 'super-series' texts, where the latter may have included thousands of problems representing a systematic exploration of ever more elaborate forms of problems bearing on a simple topics, such as the sides of a rectangle (Proust 2009, 2012). Here, Proust argues, 'The series do not reflect a simple practical classification of existing material, but rather reveal a powerful mechanism that created thousands of new problem statements. Innovative work does not involve methods for solving problems, which in any case are absent from these texts, but rather focuses on the statements themselves, and more specifically, on the mechanism of actually making lists of statements' (2012). Thus, while at least some procedure texts may have been used by students in the classroom, the catalogues and series texts indicate a diversity of Old Babylonian mathematical practice.

Another, and considerably more loquacious, category of tablets is formed by some of the 'recombination' texts. Among the most spectacular mathematical tablets, these large multi-column texts (such as BM 85194) contain compilations of problems on many different topics, presumably derived from thematic collections. The tablets are copies, not original work, and, judging by the errors, the contents were not always completely understood by the copyist. Robson (1999) identified a set of such tablets as being the work of one Iškur-mansum (probably) of Sippar. Interestingly, his is the only name known to us from a colophon of a problem text, although the names of several students appear in colophons of mathematical tables. However, we do not know to what use Iškur-mansum, and the other copyists like him put their texts.

Veldhuis has argued that, 'mathematical tables are primarily exercises in writing' and that education at Nippur was, 'not guided by the list of skills a future scribe had to master. It seems that handing down the Sumerian language and tradition as completely as possible was considered to be all important' (1997, 82). The elaborate recombination mathematical problem texts have little practical purpose, and do not fit easily into a conservative educational system emphasizing 'the heritage of Sumerian writing and Sumerian poetics' (Veldhuis 1997).

However, in his study of Sumerian lexical texts, Niek Veldhuis has suggested a category of 'extra-curricular' texts, one that were cherished by the experts, but not taught to students on a daily basis. 'Extra-curricular lists ... were collected in esthetically-pleasing sets of tablets as tokens of the intellectual capacities or pretensions of their owners' (2004, 86). While 'Curricular lists were copied and learned by every schoolboy; the extra-curricular lists were owned by the elite within the elite. The ones who collected esthetically pleasing sets of archaic knowledge were probably involved in education as teachers; they may have belonged to a group that we would call 'academics', or professors' (2004, 95). It is entirely possible that some of these 'esthetically pleasing' advanced texts with copies of a diverse collection of problems fulfilled a similar function. They were clearly written be someone with access to the daily round of advanced mathematical education, for however few people may have been involved in such an endeavor, but the tablets themselves do not belong in the classroom. Veldhuis' conclusion for the extracurricular lexical lists is that, 'For knowledge, being irrelevant may actually be some kind of a plus. The less useful knowledge is, the more easily it may be used for making precise social distinctions' (Veldhuis 2004, 102). This conclusion may well apply in mathematics as well.

Conclusions: The Legacy of Neugebauer and Thureau-Dangin

In this paper, for reasons of space, I have tried to sketch just a few of the major lines of development of the historiography of Mesopotamian mathematics since Neugebauer left the field. Much has been left unsaid, but the topics covered should be enough to show that the area has undergone radical revisioning since the work of the early pioneers. From this vantage point, how should we evaluate their legacy?

Francois Thureau-Dangin (1872–1944) was from a generation earlier than Neugebauer (1899–1990). Thureau-Dangin came to Mesopotamian mathematics as an Assyriologist and philologist. Neugebauer came to it as a mathematician, although he was able to draw on the philological expertise of Albrecht Goetze from 1930, and later also that of Abraham Sachs (see Proust (2015)). Thureau-Dangin published widely, including many short articles in *Revue d'Assyriologie* on metrological and lexical issues related to mathematics from an early date, but did not engage deeply with mathematics until Neugebauer entered the field. Neugebauer's monumental *Mathematische Keilschrifttexte* (1935–1937) was met by what many saw as Thureau-Dangin's riposte, *Textes mathématiques babyloniens* of 1938. The rival publications, and divergent interpretations, of the two leading experts on Babylonian mathematics were met with some alarm by early commentators. In *Isis*, George Sarton remarked, 'These two scholars agree in giving credit to the old "Babylonians" for a deeper mathematical knowledge than had been hitherto suspected, but they disagree as to details… When the experts disagree the non-experts are in an awful quandary' (1940, 398).

One of the points of contention was the language of the texts. While Old Babylonian scribal education was centered on learning to read and write Sumerian, then a dead language outside the academy, there was a near-universal understanding that mathematical tablets were read in Akkadian, the language of the scribes, regardless of whether the signs were written in syllabic Akkadian, Sumerian, or Sumerograms. Neugebauer chose to reflect the signs as written on the tablets in his transliteration; Thureau-Dangin translated everything into Akkadian. Sarton commented, 'The disagreements between the two experts when they occur are truly fundamental, in the sense that they concern the text itself, being due to different "readings" of it… Neugebauer reads them in the old Sumerian style, but the scribes of the clay tablets which have come to us were Babylonians who used or misused the ancient symbols for the writing of their own (Semitic) language. Thureau-Dangin concludes that we should read these tablets in the spirit in which they were written… If a text is correctly read and translated, further elucidation of it by the modern mathematician may be right or wrong; if the text is not correctly read *everything* is wrong.' (1940, 400–401). Sarton's criticism is somewhat overwrought.

Help in dealing with the awful quandary was offered by Gandz in the next paper in the same issue of the journal. Setting up Neugebauer as the mathematician and Thureau-Dangin as the philologist, Gandz argued, 'To interpret ancient literature the philologist alone is competent … The mathematician and the philologist approach reality from two different angles …However, for some reason or other, the philologist has, so far, shown only very little interest in mathematics. Hence history of mathematics, of ancient mathematics in particular, has long remained a fallow field, unsown and untilled. Especially the critical investigation of mathematical texts has long been sadly neglected' (1940, 406). After discussing the language problem, Gandz compared the translations of Neugebauer and Thureau-Dangin for a couple of problems in which Thureau-Dangin had indeed made significant improvements in interpretation (problems involving series). Gandz concluded his paper with a section titled, 'Thureau-Dangin, the Interpreter of Babylonian Mathematics'. The section opened, 'In this work of his, TH. D. reveals himself as the great expert in Assyriology, as the masterful interpreter of the old Babylonian texts', and closed 'THUREAU-DANGIN, who laid the foundation for the sound philological interpretation of Babylonian mathematics' (1940, 424–425). Thus, Gandz came down firmly on the side of the older scholar Thureau-Dangin at the expense of the newcomer Neugebauer.

Gandz also opined that, 'It is the tragic fate of pioneers in every field of human endeavor that their results are only preliminary, that they are soon superseded and replaced by the new improved results of those who come after them' (1940, 407), a fate which, ironically, seems to have befallen his own work more than that of Neugebauer and Thureau-Dangin.

This harsh early reception should not overshadow the immense achievement of Neugebauer in assembling and interpreting Old Babylonian texts in *MKT* and *MCT*. For anyone intending to study the texts available to Neugebauer in the Thureau-Dangin 1938 and 1940s, his work is still the place to start. Thureau-Dangin's immense philological expertise built on Neugebauer's publications and clarified numerous detailed problems of interpretation (*TMB* was largely a restatement and compilation of a long series of specialist articles mostly published in *Revue d'Assyriologie*). Many other scholars have contributed to refining the readings of some the tablets published by Neugebauer and Thureau-Dangin, but no one has yet felt the need for a complete new edition.

More texts are available to us now than Neugebauer had access to, and their publication has extended our understanding of the range and depth of the internal structure Old Babylonian mathematics. An important collection of tablets from Susa was published by Bruins and Rutten in 1961. Robson (2004) published the collection of mathematical tablets held by the Ashmolean Museum and included an outline of mathematics in Old Babylonian Kish as a counterbalance to the better known sources from Nippur and Ur; a variety of tablets have been excavated in the Diyala region (the various excavations and finds are summarized in Isma'el and Robson 2010); Robson (1999) and Robson (2008) contain very useful lists of texts published since Neugebauer's work; the Nippur corpus is now largely complete (see Proust 2007 and 2008 for catalog), and Friberg (2007) has published the extensive private Schøyen collection. New tablets steadily appear from all periods and the continuous accretion helps the field steadily advance.

As to the method of transliteration of Sumerograms that so exercised Sarton and Gandz, Neugebauer's cautious approach more accords with modern Assyriological style than Thureau-Dangin's (characterized by Høyrup as 'philologically inconvenient' (1996, 5). Also, given the current more nuanced view of the uses of advanced texts, the issue of Akkadian versus Sumerian is perhaps not quite so clear-cut as it used to seem.

If Neugebauer's groundbreaking text editions have largely stood the test of time, what of his methodological orientation? The history of mathematics always involves a tension between mathematics and history, between eternal verities and particular moments. Writing Neugebauer's éloge for the *Proceedings of the American Philosophical Society*, Noel Swerdlow commented, 'if the truth be told, on a deeper level Neugebauer was always a mathematician first and foremost, who selected the subjects of his study and passed judgment on them, sometimes quite strongly, according to their mathematical interest. And for this we must be grateful, for only a true mathematician would recognize and be willing to expend the effort necessary to reveal the depth of Babylonian mathematics' (1993, 142). In Swerdlow's view, it was Neugebauer's mathematical taste that was a pre-requisite for developing the field, rather than philological competence. Similarly, in their recent evaluation of Neugebauer, Rashed and Pyenson declared, 'Neugebauer cultivated a mathematical approach to history of mathematics, and he addressed his work, essentially, to mathematicians' (2012, 5). That is, Neugebauer connected his work on Babylonian mathematics to mathematicians, not Assyriologists.

Philip Davis, in a personal memoir reflecting on Neugebauer, also mentioned his textual focus and distrust of the connection of history of mathematics to wider historical currents, 'The text was the thing... he was suspicious of and had little patience with attempts to link the history of mathematics with general history' (1994, 130). One of the great developments in the historiography of Mesopotamian mathematics since the time of Neugebauer is that its administrative and educational loci have naturally led scholars to ask wider questions and to seek the connections between mathematical education, and mathematical usage and practice. Indeed, one of the most forthright champions of the wider social contextualization, as well as a wider historical view counterbalancing the over-emphasis on the Old Babylonian period as a result of the abundance of texts from that time is Robson, declaring her recent book 'A Social History' (2008).

To summarize how the field has changed since Neugebauer left it, I would emphasize first, that it now has a history. Old Babylonian mathematics did not spring fully formed from the ground in Nippur. That history gives a richness and depth to the field. The learning of metrology was integrated into the overall Sumerian curriculum. The sexagesimal system was the end product of a thousand-year journey towards an abstract concept of number, and the concept Mesopotamian scribes hit upon was more abstract in some ways than our current usages. That abstraction did not permeate their mathematics: arithmetical operations took note of what they operated on, and in many cases should be seen as geometric operations. When asked, "What did they know?", we would now respond, "What did **who** know?", for copying problems is not the same as solving them and may not imply that the mathematics represented on a particular tablet was being actively taught and learned. The topics covered here are only a small selection of the many different issues that have been explored by scholars over the past decades.

All historians of Mesopotamian mathematics owe Neugebauer an enormous debt. His work is still continually consulted and referenced. But the discipline has moved on, asking new questions and re-evaluating old evidence. It continues to evolve.

References

Bruins, E.M., and M. Rutten. 1961. *Textes mathématiques de Suse*, MDP, vol. 34. Paris: Geuthner.

Chrisomalis, S. 2010. *Numerical notation: A comparative history*. New York: Cambridge University Press.

Civil, M. 1979. *Ea A = nâqu, Aa A = nâqu, with their forerunners and related texts*, Materials for the Sumerian Lexicon, vol. 14. Rome: Pontifical Biblical Institute.

Davis, P.J. 1994. Otto Neugebauer: Reminiscences and appreciation. *American Mathematical Monthly* 101: 129–131.

Deimel, A. 1923. *Die Inschriften von Fara, II: Schultexte aus Fara*. Leipzig: J.C. Hinrichs.

Deimel, A. 1924. *Die Inschriften von Fara, III: Wirtschaftstexte aus Fara*. Leipzig: J.C. Hinrichs.

Foster, B.R., and E. Robson. 2004. A new look at the Sargonic mathematical corpus. *Zeitschrift für Assyriologie* 94: 1–15.

Friberg, J. 1978. *The third millennium roots of Babylonian mathematics. I. A method for the decipherment, through mathematical and metrological analysis, of proto-Sumerian and proto-Elamitesemipictographic inscriptions*. Preprint 78-09, Chalmers University of Technology.

Friberg, J. 1996. Pyramids and cones in cuneiform and other mathematical texts. New hints of a common tradition. *Proceedings of the Cultural History of Mathematics* 6: 80–95.

Friberg, J. 2000. Mathematics at Ur in the Old Babylonian period. *Revue d'Assyriologie et d'Archéologie Orientale* 94: 97–188.

Friberg, J. 2005. On the alleged counting with sexagesimal place value numbers in mathematical cuneiform texts from the Third Millennium BC. *Cuneiform Digital Library Journal* 2005: 2.

Friberg, J. 2007. *A remarkable collection of Babylonian mathematical texts*, Manuscripts in the Schøyen collection: Cuneiform texts I. New York: Springer.

Gandz, S. 1940. Studies in Babylonian mathematics II: Conflicting interpretations of Babylonian mathematics. *Isis* 31: 405–425.

George, A. 2005. In search of the é.dub.ba.a: The ancient Mesopotamian school in literature and reality. In *An experienced scribe who neglects nothing: Ancient Near Eastern studies in honor of Jacob Klein*, ed. Y. Sefati et al., 127–137. Bethesda: CDL Press.

Guitel, G. 1963. Signification mathématique d'une tablette sumérienne. *Revue d'Assyriologie et d'Archéologie Orientale* 57: 145–150.

Hilprecht, H.V. 1906. *Mathematical, metrological and chronological tablets from the library at Nippur*, BE 20,1. Philadelphia: Department of Archaeology: The University of Pennsylvania.

Høyrup, J. 1982. Investigations of an early Sumerian division problem. *Historia Mathematica* 9: 19–36.

Høyrup, J. 1996. Changing trends in the historiography of Mesopotamian mathematics: An insider's view. *History of Science* 34: 1–32.

Høyrup, J. 1998. The finer structure of the Old Babylonian mathematical corpus. Elements of classification, with some results. In *Assyriologica et Semitica, Festschrift für Joachim Oelsneranläßlich seines 65. Geburtstages am 18. Februar 1997*, ed. J. Marzahn and H. Neumann, 117–178. Kevelaer: Neukirchen-Vluyn.

Høyrup, J. 2002. *Lengths, widths, surfaces: A portrait of Old Babylonian algebra and its kin*. New York: Springer.

Isma'el, K.S., and E. Robson. 2010. Arithmetical tablets from Iraqi excavations in the Diyala. In *Your praise is sweet: A memorial volume for Jeremy Black from students, colleagues and friends*, ed. H.D. Baker, E. Robson, and G.G. Zólyomi, 151–164. London: British Institute for the Study of Iraq.

Jestin, R. 1937. *Tablettes sumériennes de Shuruppak au Musée de Stamboul*. Paris: E. de Boccard.

Jestin, R. 1957. *Nouvelles tablettes sumériennes de Shuruppak au Musée d'Istanbul*. Paris: Maisonneuve.

Knuth, D.E. 1972. Ancient Babylonian algorithms. *Communications for the Association of Computing Machinery* 15: 671–677.

Kramer, S.N. 1949. Schooldays: A Sumerian composition relating to the education of a scribe. *Journal of the American Oriental Society* 69(4): 199–215.
Limet, H. 1973. *Etude de documents de la période d'Agade appartenant à l'Université de Liège*. Paris: Société d'Éditions 'Les Belles Lettres'.
Martin, H. 1988. *Fara: A reconstruction of the ancient Mesopotamian city of Šuruppak*. Birmingham: Chris Martin & Associates.
Martin, H., F. Pomponio, G. Visicato, and A. Westenholz. 2001. *The Fara tablets in the University of Pennsylvania Museum of Archaeology and Anthropology*. Bethesda: CDL Press.
Melville, D.J. 2002a. Ration computations at Fara: Multiplication or repeated addition? In *Under one sky: Atronomy and mathematics in the Ancient Near East* (London, 2001), AOAT, vol. 297, ed. J.M. Steele and A. Imhausen, 237–252. Münster: Ugarit-Verlag.
Melville, D.J. 2002b. Weighing stones in ancient Mesopotamia. *Historia Mathematica* 29: 1–12.
Melville, D.J. 2005. The area and the side I added: Some Old Babylonian geometry. *Revue d'histoire des mathématiques* 11: 7–21.
Neugebauer, O. 1927. Zur Entstehung des Sexagesimalsystems. *Abhandlungen der Gesellschaft der Wissenschaften in Göttingen, Mathematisch-Physikalische Klasse* 13: 1–55.
Neugebauer, O. 1935–1937. *Mathematische Keilschrifttexte* I-III (MKT). Berlin: Springer.
Neugebauer, O., and A. Sachs. 1945. *Mathematical cuneiform texts (MCT)*, American oriental series, vol. 29. New Haven: American Oriental Society.
Nissen, H.J., P. Damerow, and R. Englund. 1993. *Archaic bookkeeping: Early writing and techniques of economic administration in the ancient Near East*. Chicago: University of Chicago Press.
Pomponio, F., and G. Visicato. 1994. *Early dynastic administrative texts of Šuruppak*. Napoli: Istituto universitario orientale di Napoli, Dipartimento di studi asiatici.
Powell, M.A. 1971. *Sumerian numeration and metrology*, Unpublished Ph.D. dissertation. University of Minnesota, Minneapolis.
Powell, M.A. 1972a. The origin of the sexagesimal system: The interaction of language and writing. *Visible Language* 6: 5–18.
Powell, M.A. 1972b. Sumerian area measures and the alleged decimal substratum. *Zeitschrift für Assyriologie und Vorderasiatische Archäologie* 62: 165–221.
Powell, M.A. 1976. The antecedents of Old Babylonian place notation and the early history of Babylonian mathematics. *Historia Mathematica* 3: 414–439.
Powell, M.A. 1990. Masse und Gewichte. In *Reallexikon der Assyriologie*, vol. 7, ed. D.O. Edzard et al., 457–530. Berlin/New York: De Gruyter.
Proust, C. 2007. *Tablettes mathématiques de Nippur*. Paris: Institut français d'études anatoliennes Georges-Dumezil.
Proust, C. 2008. *Tablettes mathématiques de la collection Hilprecht*, Texte und Materialen der Frau Professor Hilprecht Collection, vol. 8. Leipzig: HarrassowitzVerlag.
Proust, C. 2009. Deux nouvelles tablettes mathématiques du Louvre: AO 9071 et AO 9072. *Zeitschrift für Assyriologie und Vorderasiatische Archäologie* 99: 167–232.
Proust, C. 2012. Reading colophons from Mesopotamian clay tablets dealing with mathematics. *NTM Zeitschrift für Geschichte der Wissenschaften, Technik und Medizin* 20: 123–156.
Proust, C. 2015. *Mathematical and philological insights on cuneiform texts. Neugebauer's correspondence with fellow Assyriologists*, Dordrecht: Springer.
Rashed, R., and L. Pyenson. 2012. Otto Neugebauer, Historian. *History of Science* 30, preprint.
Ritter, J. 1995a. Babylon -1800. In *A history of scientific thought*, ed. M. Serres, 17–43. Oxford: Blackwell.
Ritter, J. 1995b. Measure for measure: Mathematics in Egypt and Mesopotamia. In *A history of scientific thought*, ed. M. Serres, 44–72. Oxford: Blackwell.
Ritter, J. 2004. Reading Strasbourg 368: A thrice-told tale. In *History of science, history of text*, ed. K. Chemla, 177–200. New York: Springer.
Robson, E. 1999. *Mesopotamian mathematics, 2100-1600 BC. Technical constants in bureaucracy and education)*, OECT, vol. 14. Oxford: Clarendon.

Robson, E. 2000. Mathematical cuneiform tablets in Philadelphia. I. Problems and calculations. *SCIAMVS* 1: 11–48.
Robson, E. 2001. The Tablet House: A scribal school in Old Babylonian Nippur. *Revue d'Assyriologie et d'Archéologie Orientale* 95: 39–66.
Robson, E. 2002. More than metrology: Mathematics education in an Old Babylonian scribal school. In *Under one sky: Astronomy and mathematics in the ancient Near East* (London, 2001), AOAT, vol. 297, ed. J.M. Steele and A. Imhausen, 325–365. Münster: Ugarit-Verlag,
Robson, E. 2004. Mathematical cuneiform tablets in the Ashmolean Museum, Oxford. *SCIAMVS* 5: 3–65.
Robson, E. 2008. *Mathematics in Ancient Iraq: A social history*. Princeton: Princeton University Press.
Sarton, G. 1940. Remarks on the study of Babylonian Mathematics. *Isis* 31: 398–404.
Schlimm, D., and T.R. Widom. 2012. Methodological reflections on typologies for numerical notations. *Science in Context* 25(2): 155–195.
Steinkeller, P. 1987. The administrative and economic organization of the Ur III state: The core and the periphery. In *The organization of power: Aspects of bureaucracy in the Ancient Near East*, Studies in ancient oriental civilization, vol. 46, ed. McG. Gibson and R.D. Biggs, 19–41. Chicago: Oriental Institute of the University of Chicago.
Swerdlow, N. 1993. Otto E. Neugebauer (26 May 1899–19 February 1990). *Proceedings of the American Philosophical Society* 137: 138–165.
Thureau-Dangin, F. 1928. L'Origine du système sexagésimal. *Revue d'Assyriologie et d'Archéologie Orientale* 25: 115–121.
Thureau-Dangin, F. 1929. L'Origine du système sexagésimal. Un postscriptum. *Revue d'Assyriologie et d'Archéologie Orientale* 26: 43.
Thureau-Dangin, F. 1932. *Esquisse d'une histoire du système sexagésimal*. Paris: Geuthner.
Thureau-Dangin, F. 1936. L'Équation du deuxième degré dans la mathématique babylonienne d'après une tablette inédite du British Museum. *Revue d'Assyriologie et d'Archéologie Orientale* 33: 27–48.
Thureau-Dangin, F. 1938. *Textes mathématiques babyloniens (TMB)*. Leiden: Ex Oriente Lux 1.
Thureau-Dangin, F. 1939. Sketch of a history of the sexagesimal system. *Osiris* 7: 95–141.
Tinney, S. 1998. Texts, tablets and teaching. *Expedition* 40(2): 40–50.
Tinney, S. 1999. On the curricular setting of Sumerian literature. *Iraq* 61: 159–172.
Van De Mieroop, M. 2004. *A history of the ancient Near East ca. 3000–232* bc. Oxford: Blackwell.
Veldhuis, N. 1997. *Elementary education at Nippur: The lists of trees and wooden objects*. Unpublished doctoral thesis, University of Groningen.
Veldhuis, N. 2004. *Religion, literature, and scholarship: The Sumerian composition Nanše and the birds. With a catalogue of Sumerian bird names*, Cuneiform monographs, vol. 22. Leiden: Brill Publications.
Visicato, G. 1995. *The Bureaucracy of Šuruppak*. Münster: Ugarit-Verlag.
Visicato, G. 2000. *The power and the writing*. Bethesda: CDL Press.
Woods, C. (ed.). 2010. *Visible language: Inventions of writing in the ancient Middle East and beyond*. Chicago: The Oriental Institute.

Babylonian Astronomy 1880–1950: The Players and the Field

Teije de Jong

Introduction

This essay aims at telling the story of the rediscovery of Babylonian astronomy and of the wrestling of the early pioneers with the astronomical cuneiform texts in trying to understand the ingenious Babylonian numerical schemes for the computation of the celestial positions of the Sun, Moon and planets. When Otto Neugebauer entered the stage in the early 1930s, this pioneering phase had already come to an end. While at that time the field of Babylonian mathematical astronomy had been created, it needed Neugebauer to develop it into a well-established discipline in the history of science. This he accomplished almost single-handedly by systematically analyzing all texts available to him at the time in great depth and detail, eventually resulting in the publication of his magnum opus *Astronomical Cuneiform Texts* (Neugebauer 1955; here often referred to as ACT). In this essay I will strictly limit myself to the period 1880–1950, but most of what is in ACT is previewed in papers published before 1950.

By focusing mainly on Babylonian mathematical astronomy, I will not discuss the work of early Assyriologists, such as C. Bezold (1859–1922), F. Thureau-Dangin (1872–1944), R. C. Thompson (1876–1941), and E. F. Weidner (1891–1976), who made important contributions to the understanding of the earlier phases of Babylonian astronomy and astrology as described in the omen series *Enūma Anu Enlil*, the astronomical compendium *MUL.APIN*, the letters and reports sent by Assyrian and Babylonian scholars to the Neo-Assyrian kings, and other texts from the second and the first half of the first millennium BC.

T. de Jong (✉)
Astronomical Institute 'Anton Pannekoek', University of Amsterdam, Amsterdam, The Netherlands
e-mail: T.deJong@uva.nl

A. Jones et al. (eds.), *A Mathematician's Journeys*, Archimedes 45,
DOI 10.1007/978-3-319-25865-2_9

The Beginning

In 1880 there was no field and only one player: Johann Nepomuk Strassmaier (1846–1920) (Fig. 1). Strassmaier was born in a family of simple Bavarian country folk (Pollen 1920). At the age of 19 he entered the Jesuit order and there received the traditional thorough Jesuit education, first in Germany and from 1872 onwards in England, after the Jesuits were forced to leave Germany as a consequence of the so-called 'Kulturkampf' (see for instance Gross 2004). From 1872 until 1917 the German province of the 'Societas Jesu' remained in exile in the Netherlands, Belgium and England. From 1878 until his death in 1920 Strassmaier lived in a Jesuit residence in London. He fell seriously ill in 1897, never fully recovered from his operation and never returned to active research afterwards.

During his theology education in England from 1873 to 1878 Strassmaier spent his summer vacations in London studying cuneiform tablets of the Kuyunjik (Nineveh) collection at the British Museum, from 1875 onwards authorized by the formal permission of Dr. Samuel Birch, the Keeper of Oriental Antiquities. Strassmaier was generally interested in languages and had studied Hebrew, Syriac, Arabic and Chinese, next to Latin and Greek. Now he added Akkadian to this list and had to accomplish this by self-study because in those early days a formal grammar was not available (Deimel 1920).

After having been ordained priest in 1876 and having finished his so-called Tertiary education Strassmaier moved in 1878 to the Jesuit residence in Mount Street, London—within walking distance of the British Museum—to formally start his study of cuneiform inscriptions in preparation of writing a book on the History

Fig. 1 Johann Nepomuk Strassmaier (1846–1920) (From Budge 1925)

of the Semitic Languages. As he once explained to one of his fellow Jesuits (Pollen 1920) his plan was to:

> carefully copy and systematically publish the tablets in the British Museum, and well as many as possible, not only those with historical or religious content but also the more 'boring' economic texts.

This program was apparently approved by his superiors who may have realized the potential importance of studying cuneiform texts after George Smith (1840–1876) discovered in 1872 a tablet in the British Museum collection with a text that showed parallels with the biblical story of the flood.

Strassmaier's drive and attitude may be illustrated by two anecdotes told by E. A. Wallis Budge (1857–1934) in a letter to one of Strassmaier's biographers (Pollen 1920). Budge got to know Strassmaier well, first in the 1870s when Budge was a young boy during his daily visits to the British Museum trying to master the secrets of cuneiform writing, and later from 1883 onwards when he took up employment at the Museum. Budge writes:

> For twenty years or more Strassmaier was a very familiar figure in the Students' Room of the Department of Egyptian and Assyrian Antiquities. He arrived punctually at 10 a.m. and sat there working all day without lunch, until he was turned out at 4 p.m. His skill in reading tablets was very great, and his copies were among the best and most accurate which have ever been made. From first to last he must have copied one half of the collections which were in the British Museum in his day.

And a bit further down:

> I asked him once why he did not get on with the book and he said: How can a history of these languages be written, whilst 60.000 cuneiform tablets remain uncopied and untranslated.

The First Decade

While copying and studying tablets in the British Museum, Strassmaier noticed that some contained little text but large numerical tables (e.g. in the Spartali collections that were purchased by the British Museum between 1878 and 1880). He was intrigued by these tablets but did not feel adequately equipped to try to study their contents; so he started looking around for help.

In the fall of 1880 Strassmaier visited Blijenbeek castle near Afferden in Limburg, the Netherlands, one of the manors that had been put at the disposition of the German Jesuits by members of the German/Dutch catholic nobility as a temporary residence after they were forced to leave Germany (Fig. 2). The purpose of his visit was to discuss a 'sabbatical' visit to Blijenbeek to prepare the publication of his *Alphabetisches Verzeichnis der Assyrischen und Akkadischen Wörter der Cuneiform Inscriptions of Western Asia vol. II* (Strassmaier 1886).

Fig. 2 Blijenbeek castle near Afferden, Limburg, the Netherlands. Here Epping resided from 1876, when he was appointed professor of Mathematics and Astronomy, until 1885 when the Jesuit Philosophy education was transferred to the House Exaeten near Baexem, Limburg, the Netherlands. Strassmaier stayed here from 1881 to 1884 during his study leave (Courtesy the Archiv of the North German Province of the Societas Jesu, Munich, Germany)

At Blijenbeek Strassmaier met Joseph Epping, more than 10 years his senior, who had been his teacher in mathematics and astronomy when he was a student at the Jesuit Collegium Magnum in Maria Laach, Germany in the 1860s.

Here Joseph Epping (1835–1894) enters my narrative as the second player[1] in the still not existing field. Epping was born in Bevergem, Nordrhein-Westfalen in Germany, close to the Dutch border in a middle-class family; his parents died early and he was raised by relatives. According to his biographer and fellow-Jesuit Alexander Baumgartner (1894) he was a stocky, cheerful, humorous little fellow, without much pretence. He studied mathematics at the University of Münster before entering the Jesuit order in 1859. From 1864 until 1867 he was Professor of Mathematics and Astronomy at the Jesuit Collegium Magnum in Maria Laach, Germany and then, after having finished his theology education in 1872, he volunteered to be stationed in Quito, Ecuador to become Professor of Mathematics and Astronomy at the newly founded Polytechnic Institute. When in 1875 the president of Ecuador and founder of the Institute was murdered, the ensuing political unrest made work at the Institute increasingly difficult so that in 1876 the Jesuit professors were called back to Europe. Upon his return Epping was appointed Professor of Mathematics and Astronomy at the German Jesuit College in Blijenbeek where in 1880, four years later, his former pupil Strassmaier asked him for help with the interpretation of the astronomical cuneiform texts. Within a year their collaboration

[1] In spite of several searches in different archives and enquiries at several institutions I have been unsuccessful so far in obtaining a photograph of Epping.

resulted in a first short publication in the Jesuit journal *Stimmen aus Maria Laach* (Strassmaier and Epping 1881) entitled: "Zur Entzifferung der astronomischen Tafeln der Chaldäer". Neugebauer (1975, p. 349) refers to this short paper as "a masterpiece of a systematic analysis of numerical data of unknown significance". I will come back to it below.

From 1881 to 1884, during Strassmaier's study leave in Blijenbeek, their collaboration further developed. After Strassmaier returned to London, Epping continued more or less on his own, from 1885 onwards in the house Exaeten near Baexem, Limburg, the Netherlands, where the Jesuit philosophy education had been transferred to. In 1889 Epping published a 200-page monograph, *Astronomisches aus Babylon*, in which he presented an extensive study of six texts, including the ones he had previously analyzed in the first paper. The results were spectacular: in one strike the field of Babylonian Astronomy had been created.

The next five years Epping published another ten papers, partly co-authored by Strassmaier, based on texts provided by the latter. After 1885 Epping's health steadily deteriorated and in 1894 he died, prematurely, at the age of 59.

Zur Entzifferung der astronomischen Tafeln der Chaldäer

The inconspicuous article with the above title published in 1881 in the Jesuit religious journal *Stimmen aus Maria Laach* can now be seen to be a pioneering landmark in the study of Babylonian astronomy. Strassmaier wrote the general historical and philological introduction, then Epping described his wrestling with the material: a problem at that time with no equations and many unknowns.

In his introduction Strassmaier remarks that Ptolemy uses in the *Almagest* a number of observations by the 'Chaldeans' from the eighth to the third century B.C. He speculates that the astronomical texts from Babylonia may well contain the observations quoted by Ptolemy. He also expresses his concern about the fragmentary state of the texts and their durability (my translation):

> The few fragmentary remains show us only how much we have to regret the loss of the complete collection; they are all written on unbaked clay in the cursive cuneiform Babylonian script, and as such difficult to read and to copy. A trained eye will therefore later certainly be able to see more on these inscriptions then a first copier with his copy can extract from it. Since all these inscriptions on unbaked clay will erode once exposed to the air, they will gradually become more difficult to read, so that after not even a very long time these remains of a Babylonian literature will be lost for scientific exploration, if they are not carefully studied or accurately copied before.

Apparently Strassmaier has already a good grasp of the contents and of the different classes of the astronomical texts. He continues (my translation):

> The fragments of these texts show that several works about astronomy were present: some contain long lists of numbers with headings containing astronomical terms, like for instance B.S.† 2343.[2] This class may well concern the calculation of risings and settings of the plan-

[2] See inset B.S.† 2343–A 120-year Perspective.

Fig. 3 The tablet Spartali 129, later catalogued as BM 34033 (Courtesy the Trustees of the British Museum, London)

> ets; others appear to be related to the computation of New and Full Moon; some, in which several names of stars occur accompanied by numbers, appear to contain observations and calculations of the course of the planets or of lunar eclipses.
>
> Unfortunately these valuable remains of antiquity are so broken and so fragmentary that for the most part they will remain undeciphered for ever. Only a few fairly complete tablets are preserved, which seem to form a separate class. These tablets, about five inches high and three inches wide, when fully preserved, contain, each for one year, the constellation of the planets for increasing month dates, and these must presumably be the records to which the ancient authors refer. The most complete of these is Sp 129 in the British Museum, on which only few columns are damaged.

As we shall see shortly, Sp 129 (Fig. 3) is not an observational text but a so-called Normal Star Almanac (see Hunger and Pingree 1999, p. 159).

Then it is Epping's turn; his part is entitled "Astronomische Enthüllungen" ("Astronomical Revelations"). He begins with accounting how he is persuaded by Strassmaier to work on the interpretation of the astronomical texts (my translation):

> After P. Strassmaier has presented the philological-historical aspect of the astronomical cuneiform texts, the question arises, if astronomy is able to create some light here. This task then imposed the same pater on my humble self, by handing over to me several tablets that were copied by him. Then I should indeed have seized this opportunity with both hands; since it was in no way to foresee that a precious historical treasure lay buried in these tablets. But the difficulties, which should be overcome in such work, should also not be underestimated; and with me the conditions were not adequately fulfilled. Namely, such a person should be at least somewhat familiar with cuneiform texts and at the same time be well educated in astronomy. As far as the first point is concerned, I could not remember having ever seen such hooks, and as far as the second point is concerned, astronomy is certainly not totally alien to me, but I did not believe to be such a computational artist, that I could solve an equation, that had so large a number of unknowns, and so little a number of knowns.

Epping mentions that he has in front of him two kinds of tablets: one kind with computations and one with observations; he starts with one of the first kind because "numbers are more easy to analyze". The tablet he chooses had been copied and translated by Strassmaier but had not yet been formally registered (as SH 81-7-6, 277).[3] It contains seven columns of numbers and Epping shows that these are part of a stepwise computation in the sixth column of the dates and times of New Moon in a number of subsequent months. He is impressed by the fact that the Babylonian astronomers were apparently aware of the large variations in the length of the synodic month, due to the variation in the lunar velocity, and that they were able to compute this. By comparing with modern calculations he shows that the size of these variations is of the right order of magnitude. He ends this part of his analysis by noting that the level of culture of the Chaldaeans must have been high to be able to develop such sophisticated theories, but that a lot more work has to be done to fully interpret these kinds of tablets.

He then goes on to the "observational" text Sp 129 (see Fig. 3), actually (as we know now) a computed so-called Normal Star Almanac. The text gives SE 189 as date (at that time already suspected to be equivalent to 123 B.C.). Epping computes positions of Venus (dilbat) and Jupiter (guttu, actually Mercury!) for 123 BC and several years around it to first establish the correct chronology. The positions of Venus agree for 123 BC but those of Jupiter do not. In spite of extensive computations and searches he does not make much progress and puts it aside. But urged from several quarters (Strassmaier?) he resumes his efforts a few month later in the spring of 1881 and decides to turn things around and starts with 123 BC looking for consistency in planet names in the text at computed positions. This approach is successful and results in the confirmation that Seleucid Era 189 corresponds to 123/122 B.C. and the correct identification of Jupiter (te-ut) and Mars (an).

[3] This fragment is part of BM 34580 published as ACT 122 (Neugebauer 1955), one of the most complete lunar ephemerides of system B, which consists of 9 fragments (see inset Babylonian Lunar Theory: 1880–1950). The fragment studied by Epping is the largest central piece of the ephemeris of which the text on the reverse side is best preserved. It contains columns F_1 through M_1 of a New Moon ephemeris for 13 months of the years 209 and 210 of the Seleucid Era (ACT p. 144–146).

With the correct identifications of the names of Venus and Jupiter Epping is able to also reproduce the "observations" in the text Sp 128 (also a Normal Star Almanac) for 111 B.C. And he ends as follows (my translation):

> A beginning has been made with deciphering the astronomical tablets of Babylon. If for anything, then for the explanation of unknown texts, 'all beginning is difficult'. With more ease the deciphering of other astronomical tablets will now be possible, those that have already been uncovered as well as those that will be obtained in the future; since ever more new tablets arrive at the British Museum. The profit for science will be threefold, first—as is clear by itself—for the deciphering of cuneiform texts, then for astronomy and for chronology.

Astronomisches aus Babylon

After the publication of their first little paper Epping continues his research at Blijenbeek, where Strassmaier is working on his *Alphabetisches Verzeichniss*. Strassmaier provides him with copies, transliterations and translations of two other fragments of the lunar ephemerides (SH 81-7-6, 272 and Sp 162), as well as one more Normal Star Almanac (Sp 175). In total Epping now has six texts at his disposal: three lunar ephemeris texts and three Normal Star almanacs.

After eight years of painstaking research, interrupted by periods of illness and by the move in 1885 of the Jesuit philosophy college and its professors and students from Blijenbeek Castle to the House Exaeten near Baexem, Limburg, the Netherlands (Fig. 4) the fruits of his labors are published in 1889 in a 200-page monograph, entitled *Astronomisches aus Babylon* (Fig. 5). The results are spectacular: with one strike the field of Babylonian astronomy has been put back on the intellectual world map.

Fig. 4 The House Exaeten near Baexem, Limburg, the Netherlands where Epping resided and worked from 1885 until his death in 1894. On the *right* the newly built Jesuit wing (Courtesy the Archiv of the North German Province of the Societas Jesu, Munich, Germany)

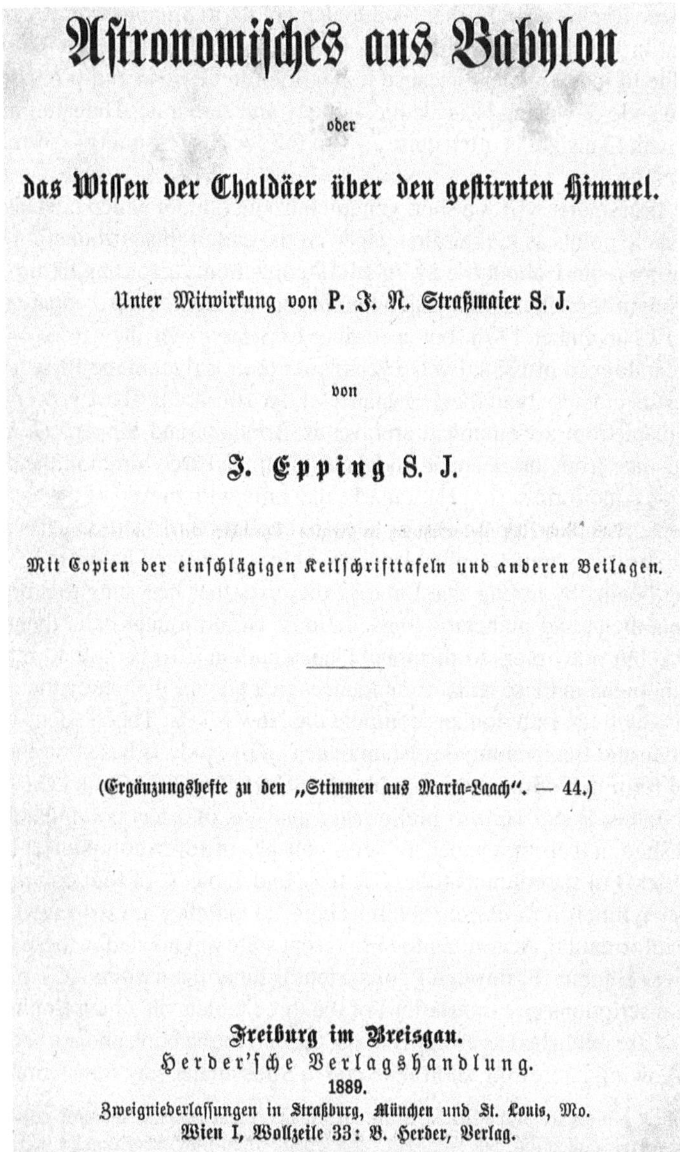

Astronomisches aus Babylon

oder

das Wissen der Chaldäer über den gestirnten Himmel.

Unter Mitwirkung von P. J. N. Straßmaier S. J.

von

J. Epping S. J.

Mit Copien der einschlägigen Keilschrifttafeln und anderen Beilagen.

(Ergänzungshefte zu den „Stimmen aus Maria-Laach". — 44.)

Freiburg im Breisgau.
Herder'sche Verlagshandlung.
1889.
Zweigniederlassungen in Straßburg, München und St. Louis, Mo.
Wien I, Wollzeile 33: B. Herder, Verlag.

Fig. 5 Title page of "Astronomisches aus Babylon" the 200-page monograph published by Epping and Strassmaier in 1889 as Ergänzungshefte 44 to the Jesuit journal *Stimmen aus Maria-Laach*. Its publication marks the birth of the discipline of Babylonian astronomy. Notice that Strassmaier is not listed as co-author but as collaborator (Courtesy the library of the Nederlands Instituut voor het Nabije Oosten, Leiden, the Netherlands)

The monograph was published as Supplement 44 to *Stimmen aus Maria-Laach*, the journal in which also the first paper had appeared. The text may be somewhat inaccessible to many readers because it is written in German and printed in gothic letters; the style of writing is academic, homely and elaborate. The latter makes that Epping's path to insight is often quite easy to follow. His reasoning is overall logical and systematic.

Again Strassmaier writes a short general introduction, in which he makes by and large the same points as in their first paper. At the end of this introduction he makes an interesting remark about the S† 76-11-17 collection. According to the date in its name acronym the tablets and fragments in this collection were apparently registered on 17 November 1876, but according to Strassmaier they were only finally properly catalogued in 1888, twelve years after their arrival in the British Museum. He further speculates about the provenance of the astronomical texts, proposing that they originate from astronomical archives in Borsippa and Sippar.[4] He notes that these texts date from the Seleucid and Arsacid era and are written in the difficult to read cursive cuneiform script. He remarks that only with the aid of the astronomical calculations by father Epping and by repeated collation of the texts, has it become possible to properly translate the planetary tablets, presented below.

Epping begins by noting that three of the texts that he wants to study contain lunar phenomena and planetary constellations. He no longer calls them observational texts but now refers to them as "Planetentafeln". To be able to interpret the lunar phenomena in these texts he announces that he will first study the other three tablets in which the Babylonians compute the New Moon. This he does in Chapter II, "Chaldäische Berechnung des Neumondes". His study is based on three tablets translated for him by Strassmaier: Tablet A (SH 81-7-6, 272), Tablet C (SH 81-7-6, 277) and Tablet B (Sp 162). A preliminary analysis of a few columns of Tablet A was published in their first paper. Tablet A consists of seven columns (a, b, c_1, c_2, d, e, m), Tablet B of six columns (d, e, f, g, h, l), and Tablet C of four columns (g, h, i, k). Similar symbols for different columns indicate that they are suspected to contain the same information. According to our present state of knowledge these three fragments cover columns F_1 through P_3 of system B lunar theory (see ACT p. 42).

The transcriptions and translations of the three tablets on which Epping's analysis is based are published as an appendix at the end of the book and are accompanied by the following interesting short note due to Strassmaier (my translation):

> These three tablets are in the British Museum, tablets A and C in the Shemtob collection, B in the Spartali collection, but were not catalogued, when they were copied in 1879, and could not be further collated, because the assistant Th.G. Pinches could not find them anymore

[4] But see Neugebauer 1975, 352.

This is curious in view of the fact that, when Abraham Sachs in the early 1950s got access to the notes and copies of Pinches[5] at the British Museum, he found copies in Pinches' beautiful hand of both ephemerides: BM 34580+ (LBAT 66; see ACT 122) consisting of eight joined fragments[6] and BM 34066+ (LBAT 60, 61, 62 and 63; see ACT 120) consisting of four fragments. Apparently the tablets turned up again later because when Kugler (1900, p. 9ff.) reanalyzed the first ephemeris ten years later he had a transcription at his disposal due to Strassmaier based on these same eight fragments.

In Chapter II Epping manages to give a first essentially correct explanation of the stepwise computation of the date and time of New Moon (midnight epoch) for consecutive months in column L_1 (his column e) starting with the duration of the synodic month (tabulated as an excess over 29 days) in column G_1 (his column b). He realizes that the auxiliary columns H_1, J_1 and K_1 (his columns c_1, c_2 and d) on which the computation of G_1 is based, are periodic without trying to identify the underlying algorithm (but see below). Finally, in this chapter he speculates that the last columns g, h, i and k are related to the duration of lunar visibility around New Moon. He comes back to this in Chapter III.

He then turns to the main theme of the book, the study of the three 'Planetentafeln' (Planetary Tables) that Strassmaier has provided him with: Sp 129 (Tablet I, dating from SE 189), Sp 128 (Tablet II, from SE 201) and Sp 175 (Tablet III, from SE 188). A first analysis of the planetary positions in Tablets I and II in their 1881 paper had resulted in the identification of the Akkadian names of Venus and Jupiter (see above, section "The Beginning"). Tablet III is a new one. Tablet I is almost undamaged and contains data for all 13 months of SE 189 (see Fig. 3), while Tablets II and III are broken fragments so that several months in the middle of the year are missing.

Before turning to the planets Epping starts out by discussing the Moon data in the Planetary Tables in Chapter III, "Chaldäische Mond-Ephemeriden". He notes that for each month the text in the upper left hand part contains lunar dates around New Moon and Full Moon and that for each date numbers are given that may be related to the duration of visibility of the Moon on these dates in SE 188, 189 and 201. He decides to study these lunar data first. The results of his analysis are impressive. He first establishes the properties of the Babylonian luni-solar calendar and the relation with the Julian calendar in the years 123, 122 and 111 BC, things that we take for granted not realizing that Epping is the pioneer who first worked it out based on the time pegs provided by the lunar data in the texts (including predictions of lunar eclipses).

Epping then shows that the data listed are (two) predictions of the duration of lunar visibility around New Moon, one during one of the last days of the month,

[5] Theophilus Goldridge Pinches (1856–1934) joined the staff of the British Museum in 1878 and worked there as assistant and later as curator until 1900, when he was fired after a feud with his superior E.A. Wallis Budge, Keeper of the Egyptian and Assyrian Antiquities in the British Museum. He was Lecturer in Assyriology at University College London until his death in 1934 (see his Obituary published in Nature 134, 16, 1934).

[6] See also inset Babylonian Lunar Theory 1880–1950.

when the Moon is visible for the last time just before the Sun rises in the morning, and one on the evening of the first day of the month, when the lunar crescent becomes visible again for the first time in the evening shortly after sunset; and (four) predictions on days around Full Moon of the time differences between all four possible combinations of moonrise and moonset and with sunrise and sunset. This type of Babylonian lunar data is presently known as the 'lunar-six', of which the earliest records date back to the seventh century BC (Huber and Steele 2007). To prove this Epping has to calculate rising and setting times of the Sun and Moon in Babylon, based on modern astronomical theory (involving a non-negligible amount of computation). Because no lunar data are missing, and based on an analysis of the errors in the data Epping is led to the correct conclusion that the lunar data in the Planetary Tables are predicted rather than observed.

Then, Epping notes the similarity of the magnitude of the duration of the lunar crescent visibility on day 1 in the Planetary Tables with the numbers in column i (= P_1 of ACT) of Tablet B and columns i and k (= P_3 of ACT) of Tablet C of the lunar ephemerides discussed in the preceding Chapter II. Extensive calculations and clever reasoning enable him to conclude that the numbers in column i of the ephemerides indeed must be the duration of lunar crescent visibility after sunset on the evening of the first day of the new month, and that column k represents the duration of lunar visibility just before sunrise a few days earlier on the day of last visibility of the waning lunar crescent. He further shows by numerical connection that Tablets A and C must be part of the same lunar ephemeris (ACT 122). Finally, he manages to explain the remaining columns. Column m on Tablet A is a misnomer and should have been named f (following e, as on Tablet B); it contains the lunar date of New Moon with time measured with respect to sunrise or sunset (column M_1 in ACT). Column g (N_1 in ACT) contains the difference in time between New Moon and sunset on the evening of first lunar crescent visibility, and h (O_1 in ACT) contains the elongation of the Moon on the evening of first visibility.

After the presentation of these quite interesting results Epping now comes to the main substance of his book in Chapter IV "Chaldäische Planeten-Ephemeriden" where he addresses the remaining parts of the text of the Planetary Tables Sp 129, Sp 128 & Sp 175. The tablets are organized month by month with one year per tablet, the first six months on the obverse, the last six months on the reverse (only Sp 129 is complete). He notes that the text is interspersed with numbers and that per month lines with increasing day numbers (up to 30) can be recognized. At the end of the chapter Strassmaier provides a full translation of the three tablets with associated explanatory philological comments. Epping first discusses the Babylonian names of the five planets (correcting some identifications that he made in their first paper eight years before). The contents of the texts as resulting from Epping's study are briefly summarized below:

- Predicted dates of conjunctions of planets with stars near the ecliptic. Epping correctly identifies 28 "Normalsterne", apart from a few weak stars in the constellation Cancer. This is why this class of texts is called Normal Star

Almanacs (Sachs 1948). For a modern version of the Normal Star list I refer to Hunger and Pingree (1999, p. 148–149).
- Predicted dates of heliacal rising, heliacal setting, acronychal rising and stationary points of the planets.
- Predicted dates of the first visibility of Sirius.
- Predicted dates of equinoxes and solstices.

Then in Chapter V "Haupt-Ergebnisse" Epping summarizes the main results of his work. At the end of this summary he comes, again in his simple homely style, to a visionary conjecture in an attempt to answer the question how the Babylonians were able to make all these fairly accurate predictions. Based on two other texts provided by Strassmaier (apparently goal-year texts; see also Fig. 6) he suggests that they used periods to transform previous observations into predictions. He mentions periods for Venus (8 years), Mercury (46 years), Jupiter (12 years) and Saturn (59 years), all well-known now[7] but in 1889 brand new.

Tablet ID	Museum #	LBAT #	Years covered	Contents	References
SH 81-7-6, 272	BM 34580+	66	104-101 BC	Moon sys B ACT 122	SE 1881, ES1889, E1890a
Sp 129	BM 34033	1055	123 BC	Normal Star Almanac	SE1881, ES1889
Sp 128	BM 34032	1059	111 BC	Normal Star Almanac	SE1881, ES1889
SH 81-7-6, 277	BM 34580+	66	104-101 BC	Moon sys B ACT 122	ES1889, E1890a
Sp 162	BM 34066	60	133-132 BC	Moon sys B ACT 120	ES1889, E1890a
Sp 175	BM 34078	1051	124 BC	Normal Star Almanac	ES1889
Rm IV 118A	BM 33562A	**1445	80 BC	Lunar eclipse report	E1889b
78-11-7, 4	BM 33066	**1477	523 BC	Excerpt text no. 55	E1890b
Rm 678	BM 92682	**1297	76 BC	Goal-year text no. 86	ES1890
S† 76-11-17, 1949	BM 32222	**1237	194 BC	Goal-year text no. 20	ES1890, ES1891a
SH 88-7-21, 9	CBS 17	**1295	87 BC	Goal-year text no. 82	ES1890, ES1891b
Rm 844			89 BC	Horoscope fragment	ES1891b
Rm 845			329 BC	Diary -328 (fragment)	ES1891b
Rm 710		**498	89 BC	Diary -88 (fragment)	ES1891b
82-7-4, 137	BM 92688+	**243	274 BC	Diary -273 (fragment)	ES1891b, ES1892
Rm IV 397	BM 33837	**284	233 BC	Diary -232 (fragment)	ES1891b, ES1892
Sp I, 131	BM 34035		138 BC	Astrological text	ES1891b
Sp II, 48	BM 34576+		405-99 BC	Saros list	S1892
Sp II, 955	BM 34576+		549-423 BC	Saros list	S1893
Sp II, 71	BM 34579	1428	401-272 BC	Saros Canon	ES1893, S1895
SH 81-7-6, 93	BM 45688	50	175-152 BC	Lunar eclipses	ES1893
(i)	(ii)	(iii)	(iv)	(v)	(vi)

Fig. 6 Texts studied and published by Epping (E) and Strassmaier (S) during their collaboration from 1881 to 1893. The acronyms SH and Sp in column (i) refer to collections bought from the antiquities dealers Shemtob and Spartali, Rm refers to collections assembled by Hormuz Rassam in assignment of the British Museum, and S† refers to the collection shipped by George Smith but arriving at the British Museum after his death. Dates in the tablet ID refer to the date that a collection arrived or was registered at the British Museum. The numbers in the Tablet ID indicate the sequence number in that collection. Strasssmaier notes about tablet SH 88-7-21, 9 that it is 'jetzt in Philadelphia, University Pennsylvania' and refers to it as the 'Amerikanishes Tablet' (presently know as CBS 17). The numbers in column (iii) are those of the hand-written copies by Pinches and Strassmaier published by Sachs (1955) in "Late Babylonian Astronomical and Related Texts". The numbers of the Diaries, the Excerpt text and the Goal-year texts in column (v) are those of the modern editions of these texts in the *Astronomical Diaries and Related Texts from Babylonia* by Hunger, Sachs and Steele (1988–2006). For the exact references of the papers in column (vi) the reader is referred to the Bibliography at the end of this paper

[7] See Hunger and Pingree 1999, 203–205.

In 1890 Epping publishes a third paper, again in *Stimmen aus Maria-Laach*, again in German gothic print, this time entitled "Die babylonische Berechnung des Neumondes" (Epping 1890a). In this paper he reports improvements in the understanding of the calculation procedure of three columns of the lunar ephemerides in Tablets A and B, which had remained elusive so far.[8] He first notes that columns b (G_1 in ACT) of Tablets A and B are "oscillierende Differenzreihe erster Ordnung" (oscillating difference sequences of the first order = Neugebauer's zigzag function in ACT). He discovers the mirroring principle near the extremes and he identifies the numerical values of its maximum and minimum and of its period of 251 synodic months. He recognizes that this period is a lunar anomalistic period, i.e. after 251 synodic months the Moon returns to its perigee/apogee. From these data he derives the Babylonian value for the mean length of the synodic month of 29;31,50,8,20 days (in sexagesimal notation), without noticing that this is exactly the value transmitted through Hipparchus to Ptolemy.[9] Finally, he explains columns c_1 (H_1 in ACT) and c2 (J_1 in ACT) as first and second order oscillating difference sequences. So after ten years of intensive study Epping has succeeded in providing correct explanations for 10 out of the 12 different columns in the lunar ephemerides of system B, available to him in Tablets A, B and C, a truly admirable achievement.

The collaboration of Epping and Strassmaier continued for four more years until Epping's passing away at the premature age of 59 on 22 August 1894. The texts that they studied and the papers written during their collaboration are listed in Fig. 6. The subject matter of these texts shows that their pioneering studies covered quite a large range of topics, including lunar eclipses and their periodicity (the Saros).

The Period 1895–1930

Shortly after Epping's death in 1894 his fellow Jesuit Joseph Hontheim (1858–1929) seems to have received the assignment to continue the work on Babylonian astronomy (see the introduction to "Die Babylonische Mondrechnung", Kugler 1900, p. vii).

Although Hontheim started to familiarize himself further with the material, his interests were more theological and soon other duties forced him to abandon the project. Then it lay unattended for a few years until Franz Xaver Kugler (who must have been a former pupil of Epping at the Jesuit college Exaeten, and who may have expressed interest) was apparently in 1897 assigned the task to continue Epping's work. Whether Strassmaier has played any role in this is not known but one would expect so although, at the time that Kugler finally took over, Strassmaier had become seriously ill.

[8] As acknowledged in a footnote these improvements were stimulated by a letter from Herr August Lorenz from Gross-Leubusch, Germany.

[9] It was F.X. Kugler who first realized this 10 years later (see BMR p. 24).

Fig. 7 Franz Xaver Kugler (1862–1929) (From Esch 1929)

Here enters the third player: Franz Xaver Kugler (1862–1929), now on a small playing field. Kugler was born in Koningsbach (Rheinpfalz, Germany) in a family of landowners (Fig. 7). He studied chemistry at the Technische Hochschule in Munich from 1880 to 1885 and received his doctorate in chemistry from the University of Basel in 1886. Although it seems that he was preparing for a career in chemistry, he entered the Jesuit society (still in exile in the Netherlands) a few months later in 1886. Following the nominal Jesuit education program, he studied philosophy in Exaeten from 1886 to 1889 (where Epping is his professor of "Mathesis et Astronomia") and theology in Ditton Hall in England from 1889 to 1992, and is ordained priest in 1893. One year later, in 1894, he is appointed Professor in Mathematics (as successor of Epping) at the just finished Collegium Magnum St. Ignatius in Valkenburg, Limburg, the Netherlands, where he lectured and worked, for more than 30 years, until his death in 1929 (Fig. 8).

In 1897 he is officially assigned to continue Epping's work. Already 3 years later—after what must have been a very intensive period of learning and study—he publishes his first book on Babylonian astronomy: *Die Babylonische Mondrechnung* (Kugler 1900; also referred to as BMR; Fig. 9). His results are impressive and constitute a major step forward compared to Epping's work. He explains the overall structure of the Babylonian lunar ephemerides and most of the columns in full detail and he recognizes that the lunar ephemerides are computed according to two different systems: system I (ACT, system B) and system II (ACT, system A).

In the introduction to his book Kugler complains that he could not get much help from Strassmaier, because of the latter's illness, and that he struggled with the reading of the texts because of his limited assyriological knowledge; he warns that philologically his book may contain errors or inaccuracies. All of the texts that he discusses are based on Strassmaier's copies, either from Epping's 'Nachlass' and/or provided by Strassmaier later.

From a few letters between Strassmaier and Kugler that I found in the Archive of the German Jesuit Province in Munich it can be seen that the relation between the

Fig. 8 The Collegium Magnum St. Ignatius in Valkenburg, Limburg, the Netherlands around 1910. The college was built in 1893/1894 and housed the philosophy and theology education of the German Jesuit Province. Here Kugler lived and worked as professor of Mathematics and Astronomy from 1894 until his death in 1929 (From *25 Jahre Ignatius Kolleg Valkenburg 1894–1919*)

two men was strained. Strassmaier apparently tried to remain involved with Kugler's work, because he was concerned about Kugler's linguistic shortcomings, but he was brutally rebuked by Kugler who was of the opinion that the main work to be done was the astronomical analysis of the texts and that Strassmaier did not know a thing about that. After the publication of *Die Babylonische Mondrechnung* in 1900 it seems that Strassmaier tried to persuade his superiors and one of his colleagues to convince Kugler that he needed his assistance and to mediate between them.

Among Kugler's papers in the Jesuit Archive I found a copy of a letter that Kugler sent on 29 October 1900 (Fig. 10), shortly after *Die Babylonische Mondrechnung* had appeared, to the colleague who had written to him about Strassmaier's dismay (probably A. Baumgartner, an internationally known scholar of comparative literature, for whom Strassmaier, still suffering from the consequences of his operation, was doing correction work). In this letter Kugler is forcefully defending himself, pointing out that Strassmaier has been bothering him with useless advice and suggestions rather than helping him with providing and collating texts in the British Museum and that the main contribution to the interpretation of astronomical texts comes from the astronomer who is figuring out what their meaning is rather than from the Assyriologist who only provides their translation.

Kugler starts his letter as follows (my translation):

> Most Reverend dear Pater.
>
> Pax Christi
>
> Your good intentions are fully acknowledged. Only in a few days you can not clearly realize, what here the correct thing is; because, since you are different at home in the astronomy, as I am, you do not know how to correctly value the difficulties that are connected with the deciphering of the Babyl. cuneiform tablets.

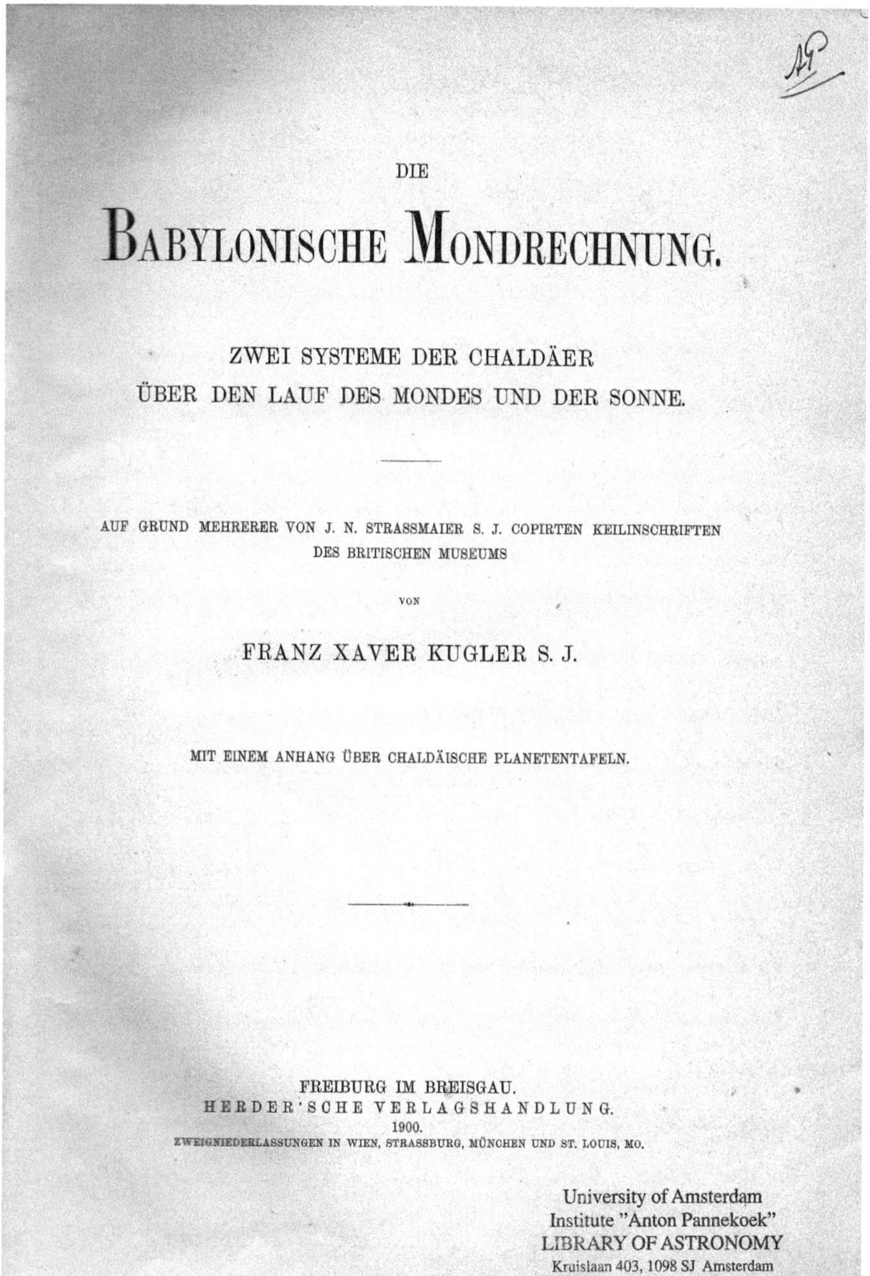

DIE

BABYLONISCHE MONDRECHNUNG.

ZWEI SYSTEME DER CHALDÄER
ÜBER DEN LAUF DES MONDES UND DER SONNE.

AUF GRUND MEHRERER VON J. N. STRASSMAIER S. J. COPIRTEN KEILINSCHRIFTEN
DES BRITISCHEN MUSEUMS

VON

FRANZ XAVER KUGLER S. J.

MIT EINEM ANHANG ÜBER CHALDÄISCHE PLANETENTAFELN.

FREIBURG IM BREISGAU.
HERDER'SCHE VERLAGSHANDLUNG.
1900.
ZWEIGNIEDERLASSUNGEN IN WIEN, STRASSBURG, MÜNCHEN UND ST. LOUIS, MO.

University of Amsterdam
Institute "Anton Pannekoek"
LIBRARY OF ASTRONOMY
Kruislaan 403, 1098 SJ Amsterdam
The Netherlands. Tel. +31 20 - 525 7489

Fig. 9 Titlepage of *Die Babylonische Mondrechnung*, the first publication of Franz Xaver Kugler on Babylonian astronomy. Notice the reference to Strassmaier in the subtitle as having provided the transcriptions. Also notice the autograph AP of Antonie Pannekoek, the original owner of the book, in the *upper right-hand corner* (see section "Epilogue") (Courtesy the University Library, University of Amsterdam, the Netherlands)

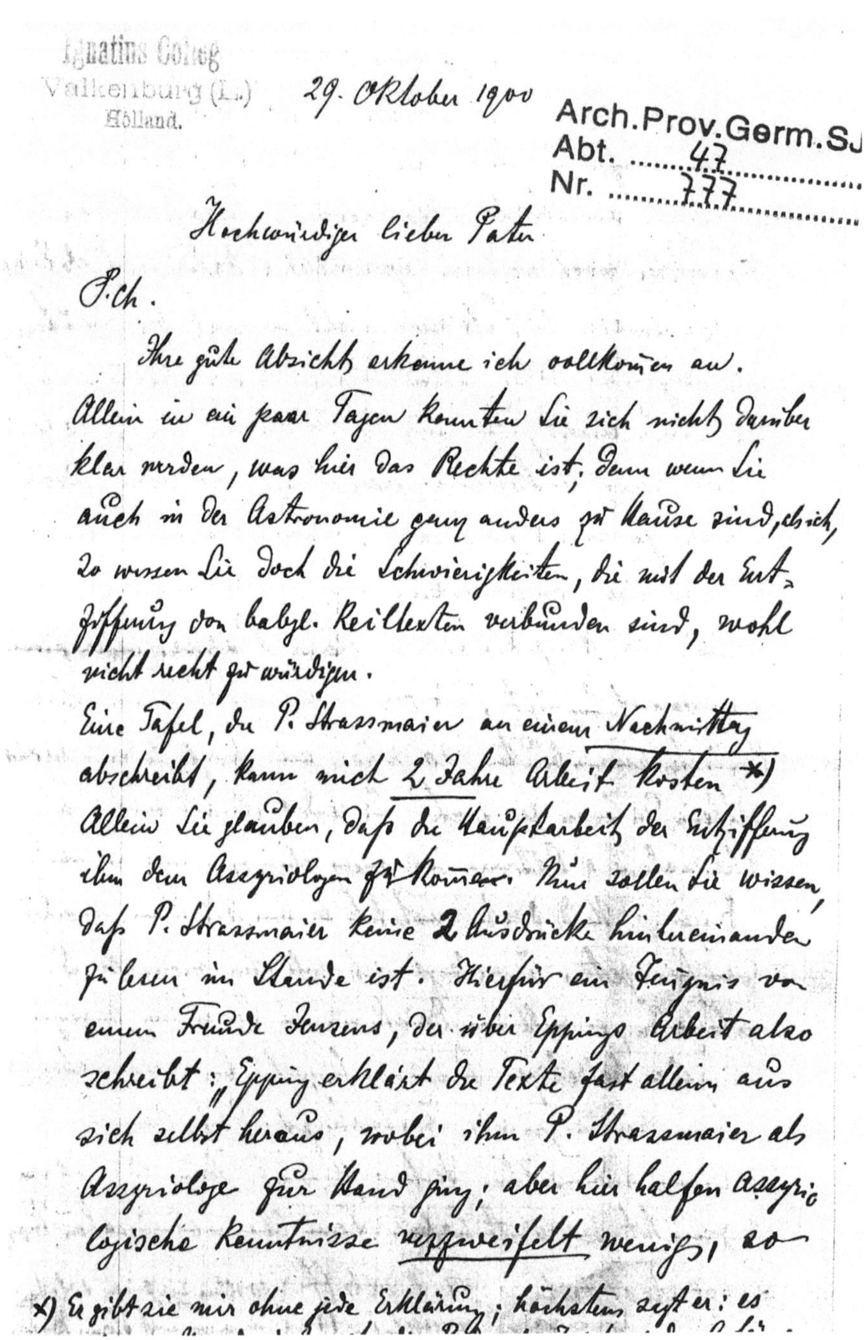
Ignatius Colleg
Valkenburg (L.)
Holland.

29. Oktober 1900

Arch.Prov.Germ.SJ
Abt. 47
Nr. 777

Hochwürdiger lieber Pater
P. Ch.

Ihre gute Absicht erkenne ich vollkommen an. Allein in ein paar Tagen konnten Sie sich nicht darüber klar werden, was hier das Rechte ist; denn wenn Sie auch in der Astronomie ganz anders zu Hause sind, als ich, so wissen Sie doch die Schwierigkeiten, die mit der Entzifferung von babyl. Keiltexten verbunden sind, wohl nicht recht zu würdigen.

Eine Tafel, die P. Strassmaier an einem Nachmittag abschreibt, kann mich 2 Jahre Arbeit kosten *)

Allein Sie glauben, daß die Hauptarbeit der Entzifferung ihm dem Assyriologen zukomme. Nun sollen Sie wissen, daß P. Strassmaier keine 2 Ausdrücke hintereinander zu lesen im Stande ist. Hierfür ein Zeugnis von einem Freunde Jensens, der über Eppings Arbeit also schreibt: „Epping erklärt die Texte fast allein aus sich selbst heraus, wobei ihm P. Strassmaier als Assyriologe zur Hand ging; aber hier halfen assyriologische Kenntnisse verzweifelt wenig, so

*) Er gibt sie mir ohne jede Erklärung; höchstens sagt er: es

Fig. 10 The first twenty lines of a letter, dated 29 October 1900, of Kugler to a fellow Jesuit, who apparently tried to mediate in his conflict with Strassmaier. Whether it is a copy of the letter that was actually sent or just a draft that was never sent is not clear. I suspect that the addressee is Dr. A. Baumgartner, whom Strassmaier was assisting with correction work for his book on Indian literature (Courtesy Archiv of the North German Province of the Societas Jesu, Munich, Germany)

A tablet, that P. Strassmaier copies in one afternoon, may cost me 2 years of work. Only you believe that the main work of deciphering belongs to him the Assyriologist. Now you should know that P. Strassmaier is not able to read 2 expressions after each other. Hereto a testimony of a friend Jensen, who writes about Epping's work as follows: "Epping explains the text almost exclusively out of itself, where Strassmaier as Assyriologist lends him a helping hand; here is Assyriological knowledge of desperately little assistance, so little that one may call Epping's work a work of deciphering in the most original meaning of the word".[10] What has been said here, holds all the more for Die Babylonische Mondrechnung, in so far as the character is even less linguistic as E.'s work.

And he ends his letter as follows:

Such is the state of affairs. I have made more than 10 attempts to show P. Str. the area where he can achieve something original; P. Provincial has also tried it; all in vain. He produced only confused ideas, that already Epping & Hontheim could not cope with, he wants work to be carried out that is an absurdity. I have done enough. If he wants to help me, then I will acknowledge everything in detail—conform the truth. He could be of great service to me by collations and at the same time put his texts right. Only I may claim then that will be said: for the deciphering of the text should this and that be correctly put. But he is afraid of that. Then I can not help him. I also want to be able to publish my work then and there, where it seems most appropriate and not leave the printing process to him. If that would have happened with Epping's book, then the army of printing errors, in particular wrong numbers, would not have occurred. I had to write an astronomer, that much in Epping's manuscript was fully correct according to my examination, but that it was often corrupted by the copyist (who did not understand the subject matter). I did not say who the copyist was; but I know it from his own mouth.

Now we have enough. Once more cordial thanks for your good intention; dear God will reward you.

With cordial greetings,

Your dear fellow-brother F.X. Kugler S.J.

Also among Kugler's papers I found a postcard from Strassmaier, sent to Kugler about one half year later (21 May 1901), apparently in response to two letters that he received from Kugler asking for help with a Jupiter text (Fig. 11). Strassmaier writes (my translation):

R.P. I have received your long letter of 17 May and of 21 May with tablet Sp II. 42+574+68+107 yesterday in good order. Since right now I have put my time at the disposition of P. Baumgartner for the correction of Indian literature and I should read all East Asian literature, I have right now not the required leisure to check all numbers, and you have to exercise some patience. From your long letter I can also not clearly recognize which freedom you allow me to make comments, propositions and suggestions and I absolutely don't want to loose time, to write something useless to your irritation and annoyance. I had no idea that you were working on Jupiter; thereto I had a lot to say, that is impossible to get from books. But before I make a proposition, I must know that you accept it. Scolding and fighting I don't want anymore: but that it is possible to find other fragments I very much

[10] This passage is a quote from the review by P. Jensen of *Astronomisches aus Babylon* published in *Zeitschrift für Assyriologie* 5 (1890), 121–133. It is part of a longer citation of Jensen's review in Kugler's introduction to *Die Babylonische Mondrechnung* (1900, v–vi).

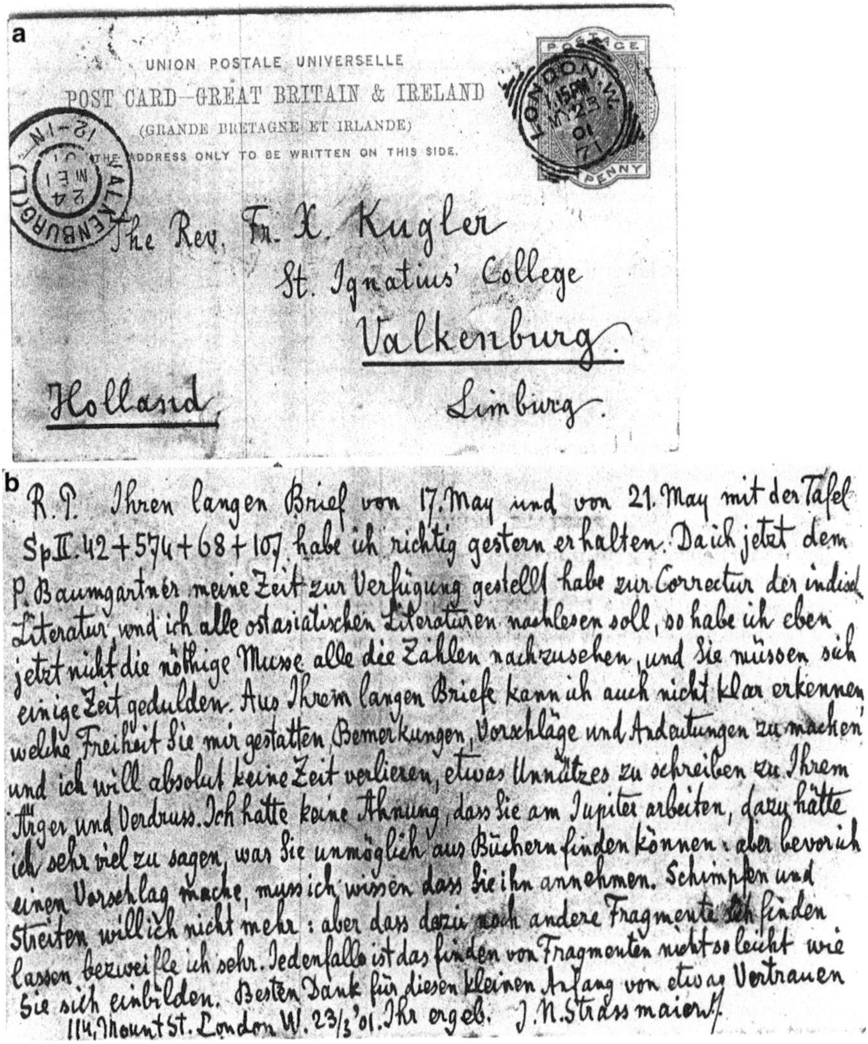
a

UNION POSTALE UNIVERSELLE
POST CARD—GREAT BRITAIN & IRELAND
(GRANDE BRETAGNE ET IRLANDE)
THE ADDRESS ONLY TO BE WRITTEN ON THIS SIDE.

The Rev. Fr. X. Kugler
St. Ignatius' College
Valkenburg.
Holland
Limburg.

b

R.P. Ihren langen Brief von 17. May und von 21. May mit der Tafel
Sp II. 42+574+68+107 habe ich richtig gestern erhalten. Da ich jetzt dem
P. Baumgartner meine Zeit zur Verfügung gestellt habe zur Correctur der indisch
Literatur, und ich alle ostasiatischen Literaturen nachlesen soll, so habe ich eben
jetzt nicht die nöthige Musse alle die Zahlen nachzusehen, und Sie müssen sich
einige Zeit gedulden. Aus Ihrem langen Briefe kann ich auch nicht klar erkennen
welche Freiheit Sie mir gestatten, Bemerkungen, Vorschläge und Andeutungen zu machen
und ich will absolut keine Zeit verlieren, etwas Unnützes zu schreiben zu Ihrem
Ärger und Verdruss. Ich hatte keine Ahnung, dass Sie am Jupiter arbeiten, dazu hätte
ich sehr viel zu sagen, was Sie unmöglich aus Büchern finden können: aber bevor ich
einen Vorschlag mache, muss ich wissen dass Sie ihn annehmen. Schimpfen und
Streiten will ich nicht mehr: aber dass dazu noch andere Fragmente sich finden
lassen bezweifle ich sehr. Jedenfalls ist das finden von Fragmenten nicht so leicht wie
Sie sich einbilden. Besten Dank für diesen kleinen Anfang von etwas Vertrauen
114, Mount St. London W. 23/5'01. Ihr ergeb. J.N. Strassmaier

Fig. 11 (**a**) and (**b**) Postcard sent on 21 May 1901 by Strassmaier to Kugler (Courtesy Archiv of the North German Province of the Societas Jesu, Munich, Germany)

doubt.[11] In any case, finding fragments is not as easy as you envisage. Thank you for this small beginning of some trust.

114 Mount St., London W 23/5'01 Your devoted, J.N Strassmaier

[11] In Neugebauer's edition of this text (ACT 611) there are several more fragments that are joined to the tablet: Sp II 876+VAT 1753+VAT 1755. Apparently Sp II, 876 is already present in Pinches copy LBAT 119.

It must have been quite difficult for Strassmaier, the (sick) man who started it all, to be told to back off, because he did not know what he was talking about, and to just provide copies of texts when he was asked.

During his impressive and extremely productive career Kugler published most of his results in the form of books; in addition he published some twenty articles in *Zeitschrift für Assyriologie*, *Stimmen aus Maria-Laach* and a few other journals (see the obituary by Esch 1929). His first book *Die Babylonische Mondrechnung* (1900) was followed by *Sternkunde und Sterndienst in Babel* (SSB) of which volume I was published in 1907, the first two parts of vol. II in 1909/10 and 1912, with two supplements in 1913 and 1914; the last part of vol. II was finally published in 1924. In the mean time had appeared *Im Bannkreis Babels* (1910) and *Von Moses bis Paulus* (1922). His last book was a more religious inclined work entitled: *Sibyllinische Sternkampf und Phaeton in naturgeschichtlicher Beleuchtung* (1927). For full references of these publications see the Bibliography.

In his publications Kugler pioneered and contributed to virtually every aspect of Babylonian astronomy and every discipline in which Babylonian astronomy played a role, e.g.:

- Lunar and planetary theory; the discovery of two systems of computation, system I (system B in ACT) and II (system A in ACT) (BMR and SSB I).
- Early non-mathematical astronomy, astrological texts, correspondence of Esarhaddon, star names, etc. (SSB II).
- Mesopotamian chronology (old-Babylonian, Assyrian, & late-Babylonian calendars (SSB I & II, *Im Bannkreis Babels*). His most famous chronological discovery is the reference to the year of "the golden throne" the formula for year 8 of the old- Babylonian king Ammisaduqa appearing in one of the Venus observations in tablet 63 of the omen series Enuma Anu Enlil (SSB II, p. 257–311).
- Biblical chronology (*Von Moses bis Paulus* 1922).

It is of interest to also mention the fierce polemic with the proponents of the so-called Panbabylonistic movement in which Kugler became involved after publication of the first volume of *Sternkunde und Sterndienst in Babel* in 1907. Panbabylonism was a school of thought within Assyriology and Religious Studies that considered the Hebrew Bible and Judaism as directly derived from Babylonian culture and mythology. It appeared in the late nineteenth century and gained popularity in the early twentieth century, advocated notably by Hugo Winckler (1863–1913) and Alfred Jeremias (1864–1935) (see Surhone et al. 2010). The direct cause of this polemic was Kugler's refutation (SSB I, p. 215–225) of the hypothesis, first proposed by Fritz Hommel (1854–1936) and taken over by Winckler, that in late-Babylonian times the names for the planets Mercury and Jupiter and for Mars and Saturn had become interchanged compared to the original usage during the old-Babylonian period. When Jeremias went even further by proclaiming that the "wissenschaftliche Babylonische Astronomie" dated from the third millennium B.C. they were demolished by the chronological arguments of Kugler, served with biting sarcasm, in his paper "Auf den Trümmern des Panbabylonismus" [On the Ruins of Panbabylonism] (1909) and in his monograph *Im Bannkreis Babels* [Under the

Spell of Babylon] (1910). The subtitle of the latter book is illustrative of Kugler's polemic style of writing: "Panbabylonische Konstruktionen und Religionsgeschichtliche Tatsachen". Ernst Weidner (1891–1976) who was an early supporter of the Panbabylonistic doctrine complains in his paper "Zum Alter der babylonischen Astronomie" (1912) about Kugler's language where he characterizes Weidner's attempts to keep the idea of a high age for Babylonian astronomy alive as "galvanische Zuckungen" (galvanic convulsions) and he asserts "I will resolutely go my way, without letting myself be disturbed by the unbridled outbursts of Pater Kugler".

When Kugler started his study of Babylonian astronomy in 1897 his knowledge of cuneiform writing and of the Akkadian language was virtually non-existing. Three years later, in the Introduction to *Die Babylonische Mondrechnung* (1900) he warns the reader against possible linguistic mistakes of his, on the other hand pointing out that most of his results are based on numerical analysis and that quite often the translation of certain Akkadian terms is supported by astronomical rather than by linguistic arguments. There is evidence that Kugler tried to improve his knowledge of Akkadian and related subjects. Among his papers I found a certificate from Heidelberg University that in the summer semester of 1903 he attended lectures of Prof. Carl Bezold (1859–1922) on Assyrian paleography (1 h), on Continuation of Arabic (2 h) and on the Interpretation of the prism inscription of Esarhaddon (2 h) and by Dr. Becker on Syriac (2 h); for this he was charged the total sum of 42 Deutschmark (equivalent to about 10 US dollars in around 1910).

Kugler's method of research is inductive, rather than deductive; letting the reader share in the experience of the unfolding insight, reasoning by way of worked out cases and examples rather than by theorems. This is well illustrated by his first book *Die Babylonische Mondrechnung*. He starts out by analyzing all columns as they appear one by one in the New Moon tablet Nr. 272 (81-7-6).[12] This large tablet consisted of eight different fragments joined by Strassmaier, two of which had been studied before by Epping and Strassmaier (see above and the inset Babylonian Lunar Theory 1880–1950). Kugler step by step unraveled the way in which the ephemeris is constructed, using a few other tablets also put at his disposal by Strassmaier to support his argument. He finishes that part by summarizing the computational procedure of what he calls system I (ACT system B), adding a graphical representation of the connections between different columns. He then proceeds to system II (ACT system A) which is even more difficult to reconstruct not only because it is virgin territory but also because he has only a dozen (sometimes tiny) fragments of tablets at hand. After an admirable 'tour de force' of astute reasoning he also cracks system II (ACT system A), again finishing with a 'rückblick' and a figure illustrating the connection between the different columns. The absence of an

[12] Since I will come back to this later when discussing the work of Paul Schnabel it is interesting to note that Kugler (or Strassmaier) here is the first to draw attention to the colophon which mentions that this ephemeris contains the 'tersitum ša Kidin(nu)', the computational table of Kidinnu, who was later identified with Kidenas, the Chaldaean astronomer mentioned by the Greek and Roman writers Strabo and Plinius.

index in this and most later books written by Kugler forms a severe handicap in digesting and cross correlating his results, all the more so because in later books he often comes back to and corrects results published earlier.

The inductive method followed by Kugler in his research is, of course, quite suitable in the pioneering phase of exploring a new subject. However, the difference in style and method with Neugebauer's ACT published more than 50 years later is striking. Neugebauer starts with an overview of Babylonian lunar theory and then discusses all available texts one by one individually, pointing out variants of method and errors in each text: the deductive method. At the end of ACT he includes almost 60 pages with Indices and Bibliography, a blessing for any student of Babylonian mathematical astronomy.

Kugler (1922) described his own approach in the Introduction to *Von Moses bis Paulus* by pointing out that trying to read everything that is known about a certain subject usually stands in the way of making one's own independent judgment, and that for someone (like himself) "whose combinatorial gift is much larger than his storage capacity" this is a poor method. He continues by stating that he usually went his own way, with the danger of even sometimes making mistakes or to repeat what is already known. He adds that the latter should never lead to belittlement of previous work but to its confirmation.

Kugler had the reputation of being a difficult man. In his paper "Drei babylonische Planetentafeln der Seleukidenzeit" Schaumberger (1933) writing about Kugler's 'Lebenswerk' makes some personal observations. He mentions Kugler's victories on the Panbabylonistic battlefield around 1910 and his often somewhat strained relations with colleagues. Apparently Kugler was quite competitive; we have seen an example of this in his correspondence with Strassmaier. Among his papers there is also a letter from the German assyriologist Carl Bezold asking why Kugler is behaving so unfriendly and what has come in the way of their previously pleasant relationship.

Kugler was aware of the fact that his style of writing had often caused some irritation. In the Introduction to *Von Moses bis Paulus* (1922)—at the mature age of 60—he expresses remorse "that at previous occasions he had made all too elated use of the weapon of sarcasm". However, two sentences later he defends his combative and occasionally sharp style as sometimes necessary to be sufficiently clear and to serve the truth. He never fought the person but always the case.

Whatever his shortcomings may have been, when Kugler died in 1929 his contribution to the field of Babylonian astronomy had been monumental. With his death the pioneering phase of the subject may be considered over and the field of Babylonian mathematical astronomy well established.

SH 81-7-6, 272: Babylonian Lunar Theory 1880–1950

In their pioneering paper Strassmaier and Epping (1881) studied three (fragments of) tablets, one of which was shown to contain a column with computed dates and times of the New Moon for 13 consecutive months in the Babylonian calendar. While they were writing their paper this tablet had not yet been catalogued. But when Epping and Strassmaier (1889) published their more elaborate study eight years later it had in the mean time been assigned the catalogue nr. SH 81-7-6, 272, indicating that it was part of a collection of (fragments of) tablets acquired by the British Museum from the antiquities dealer Joseph Shemtob, dated to 6 July 1881, and catalogued as nr. 272 of that collection. In *Astronomisches aus Babylon* (1889) they were then able to show that another tablet in the same collection (SH 81-7-6, 277) was part of the same lunar ephemeris as nr. 272. Epping and Strassmaier (1989) and Epping (1890a) managed to correctly interpret the numbers in 10 of the 12 columns of the two incomplete lunar ephemerides available to him at the time (columns G_1 through P_3 of system B; see ACT p. 42–43): an astounding performance.

When Franz Xaver Kugler published *Die Babylonische Mondrechnung* in 1900 his analysis of the ephemeris SH 81-7-6, 272 was based on 8 fragments, joined for him by Strassmaier (see Fig. 12). Kugler analyzed columns A through M in depth and was able to explain most elements of Babylonian lunar theory, building on Epping's pioneering studies. He identified two different systems for the computation of lunar and planetary ephemerides (systems I and II, now called systems B and A).

In 1935 Schaumberger (in "3. Ergänzungsheft zum ersten und zweiten Buch" of Kugler's *Sternkunde und Sterndienst in Babel*), confirming Epping and Strassmaier's (1989) early results, gave a detailed explanation of the computation of the duration of the lunar visibility and of the lunar elongation at New and Full Moon in columns N through R of the ephemeris. His results were applauded by Neugebauer (1936a) in his review as "den ersten wesentlichen Fortschritt für das Verständnis der mathematischen Astronomie Babylons über die Arbeiten von Kugler hinaus". Schaumberger's analysis of the last columns of tablet BM 34580+, the British Museum number under which the ephemeris had then become known, was based on Strassmaier's transcription of the text and a drawing of the full tablet indicating all the different joined fragments provided to him by Neugebauer (private communication, see Fig. 12).

Finally, in a lengthy paper Neugebauer (1938c) solved the only remaining open issue in Babylonian lunar theory by giving the correct interpretation of functions E (lunar latitude) and Ψ (eclipse magnitude) in both lunar systems.

When A.J. Sachs in the early 1950s was allowed access to copies of texts at the British Museum made by Th.G. Pinches in the years before 1900, it turned out that copies of the text of BM 34580+ were among them (LBAT 66 in Sachs 1955).

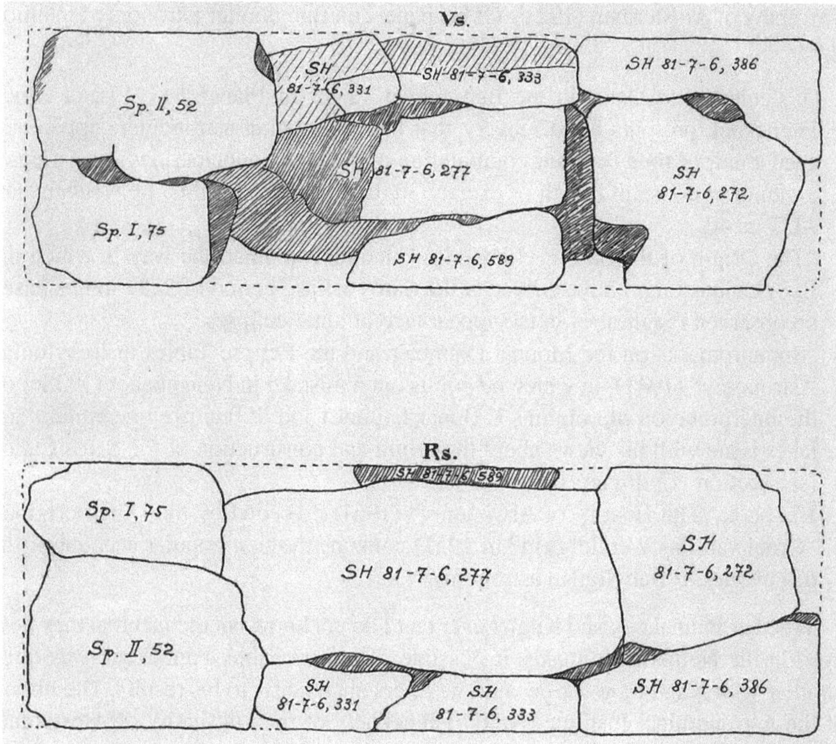

Fig. 12 Sketch by Otto Neugebauer of the eight fragments joined by Strassmaier in the 1890s of one of the most complete lunar ephemerides of system B (presently known as BM 34580+ or ACT 122). The two texts studied by Epping and Strassmaier in the 1880s are located on the reverse sides of fragment nrs. 272 and 277. Kugler (1900) already had the full text of the ephemeris as displayed here at his disposal. In ACT Neugebauer (1955) further completed the tablet by adding a ninth fragment 81-7-1, 454. Notice that Sp. I, 75 should be Sp. II, 75 (From Schaumberger 1935)

Other Early Players (1910–1935)

During the 30 years of Kugler's productive career and just before the appearance of Neugebauer on the field, only three other players can be identified who made contributions worth mentioning to Babylonian Mathematical Astronomy: Antonie Pannekoek, Johann Schaumberger and Paul Schnabel.

Antonie Pannekoek (1873–1960) was an internationally known Dutch astronomer and political activist; a prolific writer, both on socialist and Marxist themes, as well as on astronomy (see Minnaert 1974). In addition to a large number of journal and newspaper articles—most in German—on socialism and politics and over a hundred astronomical publications, he wrote several important early papers on Babylonian astronomy and a well-known book on the history of astronomy. Pannekoek is the founder of and first professor at the Astronomical Institute of the

University of Amsterdam (1921). Of his papers on Babylonian astronomy I mention here:

- "Calculation of Dates in the Babylonian Tables of Planets" (1917), in which Pannekoek presents his discovery that the Babylonian astronomers apparently used a unit of time for their computation of planetary ephemerides consisting of a month of constant length of exactly 30 days (baptized 'tithis' by Neugebauer; ACT, p. 40).
- "The Origin of the Saros" (1918), in which he explained the way in which the Babylonians might have arrived at the Saros eclipse period of 223 months based on observed regularities in the appearance of lunar eclipses
- "Some remarks on the Moon's Diameter and the Eclipse Tables in Babylonian Astronomy" (1941), in which he points out a mistake in Neugebauer (1938c) on the interpretation of columns E (lunar latitude) and Ψ (eclipse magnitude) and takes issue with his views about the origin and construction of the Saros Canon (see section "Epilogue" below)
- His book "The History of Astronomy" (1961; preceded by the Dutch original: "Groei van ons Wereldbeeld" in 1951) contains the first popular account of the discoveries of Babylonian astronomy[13]

Whether Pannekoek and Kugler ever met I do not know but the fact that they both lived in the Netherlands makes it possible, if not probable. Pannekoek was quite familiar with Kugler's work because he refers abundantly to his results. The library of the Astronomical Institute 'Anton Pannekoek' of the University of Amsterdam has in its possession Pannekoek's personal exemplar of *Die Babylonische Mondrechnung*, autographed AP (see Fig. 9) containing handwritten notes in pencil in the margin on numerous pages.

Johann Baptist Clemens Schaumberger (1885–1955) was a German catholic priest (in the Redemptorist order) who studied cuneiform writing with A. Deimel at the Pontificium Institutum Biblicum in Rome from 1909 to 1912. In the following year he was appointed as Professor für Bibelwissenschaft at the Ordenshochschule of the Redemptorist order in Gars am Inn, Austria where he lived and worked for more than 40 years until his death. Schaumberger finished the planned "3. Ergänzungsheft zum ersten und zweiten Buch" to "Sternkunde und Sterndienst in Babel" (1935) after Kugler's death and he saw himself as Kugler's successor in the so-called Babel-Bibel controversy. For short biographies of Schaumberger and his bibliography the reader is referred to Weiss (1995) and to Weidner (1955). His main publications relevant to Babylonian astronomy are:

- Several papers on the "Stella Magorum" (1925–1943; see bibliography in Hunger and Pingree 1999)
- 3. "Ergänzungsheft to Sternkunde unde Sterndienst in Babel" (Schaumberger 1935). This third supplement to Kuglers SSB is praised by Neugebauer (1936a) in his review as "the first essential progress especially for the understanding of

[13] Neugebauer 1975, 17.

the mathematical astronomy of Babylonia since the work of Kugler". Neugebauer is particularly enthusiastic about Schaumberger's explanation of the last columns in Babylonian lunar ephemerides in which the duration of lunar visibility on dates around New Moon and Full Moon (observed as the 'lunar-six') is computed (see also inset Babylonian Lunar Theory 1880–1950)

- Two papers on ziqpu-stars[14] (Schaumberger 1952, 1955; see bibliography in Hunger and Pingree 1999)

Paul Schnabel (1887–1947) studied ancient history and classical philology in Leipzig and Jena and obtained his doctorate in 1911 in Jena based on a study of "Die babylonische Chronologie in Berossos' Babyloniaka". After having fulfilled his military duties 1914–1918 during World War I he was appointed in 1920 as privat-dozent at the University of Halle-Wittenberg and in 1926 as extra-ordinarius, followed in 1934 by his appointment to full Professor in the History of the Ancient Orient. As a consequence of a permanent neurological problem following a malaria infection in 1937 he was forced to terminate his professional activities and died 10 years later in a sanatorium.

Schnabel's work on Babylonian astronomy is limited to the period 1923–1927. In his book *Berossos und die babylonisch-hellenistische Literatur* Schnabel (1923) published two lunar ephemeris texts in transcription, one of system A (VAT 209) and one of system B (VAT 7809) both from the collection of the Vorderasiatisches Museum in Berlin, Germany. Based on these texts he proposed Nabu-rimanni and Kidinnu as the founders of system A and system B and he further suggested that the Babylonians were aware of the precession of the equinoxes. His identifications of Kidinnu and Nabu-rimanni with Kidenas and Nabourianos, both known from Greek and Roman writers in late antiquity, were accepted but his suggestion of a Babylonian discovery of the precession was sharply rebutted by Kugler (1924) in *Sternkunde und Sterndienst in Babel* II (p. 382 ff.). Schnabel (1927) reacted by publishing a paper entitled "Kidenas, Hipparch und die Entdeckung der Präzession" which he concluded with the sentence that Kidinnu's discovery of precession (dated by him to 379 BC) was 'endgültig festgestellt'. This paper was often quoted in the literature in the 1930s and 1940s until the notion that the Babylonians had discovered the precession of the equinoxes was demolished by Neugebauer (1950).

In his Kidinnu paper Schnabel (1927) published transcriptions of another six lunar tablets from the Berlin collection (VAT), one with an ephemeris, five with lunar auxiliary functions and one with the daily motion of the Sun, all for lunar system B. According to Neugebauer (in ACT) these transcriptions are marred with errors. More useful was Schnabel's (1924) discussion of another three tablets: a Mars ephemeris of system A (AO 6481), a Saturn ephemeris of system B (VAT 7819), and a procedure text for Mercury and Saturn (AO 6477) (see Neugebauer ACT 501, 702 and 801).

[14] Ziqpu stars are culminating simultaneously with the rising or setting of certain constellations and may have been used for time keeping at night (Hunger and Pingree 1999, 68ff.).

Otto E. Neugebauer—The Star Player

When Neugebauer entered the scene of Babylonian astronomy in the early 1930s the playing field had been laid out but there were only few and not very strong players. One might say that he started out as running back but quickly developed other roles, those of quarter-back and referee at the same time, dominating the field for about half a century.

Otto Neugebauer (1899–1990) was born in Innsbruck, Austria. He graduated in 1917 from the Akademisches Gymnasium in Graz and enlisted in the Austrian army serving as an artillery lieutenant on the Italian front. Following his discharge from a prisoner of war camp in Italy in 1919 he studied electrical engineering and physics in Graz (1919–1921), mathematics and physics in Munich (1921–1922) and then finally settled for mathematics in Gottingen (1922–1926). His interests shifted to history of mathematics and he wrote a doctoral thesis entitled "Die Grundlagen der ägyptischen Bruchrechnung" in 1926. He was appointed Privatdozent in Göttingen in 1927 and began lecturing on mathematics and on the history of ancient mathematics. After he was forced to resign in 1933 he moved to the University of Copenhagen where he was offered a 3-year appointment. For more detailed biographical information see Swerdlow (1998).

Neugebauer spent the first 10 years of his career creating order in Egyptian mathematics and Babylonian mathematics (see his bibliography in Sachs and Toomer 1979). Quite soon after turning to the study of Babylonian astronomy, in a seminal paper entitled "Untersuchungen zur antiken Astronomie I" (Neugebauer 1938a), he defined his future research plan (which would take him almost two decades to execute). At that occasion, he also explicitly and even in a slightly emotional way expressed his indebtedness to his direct predecessor Franz Xaver Kugler as follows (my translation from German):

> I must end with a few personal remarks. Whoever gets involved with Babylonian astronomy has—one way or the other—to use Kugler's pioneering studies. With all possible emphasis I would like to stress that whatever will be said in what follows about Babylonian astronomy, were unthinkable without Kugler's work, even without my continuous citing it. It is namely not always so easy to present Kugler's results, also where they are directly used, as proof because he has almost never given them in the form of explicit formulae, but mostly explained them only on the basis of examples. However, since it is my purpose here to use the regularities in the number sequences in the astronomical cuneiform texts as generally as possible, I had to cloth many of Kugler's results in a seemingly different form, so that a citation makes little sense; it requires often an appreciable list of conclusions to derive from Kugler's presentation the general formulated law. But just this lies at the basis of the progress of scientific work, that often, that which has been gained along an extensive detour, eventually can be organized and formulated in such a way that it becomes a self-explanatory and adequate starting point for further work. In that way these studies will hopefully in the course of time contribute to the integration of Kugler's wonderful results into the history of ancient astronomy, as they have deserved to be for a long time.

In spite of the fact that Neugebauer's career is the topic of this collection I will not try to summarize Neugebauer's achievements in Babylonian astronomy in the last two decades of the period under review here. This would be a formidable task and much less inspiring for the reader than directly reading his papers. Actually, quite a good impression of the main results of his work can be obtained by just going through the titles of his research papers in that period (Sachs and Toomer 1979). In addition, much of it may be viewed—in retrospect—as preparations for his magnum opus "Astronomical Cuneiform Texts" (Neugebauer 1955).

Below I list Neugebauer's main early contributions to Babylonian mathematical astronomy in the period 1930–1950:

- Reviews of papers on the chronology of the Hammurabi age (1929, 1939 and 1941). In these reviews Neugebauer time and again warns against overestimating the importance of astronomical chronology, emphasizing that astronomical observations can provide useful chronological constraints but that independent historical information is required to choose between the astronomically possible candidate chronologies. The prime example of this is formed by the Venus observations of Ammisaduqa.
- Review of J. Schaumberger, "Drittes Ergänzungsheft zu F.X. Kugler Sternkunde und Sterndienst in Babel" (1936a) in which Neugebauer compliments Schaumberger for finally correctly explaining the last columns of Babylonian lunar theory (see inset Babylonian Lunar Theory 1880–1950).
- "Uber eine Untersuchungsmethode astronomischer Keilschrifttexte" (1936b) in which Neugebauer introduces the use of diophantine equations for the analysis of Babylonian 'zigzag' functions. This technique has powerful applications in the dating and connecting of (in particular) lunar ephemerides.
- "Jahreszeiten und Tageslängen in der babylonischen Astronomie" (1936c) in which the modeling of solar motion in Babylonian lunar theory is further clarified.
- "Untersuchungen zur antiken Astronomie I—III". Definition of research program (I, Neugebauer 1938a), further development of diophantine equations (II, Neugebauer 1938b), explanation of lunar latitude models and eclipse theory (III, Neugebauer 1938c).
- "Studies in Ancient Astronomy VI—VIII". Correction of his earlier erroneous interpretation of column E in Babylonian lunar theory (VI, Neugebauer 1946), Babylonian calendar cycles and intercalation patterns (VII, Neugebauer 1945), and the use of the water clock in Babylonian astronomy (VIII, Neugebauer 1947)
- "Solstices and Equinoxes in Babylonian Astronomy during the Seleucid Period" (Neugebauer 1948), showing that all Seleucid records of solstice and equinox dates are computed rather than observed.
- "The Alleged Babylonian Discovery of the Precession of the Equinoxes". In this paper Neugebauer (1950) once and for all did away with the speculation that the Babylonian scholars had discovered the precession of the equinoxes.

In 1939 Neugebauer moved to Brown University, Providence, Rhode Island, USA and in 1947 the Department of the History of Mathematics was founded that turned Brown into the leading institution for the study of the history of the exact sciences (Swerdlow 1998). One of his colleagues at Brown, the mathematician P. J. Davis, gives a lively account of the social and scientific atmosphere at the Department and about Neugebauer's personality in his amusing book *Ancient Loons* (2012). He writes that Neugebauer

> had that soft appearance and low-decibel manner that I associate with Austrians. Inwardly he held firm opinions and prejudices, and would occasionally burst out in anger and irritation using English swear words that I felt were an uncomfortable translation from German language originals. Not unlike Mark Twain, he was a misanthrope; he perceived the human world as consisting largely of fools, knaves, and dupes …

And he describes the discussions at the Brown cafeteria where he often ate lunch with Neugebauer, not infrequently joined by other members of the Department or their visitors:

> The conversation was relaxed but lively, scholarly but usually very general, and was terminated when the last person finished his lunch. Neugebauer was not one to twiddle his spoon leisurely in a second cup of coffee.

He tells that Neugebauer had his roster of the Greats in his profession so that

> anyone who ate lunch with him would find out within a week which of the great names were really great and which were intellectual asses. As regards the past he thought that Copernicus was overrated—he called him Koppernickel. Kepler was much better, and he loved Arthur Koestler's popularization of Kepler in *The Sleepwalkers*. Claudius Ptolemy was a great hero. As regards contemporaries, he expressed his views candidly. I was occasionally shocked and have no desire to go public with them.

And he finally quotes from the preface to *Astronomical Cuneiform Texts* (1955) where Neugebauer expresses his respect to the shades of the scribes of *Enuma Anu Enlil*:

> By their untiring efforts they built the foundations for the understanding of the laws of nature which our generation is applying so successfully to the destruction of civilization. Yet they also provided hours of peace for those who attempted to decode their lines of thought two thousand years later.

Contemporaries and Juniors

Because of his singular talent, his drive and his high standard of research Neugebauer has dominated the field of Babylonian Mathematical Astronomy for about half a century. There was little room for others. From his contemporaries I want to mention the mathematician Bartel van der Waerden (1903–1996) and from his juniors Neugebauer's dedicated collaborator, the Assyriologist Abraham Sachs (1915–1983).

The Dutch mathematician Bartel Leendert van der Waerden (1903–1996), whose name is associated with "van der Waerden's theorem" and the "van der Waerden number", developed an early interest in the history of mathematics and astronomy. He got to know Neugebauer when they both were in Göttingen in the late 1920s and he attended Neugebauer's first lecture course on the history of ancient mathematics in 1927.[15]

After having established himself as one of the leading mathematicians of his generation van der Waerden turned to studying ancient mathematics and astronomy. He wrote a few interesting papers on Babylonian astronomy in the 1940s which were later included and updated in his very clear, stimulating and occasionally speculative book *Anfänge der Astronomie* (1956, in German), published in English as *Science Awakening II: The Birth of Astronomy* (1974).

His first paper on Babylonian astronomy was entitled "Zur babylonischen Planetenrechnung" (1941). This paper was published in the first and only issue of the journal *Eudemus* edited by Neugebauer and Archibald. In it van der Waerden clarifies the way in which the Babylonians refined the mean synodic month as their unit of time in planetary ephemerides by dividing it in 30 "theoretical days" (later named "tithis" by Neugebauer adopting the term from Indian Astronomy). As mentioned before, this had independently already been discovered by Pannekoek (1917).[16]

Van der Waerden (1942, 1946) also wrote two interesting papers on the computation of first and last visibilities of the Moon and planets. In these papers he developed an elegant method to compute the dates of first and last appearances of the planets, and then applied it to the interpretation of the Venus observations of Ammisaduqa which led him to conclude that of the four possible Old Babylonian Chronologies allowed by the Venus observations the solution Ammisaduqa 1 = 1582 BC (later baptized the "Short Chronology") is the astronomically preferred one.

As mentioned above, van der Waerden and Neugebauer knew each other from their Göttingen days in the late 1920s. Both were protégés of Richard Courant, one of the leaders of the Göttingen mathematicians (with David Hilbert and Emmy Noether). Although Neugebauer clearly respected van der Waerden's work, he was also quite critical about his sometimes unfounded speculations. To illustrate this I quote from HAMA (Neugebauer 1975, p. 464, n. 10)) where he takes issue with van der Waerden's discussion of Babylonian Venus ephemerides:

> It seems to me not only pointless, but seriously misleading to readers who are not in a position to control the primary sources to make such utterly fragmentary material the basis for far reaching historical conclusions and to formulate them as if they were established results.

Their personal relation had become somewhat strained after van der Waerden decided to remain in Germany and keep his position at the University of Leipzig during the Nazi regime (see Soifer 2009, p. 367–483).

In 1941 Neugebauer met the young assyriologist Abraham J. Sachs (1915–1983) and persuaded him to come to Brown as a Rockefeller foundation fellow. Sachs joined the faculty in 1947, and was appointed associate professor in 1949. He played

[15] This is mentioned by van der Waerden himself in an interview given in 1993 and published in 1997 in Notices of the American Mathematical Society (Dold-Samplonius 1997, 319).

[16] Of the two papers contained in this only issue of *Eudemus* the first was written by another Dutchman; it is a curious coincidence that this happened to be Pannekoek.

an important role in assembling, categorizing and selecting many texts in the British Museum and elsewhere that were eventually included and analyzed by Neugebauer in ACT. In the period under review here Sachs (1948) published an important paper on the classification of astronomical texts.

B S.† 2343—A 120-Year Perspective

In the pioneering paper by Strassmaier and Epping (1881) "Zur Entzifferung der astronomischen Tafeln der Chaldäer" the former, in his historical and philological introduction, mentions the existence of tablets with astronomical texts, some of which contain long lists of numbers. As a representative example he mentions the tablet B.S.† 2343. In the second part of the paper Epping attempts to interpret another such text Sp 129 (the best preserved one according to Strassmaier) but B.S.† 2343 is never mentioned again; it turns out for good reasons because it took more than a century to fully understand its contents.

The tablet that Strassmaier in 1881 refers to as B.S.† 2343 is part of a collection that was shipped in 1876 from Iraq to London by George Smith (1840–1876), assistant in the Assyrian Department of the British Museum. The shipment arrived at the British Museum a few months after Smith had died in Aleppo of dysentery. It was recorded as arriving at the British Museum on 17 November 1876. The tablet registered as nr. 2343 in that collection, is presently known as BM 32599 (Fig. 13), and was included in Neugebauer (1955) as ACT 1050 (the one but last text in ACT). Neugebauer mentions that Strassmaier, according to his notebook, thought that "it probably concerned rising and setting of the Moon", while Kugler added remarks to the effect that it should be completed to 12 columns "thus referring to 12 months". Based on his analysis of the arithmetical structure of the text (see Fig. 14) Neugebauer suggested that it contained longitudes at rising and setting of

Fig. 13 Photograph of tablet BM 32599 (= S† 76-11-17, 2343) (Courtesy the Trustees of the British Museum, London)

some planet, probably Mercury although some of the parameters could apply to Venus as well.

It took more than 20 years before the text was correctly interpreted when Norman Hamilton told Asger Aaboe that he had identified ACT 1050 as a text giving longitudes of synodic phenomena of Venus computed according to system A (the first and only one of its kind), and another 20 years before Aaboe published the results (Hamilton and Aaboe 1998). Norman T. Hamilton (1927–1996) was a gifted mathematician who had introduced himself to Aaboe after a lecture in Chicago in 1978 with the words: "I am not a crank, but I have cracked ACT No. 60"; Aaboe notes "He certainly wasn't, and he certainly had".

Shortly after Hamilton and Aaboe's paper was published, John Britton (1938–2010) during a visit to the British Museum reexamined the tablet and corrected a misreading in Neugebauer's transcription of the tablet, so that the only missing parameter in the computational scheme could be determined. With all parameters known the text could now be completely reconstructed (Britton 2001).

Thus S† 76-11-17, 2343 is the remaining right-hand side of either one large tablet, or of the second one of a set of two tablets, containing the longitudes of all four synodic phases (Evening Last, Morning First, Morning Last and Evening First, in that order) of Venus covering a period of 230 years, from 419 to 189 BC or from 184 BC to 47 AD according to Britton (2001).

This case is a fine illustration of how much our knowledge of Babylonian astronomy has increased in the past 120 years, from the moment that Strassmaier first identified the tablet as an astronomical text in the late 1870s to its definitive interpretation in 2001.

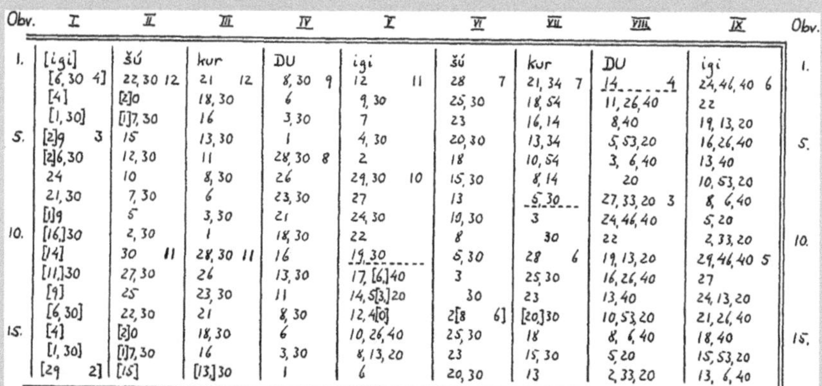

Obv.	I	II	III	IV	V	VI	VII	VIII	IX	Obv.
1.	[igi]	šú	kur	DU	igi	šú	kur	DU	igi	1.
	[6,30 4]	22,30 12	21 12	8,30 9	12 11	28 7	21,34 7	14 - - - - 4	24,46,40 6	
	[4]	[2]0	18,30	6	9,30	25,30	18,54	11,26,40	22	
	[1,30]	[1]7,30	16	3,30	7	23	16,14	8,40	19,13,20	
5.	[2]9 3	15	13,30	1	4,30	20,30	13,34	5,53,20	16,26,40	5.
	[2]6,30	12,30	11	28,30 8	2	18	10,54	3,6,40	13,40	
	24	10	8,30	26	29,30 10	15,30	8,14	20	10,53,20	
	21,30	7,30	6	23,30	27	13	5,30 - - - -	27,33,20 3	8,6,40	
	[1]9	5	3,30	21	24,30	10,30	3	24,46,40	5,20	
10.	[16,]30	2,30	1	18,30	22	8	30	22	2,33,20	10.
	[14]	30 11	28,30 11	16	19,30 - - - -	5,30	28 6	19,13,20	29,46,40 5	
	[11,]30	27,30	26	13,30	17,[6,]40	3	25,30	16,26,40	27	
	[9]	25	23,30	11	14,5[3,]20	30	23	13,40	24,13,20	
	[6,30]	22,30	21	8,30	12,4[0]	2[8 6]	[20,]30	10,53,20	21,26,40	
15.	[4]	[2]0	18,30	6	10,26,40	25,30	18	8,6,40	18,40	15.
	[1,30]	[1]7,30	16	3,30	8,13,20	23	15,30	5,20	15,53,20	
	[29 2]	[15]	[13,]30	1	6	20,30	13	2,33,20	13,6,40	

Fig. 14 Transcription of the obverse of tablet S† 76-11-17, 2343 (= BM 32599=ACT 1050) in Neugebauer's characteristic precise and clear hand writing (From Neugebauer 1955)

Epilogue

In 1941 the Dutch astronomer and historian of astronomy Anton Pannekoek published a paper in the first and only issue of the journal *Eudemus* (of which Neugebauer was one of the editors) entitled: "Some remarks on the Moon's Diameter and the Eclipse Tables in Babylonian Astronomy" (Pannekoek 1941). In this paper he identifies a mistake—a rare occasion—in Neugebauer's lengthy 1938 paper on the Babylonian theory of the lunar latitude and eclipse magnitudes[17] and he is critical about Neugebauer's modern anachronistic view of the development of Babylonian astronomy. He first quotes Neugebauer and then expresses his own views. I present this here because Pannekoek's point of view has a great deal of actuality and at the same time emphasizes the strength and the limitations of the approach of Neugebauer. The passages in Neugebauer (1938c) that Pannekoek takes issue with read:

> Hence the grouping of eclipses, given by Pannekoek, is seen to be an automatic consequence of the computing rule of E″, so that it is not necessary to consider the list of text 200 [= Sp. II, 71 = BM 34597] as a protocol of observed and non-observed eclipses, and it is no 'saros-canon' but simply represents the computed column T (year and month) of an eclipse text of the same type as text 107 [SH 81-7-6, 93 = BM 45688 = ACT 60]. Thus the only text that could be looked at as a witness of the first stage of the saros-method, has been entirely inserted into the context of the well-known theory of system II.

and a bit later:

> The situation, hence, in Babylonian astronomy is the same as with Ptolemy, who also first develops the theory of the moon's motion and then its consequence determines the possible distances between succeeding eclipses.

Pannekoek, who as an astronomer, knows that the knowledge of the skies should ultimately be based on observation, takes issue with these statements. He writes:

> It seems to me that the basis of our differences about the meaning of text 200 and the development of Chaldaean astronomy lies in a different view on the mentality of the early astronomers. Neugebauer's argumentation is logical, if we consider them solely as primitive scientists, with minds and a mode of thinking analogous to ours, qualitatively the same, but quantitatively only in its beginnings, so that we see here the first origin and gradual growth of that which is familiar as scientific spirit and methods.

and a bit further on:

> In these astronomical observations there is no trace of what we call a scientific view or aim; at that time as in the whole of antiquity the phenomena of nature are not treated with the concept of causality but of finality. They are not considered as cause and result, but as sign, omen and meaning.
>
> Hence no regularity or law is sought for. But regularities impose themselves without giving surprise. The regular returns of the lunar aspects were known of old. Now, during the high tide of astrological observation, new regularities, in the planets, in the eclipses, gradually fix themselves in the consciousness.

[17] Acknowledged and corrected in Neugebauer 1945.

And he goes on sketching the gradual development of Babylonian astronomy, all the way to its most sophisticated end product: the ephemerides, the analysis and interpretation of which became Neugebauer's 'lebenswerk'.

This appeal by Pannekoek reminds us that Neugebauer was first and foremost a mathematician and much less of an astronomer.

It seems to me that nowadays, looking back over 130 years of research in Babylonian Mathematical Astronomy, there is more room for Pannekoek's evolutionist's views than in the second half of the twentieth century, the period that was dominated by Neugebauer's scholarship.

Acknowledgements I would like to thank Dr. Clemens Brodkorb, Archivleiter of the Archiv der Norddeutsche Provinz der Societas Jesu in Munich for his assistance and the access to relevant materials in the Archiv. I further thank Peter Huber, John Steele and Christopher Walker for their critical reading of the paper and numerous useful suggestions for its improvement. Part of the research for this paper was carried out during a 3-month visit to the Department of Egyptology and Ancient West Asian Studies at Brown University. I would like to thank Professor John Steele and the staff of the Department for their hospitality and support, and the Netherlands Organization for Scientific Research NWO and the Leids Kerkhoven-Bosscha Fonds for financially supporting my visit.

Bibliography

Anon. 1919. *25 Jahre Ignatiuskolleg Valkenburg 1894–1919*. Freiburg: Buchdruckerei der Herdersche Verlagshandlung.

Baumgartner, A. 1894. Joseph Epping †. *Zeitschrift für Assyriologie* 9: 427–433.

Britton, J.P. 2001. Remarks on a system A text for Venus: ACT 1050. *Archive for the History of Exact Sciences* 55: 525–554.

Davis, P.J. 2012. *Ancient loons*. Boca Raton: CRC Press.

Deimel, A. 1920. P. Johann Nepomuk Strassmaier S.J. †. *Orientalia* 1: 5–10.

Dold-Samplonius, Y. 1997. Interview with Bartel Leendert van der Waerden. *Notices of the American Mathematical Society* 44: 313–320.

Epping, J. 1889. Aus einem Briefe des Herrn Professor J. Epping. *Zeitschrift für Assyriologie* 4: 76–82.

Epping, J. 1890a. Die babylonische Berechnung des Neumondes. *Stimmen aus Maria Laach* 39: 225–240.

Epping, J. 1890b. Sachliche Erklärung des Tablets No. 400 der Cambyses-Inschriften. *Zeitschrift für Assyriologie* 5: 281–288.

Epping, J., and J.N. Strassmaier. 1889. Astronomisches aus Babylon. *Stimmen aus Maria-Laach, Ergänzungshefte* 44.

Epping, J., and J.N. Strassmaier. 1890. Neue Babylonische Planeten-Tafeln I. *Zeitschrift für Assyriologie* 5: 341–366.

Epping, J., and J.N. Strassmaier. 1891a. Neue Babylonische Planeten-Tafeln II. *Zeitschrift für Assyriologie* 6: 89–102.

Epping, J., and J.N. Strassmaier. 1891b. Neue Babylonische Planeten-Tafeln III. *Zeitschrift für Assyriologie* 6: 217–244.

Epping, J., and J.N. Strassmaier. 1892. Babylonische Mondbeobachtungen aus den Jahren 38 und 79 der Seleuciden-Aera. *Zeitschrift für Assyriologie* 7: 220–254.

Epping, J., and J.N. Strassmaier. 1893. Der Saros-Canon der Babylonier. *Zeitschrift für Assyriologie* 8: 149–178.
Esch, M. 1929. Franz Xaver Kugler. *Vierteljahsschrift der Astronomischen Gesellschaft* 64: 294–301.
Gross, M.B. 2004. *The war against Catholicism: Liberalism and the anti-Catholic imagination in nineteenth-century Germany*. Ann Arbor: University of Michigan Press.
Hamilton, N.T., and A. Aaboe. 1998. A Babylonian Venus text computed according to System A: ACT No. 1050. *Archive for the History of Exact Sciences* 53: 215–221.
Huber, P.J., and J.M. Steele. 2007. Babylonian lunar six tables. *SCIAMVS* 8: 3–36.
Hunger, H., and D. Pingree. 1999. *Astral sciences in Mesopotamia*. Leiden: Brill.
Hunger, H., A. J. Sachs and J. M. Steele. 1988–2006. *Astronomical diaries and related texts from Babylonia*, Vols. I–III, V and VI. Vienna: Verlag der Österreichischen Akademie der Wissenschaften.
Kugler, F.X. 1900. *Die Babylonische Mondrechnung*. Freiburg im Breisgau: Herder'sche Verlagshandlung. [BMR].
Kugler, F.X. 1907. *Sternkunde und Sterndienst in Babel, I. Buch: Babylonische Planetenkunde*. Münster in Westfalen: Aschendorfsche Verlagsbuchhandlung. [SSB I].
Kugler, F.X. 1909. Auf den Trümmern des Panbabylonismus. *Anthropos* 4: 477–499.
Kugler, F.X. 1909/10. *Sternkunde und Sterndienst in Babel, II. Buch: Natur, Mythus und Geschichte als Grundlagen babylonischer Zeitordnung nebst Untersuchungen der älteren Sternkunde und Meteorologie*, I. Teil. Münster in Westfalen: Aschendorfsche Verlagsbuchhandlung. [SSB II-1].
Kugler, F.X. 1910. *Im Bannkreis Babels*. Münster in Westfalen: Aschendorfsche Verlagsbuchhandlung.
Kugler, F.X. 1912. *Sternkunde und Sterndienst in Babel, II. Buch: Natur, Mythus und Geschichte als Grundlagen babylonischer Zeitordnung nebst Untersuchungen der älteren Sternkunde und Meteorologie*, II Teil, 1 Heft. Münster in Westfalen: Aschendorfsche Verlagsbuchhandlung. [SSB II-2.1].
Kugler, F.X. 1913. *Sternkunde und Sterndienst in Babel, Ergänzungen zum ersten und zweiten Buch, I Teil, 1-VIII Abhandlung: Astronomie und Chronologie der älteren Zeit*. Münster in Westfalen: Aschendorfsche Verlagsbuchhandlung. [SSB Erg. 1].
Kugler, F.X. 1914. *Sternkunde und Sterndienst in Babel, Ergänzungen zum ersten und zweiten Buch, II Teil, IX-XIV Abhandlung: Sternkunde und Chronologie der älteren Zeit*. Münster in Westfalen: Aschendorfsche Verlagsbuchhandlung. [SSB Erg. 2].
Kugler, F.X. 1922. *Von Moses bis Paulus; Forschungen zur Geschichte Israels*. Münster in Westfalen: Aschendorfsche Verlagsbuchhandlung.
Kugler, F.X. 1924. *Sternkunde und Sterndienst in Babel, II. Buch: Natur, Mythus und Geschichte als Grundlagen babylonischer Zeitordnung nebst Untersuchungen der älteren Sternkunde und Meteorologie, II Teil, 2 Heft*. Münster in Westfalen: Aschendorfsche Verlagsbuchhandlung. [SSB II-2.2].
Kugler, F.X. 1927. *Sibyllinische Sternkampf und Phaeton in naturgeschichtlicher Beleuchtung*. Münster in Westfalen: Aschendorfsche Verlagsbuchhandlung.
Minnaert, M. 1974. Antonie Pannekoek. *Dictionary of Scientific Biography* 10: 289–291.
Neugebauer, O. 1929. Zur Frage der astronomischen Fixierung der babylonischen Chronologie. *Orientalistische Literatur Zeitung* 32: 914–922.
Neugebauer, O. 1936a. J. Schaumberger, Drittes Ergänzungsheft zu F.X. Kugler, S.J., Sternkunde und Sterndienst in Babel. *Quellen und Studien zur Geschichte der Mathematik* B3: 271–286.
Neugebauer, O. 1936b. Über eine Untersuchungsmethode astronomischer Keilschrifttexte. *Zeitschrift der Deutschen Morgenländischen Gesellschaft* 90: 121–134.
Neugebauer, O. 1936c. Jahreszeiten und Tageslängen in der babylonischen Astronomie. *Osiris* 2: 517–550.
Neugebauer, O. 1938a. Untersuchungen zur antiken Astronomie I. *Quellen und Studien zur Geschichte der Mathematik* B4: 29–33.

Neugebauer, O. 1938b. Untersuchungen zur antiken Astronomie II. Datierung und Rekonstruktion von Texten des Systems der Mondtheorie. *Quellen und Studien zur Geschichte der Mathematik* B4: 34–91.
Neugebauer, O. 1938c. Untersuchungen zur antiken Astronomie III. Die babylonische Theorie der Breitenbewegung des Mondes. *Quellen und Studien zur Geschichte der Mathematik* B4: 193–346.
Neugebauer, O. 1939. Chronologie und babylonischer Kalender. *Orientalistische Literatur Zeitung* 42: 404–414.
Neugebauer, O. 1941. The chronology of the Hammurabi age. *Journal of the American Oriental Society* 61: 58–61.
Neugebauer, O. 1945. Studies in ancient astronomy VII. Magnitudes of lunar eclipses in Babylonian mathematical astronomy. *Isis* 36: 10–15.
Neugebauer, O. 1946. Studies in ancient astronomy VI. The 'Metonic Cycle' in Babylonian Astronomy. In *Studies and essays in the history of science and learning offered in homage to George Sarton*, 435–448.
Neugebauer, O. 1947. Studies in ancient astronomy VIII. The water clock in Babylonian astronomy. *Isis* 37: 37–43.
Neugebauer, O. 1948. Solstices and equinoxes in Babylonian astronomy during the Seleucid period. *Journal of Cuneiform Studies* 2: 209–222.
Neugebauer, O. 1950. The alleged Babylonian discovery of the precession of the equinoxes. *Journal of the American Oriental Society* 70: 1–8.
Neugebauer, O. 1955. *Astronomical Cuneiform texts*, 3 vols. London: Lund Humphries. [ACT].
Neugebauer, O. 1975. *A history of ancient mathematical astronomy*, 3 vols. Berlin: Springer-Verlag. [HAMA].
Pannekoek, A. 1917. Calculation of dates in the Babylonian tables of planets. *Proceedings Koninklijke Akademie van Wetenschappen, te Amsterdam* 19: 684–703.
Pannekoek, A. 1918. The origin of the Saros. *Proceedings Koninklijke Akademie van Wetenschappen, te Amsterdam* 20: 943–955.
Pannekoek, A. 1941. Some remarks on the Moon's diameter and the eclipse tables in Babylonian astronomy. *Eudemos I* 9–22.
Pannekoek, A. 1961. *A history of astronomy*. New York: Interscience Publishers, Inc.
Pollen, J.H. 1920. Father John Strassmaier, S.J., Assyriologist. *The Month* 135: 137–145.
Sachs, A. 1948. A classification of the Babylonian astronomical tablets of the Seleucid period. *Journal of Cuneiform Studies* 2: 271–290.
Sachs, A. 1955. *Late Babylonian astronomical and related texts*. Providence: Brown University Press. [LBAT].
Sachs, J., and G.J. Toomer. 1979. Otto Neugebauer, bibliography 1925–1979. *Centaurus* 22: 257–280.
Schaumberger, J. 1933. Drei babylonische Planetentafeln der Seleukidenzeit. *Orientalia* 7: 97–116.
Schaumberger, J. 1935. *Sternkunde und Sterndienst in Babel, 3. Ergänzungsheft zum ersten und zweiten Buch*. Münster in Westfalen: Aschendorfsche Verlagsbuchhandlung. [SSB Erg. 3].
Schnabel, P. 1923. *Berossos und die babylonisch-hellenistische Literatur*. Leipzig: Teubner.
Schnabel, P. 1924. Neue babylonische Planetentafeln. *Zeitschrift für Assyriologie* 35: 99–112.
Schnabel, P. 1927. Kidenas, Hipparch und die Entdeckung der Präzession. *Zeitschrift für Assyriologie* 37: 1–60.
Soifer, A. 2009. *The mathematical coloring book*. Berlin: Springer.
Strassmaier, J.N. 1886. *Alphabetisches Verzeichniss der Assyrischen und Akkadischen Wörter der Cuneiform Inscriptions of Western Asia vol. II*. Leipzig: Hinrichs'sche Buchhandlung
Strassmaier, J.N. 1888. Arsaciden Inschriften. *Zeitschrift für Assyriologie* 3: 129–158.
Strassmaier, J.N. 1890. *Inschriften von Cambyses, König von Babylon*. Leipzig.
Strassmaier, J.N. 1892. Einige chronologische Daten aus astronomischen Rechnungen. *Zeitschrift für Assyriologie* 7: 197–204.

Strassmaier, J.N. 1893. Zur Chronologie der Seleuciden. *Zeitschrift für Assyriologie* 8: 106–113.
Strassmaier, J.N. 1895. Der Saros-Canon Sp. II, 71. *Zeitschrift für Assyriologie* 10: 64–69.
Strassmaier, J.N., and J. Epping. 1881. Zur Entzifferung der astronomischen Tafeln der Chaldäer. *Stimmen auss Maria-Laach* 21: 277–292.
Surhone, L.M., M.T. Tennoe, and S.F. Henssonow. 2010. *Panbabylonism*. Saarbrücken: VDM Verlag Dr. Müller AG & Co.
Swerdlow, N.M. 1998. *Otto E. Neugebauer 1899–1990, A biographical memoir*. Washington, DC: National Academies Press.
van der Waerden, B.L. 1941. Zur babylonischen Planetenrechnung. *Eudemus* 1: 23–48.
van der Waerden, B.L. 1942. Die Berechnung der ersten und letzten Sichtbarkeit von Mond und Planeten und die Venustafeln von Ammisaduga. *Berichte der Sächsische Akademie derWissenschaften*, math.-phys. Kl. 94: 23–56.
van der Waerden, B.L. 1946. On Babylonian astronomy I. The Venus tablets of Ammisaduqa. *Jaarbericht Ex Oriente Lux* 10: 414–424.
van der Waerden, B.L. 1974. *Science awakening II. The birth of astronomy*. Leyden: Noordhoff International Publishing.
Wallis Budge, E.A.W. 1925. *The rise and progress of Assyriology*. London: Martin Hopkinson Ltd.
Weidner, E. 1912. Zum Alter der babylonischen Astronomie. *Babylonaica* 6: 129–133.
Weidner, E. 1955. Johann Schaumberger. *Archiv für Orientforschung* 17: 490–491.
Weiss, O. 1995. Johann Baptist Clemens Schaumberger. *Biographisch-Bibliographishes Kirchenlexikon* IX, spalten 22–23.

Neugebauer's *Astronomical Cuneiform Texts* and Its Reception

John M. Steele

Introduction

Neugebauer's *Astronomical Cuneiform Texts*, published in 1955, defined the field of Babylonian astronomy for most of the second half of the twentieth century. *Astronomical Cuneiform Texts*, or ACT as it is generally referred to, contains editions of more than three hundred cuneiform tablets dealing with mathematical astronomy, each accompanied by a detailed commentary. In addition, the book contains a historical investigation of the date and provenance of the tablets, the scribes mentioned in colophons, and an extensive mathematical introduction to the lunar and planetary schemes found on the tablets. In his review of ACT, the Assyriologist A. Leo Oppenheim wrote that he "can only pay homage in a few trite phrases to the amount of devotion, patience, and scholarship which has gone into this difficult work, to which the author dedicated twenty years of his life", and that the book "ushers in the second phase in the development of our understanding of Babylonian astronomy".[1] This paper begins by tracing the history of ACT from its conception in Copenhagen during the mid-1930s to its publication two decades later by which time Neugebauer had crossed the Atlantic and was well established at Brown University. The second part of the paper discusses the reception of ACT among historians of science and Assyriologists and its impact upon the study of Babylonian astronomy.

[1] Oppenheim 1958, 157.

J.M. Steele (✉)
Department of Egyptology and Assyriology, Brown University,
1899, Providence, RI 02912, USA
e-mail: john_steele@brown.edu

A. Jones et al. (eds.), *A Mathematician's Journeys*, Archimedes 45,
DOI 10.1007/978-3-319-25865-2_10

The Twenty-Year Journey to *Astronomical Cuneiform Texts*

By the mid-1930s Neugebauer was well established as the leading figure in the study of Babylonian mathematics. He had published a number of detailed papers on the topic in the "B" series of *Quellen und Studien zur Geschichte der Mathematik, Astronomie und Physik* and was finalizing the manuscript of *Mathematische Keilschrift-Texte* (MKT), containing editions of all mathematical texts known to him with detailed commentaries, which would appear between 1935 and 1937 in three parts as volume 3 of the "A" series of *Quellen und Studien*. At that time, Neugebauer was in Copenhagen, and he gave a series of lectures in 1934 on Babylonian and Egyptian mathematics, based upon his work for MKT and his earlier studies of Egyptian mathematics.[2] These lectures would form the basis for a short book on "vorgriechische Mathematik" that Neugebauer wrote that year. The full title of the book was *Vorlesungen über Geschichte der antiken mathematischen Wissenschaften, Erster Band: Vorgriechische Mathematik*, and Neugebauer explained in the preface that the *Vorlesungen* would extend to three volumes: the present volume on pre-Greek mathematics, a second volume on Greek mathematics, and the third on astronomy. He describes his intention for the astronomy volume as follows:

> The third volume will deal with exact astronomy, that is above all with the fundamental and not highly enough regarded work of Ptolemy on the one hand and with the more difficult and less accessible, although relatively late, Babylonian astronomy.[3]

Neugebauer had not previously published anything relating to the history of astronomy except for a review of Langdon, Fotheringham and Schoch's *The Venus Tablets of Ammizaduga* (an attempt to use a tablet containing a series of celestial omens to date the Old Babylonian period of which Neugebauer was rightly skeptical), and a short response appended to an article by Fotheringham entitled "The Indebtedness of Greek to Chaldaean Astronomy" published in volume 2 of series B of the *Quellen und Studien* in 1933. Nevertheless, it is clear that Neugebauer was already well acquainted with the work of Epping and Kugler on Babylonian astronomy. In his *3. Ergänzungsheft zum ersten und zweiten Buch* of Kugler's *Sternkunde und Sterndienst in Babel*, published in 1935, Schaumberger thanked Neugebauer for providing a collation of the first ten columns of the lunar ephemeris SH 272 = BM 34580 and a sketch of the placement of the fragments of this tablet.[4] Neugebauer subsequently wrote a long and detailed review of Schaumberger's book which shows that he was fully conversant with Babylonian astronomy.[5]

---

[2] See the chapter "Otto Neugebauer's Visits to Copenhagen and His Connection to Denmark" by Brack-Bernsen in this volume.

[3] Neugebauer 1934, viii.

[4] Schaumberger 1935, 375–376. Neugebauer's sketch is published as plate 17. It is not clear whether Neugebauer's collation was based upon direct inspection of the tablet or (probably more likely) on a photograph.

[5] Neugebauer 1936a.

In 1936 Neugebauer gave a series of lectures on Babylonian mathematical astronomy.[6] Several copies of his lecture notes, handwritten in German, still exist including a set held in the Science Library of Brown University. The "Vorlesungen über babylonische Astronomie" is split into two parts, the first on the calculation of the new moon and the second on eclipse theory. Planetary theory is not discussed in these notes. The "Vorlesungen über babylonische Astronomie" are based upon Epping, Kugler and Schaumberger's publications, but Neugebauer introduced his own mathematical approach to the material such as drawings of zigzag and step functions and generalized algebraic presentations of period relations, zigzag functions and step functions. As he mentioned in the preface to the first volume of the *Vorlesungen*, his interest is in the "exact astronomy", exact in the sense of "exact science", in other words mathematical astronomy. It is clear from these first notes on Babylonian astronomy that for Neugebauer the history of astronomy is essentially the history of mathematics—it is the *mathematical* structure of astronomical theories that he will investigate rather than their *astronomical* basis, an approach which he followed in almost all of his investigations of the history of astronomy throughout his career.

Neugebauer evidently saw the potential for the type of mathematical analysis of cuneiform astronomical texts which he could undertake and turned all his attention towards it. Once MKT was seen through the press, Neugebauer published nothing further on Babylonian mathematics for almost a decade except for a few general reviews, and apparently did little or no work on Greek mathematics, which was the intended subject of the second volume of the *Vorlesungen*. Almost all of his research energy was focused on Babylonian astronomy. This began with two papers published in 1936: a short paper in the *Zeitschrift der Deutschen Morgenländischen Gesellschaft* with the title "Über eine Untersuchungsmethode astronomische Keilschrifttexte",[7] which set out the mathematical techniques he had developed for analyzing Babylonian mathematical astronomical texts, and a much longer paper in *Osiris* on "Jahreszeiten und Tageslängen in der babylonischen Astronomie",[8] which investigated the functions for length of daylight in the two lunar systems. Both papers used many of the same techniques—and some of the same diagrams—that appear in the "Vorlesungen über babylonische Astronomie" notes.

By 1938 Neugebauer had decided that in order to make progress in the study of Babylonian astronomy it would be necessary to produce the equivalent of MKT for the astronomical material. Epping, Kugler and Schaumberger's work on Babylonian astronomy had been groundbreaking, but their work proceeded from the small number of astronomical cuneiform tablets that had been identified by Strassmaier, which included observational texts, texts containing predictions made using the goal-year periods, and astrological texts as well as the texts of mathematical astronomy. The pioneers of the study of Babylonian astronomy were faced with understanding and

[6] See further the chapter "Otto Neugebauer's Visits to Copenhagen and His Connection to Denmark" by Brack-Bernsen in this volume.

[7] Neugebauer 1936b.

[8] Neugebauer 1936c.

classifying all of these different types of text and so it should not be surprising that their works were not presented in a systematic fashion. But now that the known texts had been published and analyzed for the first time, it would be desirable to restudy and reedit all of the texts within each category of astronomical text. This work could build on the foundations laid by Epping, Kugler and Schaumberger and benefit from the techniques of mathematical analysis that Neugebauer had devised. Writing in 1975 Neugebauer described the task that faced him when he started to work on the astronomical texts in the mid-1930s:

> When I started work on the astronomical texts, I had, of course, a general knowledge of Kugler's great work. I considered it as my main, and comparatively simple, task to extract from Kugler's material the mathematically oriented texts, i. e. the ephemerides and the procedure texts, supplemented by texts from Uruk which had more recently become accessible by the publications of Thureau-Dangin (texts in the Louvre) and Schnabel (Berlin Museum).
>
> It was evident from the outset that any summary of Kugler's work had to present his brilliant results in a form mathematically more concise than in the original publication which preserved many of the often involved, however ingenious, ways of actual discovery. Also the practical task of restoring damaged passages and sections of texts made it imperative to operate as systematically as possible. Consequently I developed checking methods for all numerical columns in ephemerides, based upon the simple idea of representing periodic sequences of fixed amplitude by monotone sequences in infinitely many strips. This method provided, at the same time, information about the size of the gap between related fragments of ephemerides.[9]

Neugebauer set out his manifesto for the study of Babylonian astronomy in a short article entitled "Untersuchungen zur antiken Astronomie I", which appeared in volume 4 of *Quellen und Studien* series B in 1938. Two tasks were required: a systematic approach to the study of the astronomical texts, and the preparation of editions of the texts with a rigour comparable to that found in the editions of Greek astronomical texts by Hultsch, Heiberg, Manitius and Tannery. Neugebauer himself would undertake the preparation of the editions of the mathematical astronomical texts, but this was only part of a planned bigger project to publish editions of all types of Babylonian astronomical texts:

> This edition of the astronomical texts of mathematical character is subordinated within a broader plan. Herren L. Hartmann, J. Schaumberger and A. Schott and if necessary other colleagues will edit and publish all the other classes of available astronomical texts, observational texts to astronomical-astrological texts, so we hope that together, finally, a fully complete collection of source material of Babylonian astronomy can be presented.[10]

From the beginning, therefore, Neugebauer saw the importance of the non-mathematical astronomical texts. He never, I believe, lost sight of this, but rather saw that it was in the study of the mathematical astronomical texts where he could make a meaningful contribution.

The plan described by Neugebauer would see him take responsibility for the mathematical astronomical texts while Hartmann, Schaumberger and Schott dealt

---

[9] Neugebauer 1975, 432.

[10] Neugebauer 1938, 30.

with the remainder of the astronomical and astrological material. Neugebauer did not give any detail about the division of labours between his colleagues. Some suggestions can be made, however. Schaumberger had continued Kugler's programme of research on Babylonian astronomy, in particular focusing on the Babylonian observational and predictive texts such as the Astronomical Diaries and the Goal-Year Texts as well as on the mathematical astronomical texts. Since this latter group of texts were to be edited by Neugebauer, it is likely that Schaumberger was planning to take responsibility for the observational and predictive texts. Albert Schott's contributions to the study of Babylonian astronomy are now more or less forgotten, but in the 1930s and 1940s he published two papers with Schaumberger on astronomical records on the Neo-Assyrian letters, a paper on the development of positional astronomy from Assyrian times down to the Hellenistic period and some short notes on terminology. Schott was a Professor of Assyriology at Bonn who published an important study on Gilgamesh and had worked with Julius Jordan at the site of Uruk. At Bonn, Schott participated in the history of mathematics seminar founded by Otto Toeplitz (one of Neugebauer's co-editors of *Quellen und Studien*).[11] Given Schott's interest in the Assyrian material, it seems reasonable to suppose that he was to work on the Neo-Assyrian letters and reports, and perhaps on other Assyrian texts such as the many celestial omen texts from Nineveh. The identity and planned work of L. Hartmann remains uncertain.[12] Neugebauer was the only one of the scholars mentioned in this manifesto to complete their assignment.

Neugebauer continued the manifesto by explaining that his study of Babylonian mathematical astronomy would have three strands. First, there was the edition of the texts, which would be published in two parts, one containing the texts, the second containing a commentary. This publication was to be called "mathematische-astronomischen Keilschrifttexte", to be abbreviated to MAKT. Second were to be a series of papers, published under the general title "Untersuchungen zur antiken Astronomie" which would contain detailed studies of particular problems of mathematical astronomy.[13] Finally, there would be the overall discussion of Babylonian astronomy in the planned third volume of the *Vorlesungen*. The second and third volumes of the *Vorlesungen* were never written, although something of the intention of the third volume can be seen in Neugebauer's 1975 *A History of Ancient Mathematical Astronomy* (HAMA), although this work is larger in scale than what he had planned for the *Vorlesungen*.

[11] Neuenschwander 1993, 387.

[12] I am unable to positively identify this man, although I think it is probable that it is Louis F. Hartman (note the single n at the end of the name), a biblical scholar and Assyriologist who collaborated with A. Leo Oppenheim on the publication of several cuneiform texts. A "Hartman" is also thanked for a collation of a Neo-Assyrian letter in the British Museum by Schott and Schaumberger 1941, 156.

[13] Five articles in this series (including the manifesto as paper I) were published in *Quellen und Studien*. Four further papers, written in English under the title "Studies in Ancient Astronomy" but continuing the numbering of the German papers, were published in a memorial volume for George Sarton and in the journal *Isis*.

Neugebauer published three further papers on Babylonian astronomy in the "Untersuchungen zur antiken Astronomie" series in the fourth volume of series B of the *Quellen und Studien*. All three papers were written in Copenhagen in 1938 and dealt with aspects of the two Babylonian lunar theories named by Kugler "System I" and "System II". Following his move to the USA early in 1939, Neugebauer began publishing his work in English. His first few English-language papers included general surveys and methodological statements about the study of the ancient exact sciences, as well as short notes on Old Babylonian chronology and the use of the cuneiform "zero" sign in astronomical texts. But it was not until 1945 that Neugebauer published his next major study of Babylonian astronomical texts, an examination of a column found in both lunar systems which relates to the magnitude of lunar eclipses.

Several important changes to his understanding of Babylonian mathematical astronomy and his approach to writing about it happened around the same time of Neugebauer's move to the USA and his change from writing in German to English. First and foremost, the change of language required both a change of title and of acronym for his planned edition of the texts: "Mathematische-astronomischen Keilschrifttexte" (MAKT) became "Astronomical Cuneiform Texts" (ACT). Consciously or not, Neugebauer had changed the title of his work from referring to "mathematical astronomy" to just "astronomy", a change which might suggest to the uninformed reader that what was contained in Neugebauer's book was *all* of Babylonian astronomy, not just Babylonian mathematical astronomy. The title "Astronomical Cuneiform Texts" appears already in 1941 on the inside front cover of the first volume of the journal *Eudemus* among a list of forthcoming publications.[14]

A second major change which occurred between Neugebauer's German period and English period was his abandonment of Kugler's "System I" and "System II" in favour of new designations "System A" and "System B", where Neugebauer's System A corresponds to Kugler's System II and Neugebauer's System B corresponds to Kugler's System I. Neugebauer first used these new names for the two systems in a paper entitled "Some Fundamental Concepts in Ancient Astronomy" presented at the University of Pennsylvania's Bicentennial Conference and

[14] *Eudemus* ceased publication after one volume, despite Neugebauer's later attempts to revive the journal. The first volume contains two papers on Babylonian astronomy, one by van der Waerden and the other by Pannekoek. It is clear from the inside front cover of this issue, however, that Neugebauer and his co-editor Raymond Clare Archibald, intended *Eudemus* to be the successor to *Quellen und Studien* and like its predecessor to publish both studies and sources. Five works (all by Neugebauer and his colleagues or his student) are listed as being planned for publication by the journal: The Anaphorikos of Hypsikles by M. Krause, V. De Falco and O. E. Neugebauer (published with a revised author order in the series *Abhandlungen der Akademie der Wissenchaften in Göttingen* in 1966), Egyptian Planetary Texts by Neugebauer (published by the American Philosophical Society in 1942), Mathematical Cuneiform Texts From American Collections (published as *Mathematical Cuneiform Texts* by Neugebauer and A. Sachs in the American Oriental Society Series in 1945), Studies in Greek Spherics by O. Schmidt (Schmidt's Brown University PhD dissertation which remains unpublished to this day), as well as Astronomical Cuneiform Texts.

published in 1941 in the volume *Studies in the History of Civilization*. Discussing the calculation of the length of daylight by using the rising arcs of the signs of the zodiac he begins:

> It is now a very natural question to ask about the corresponding theory in Babylonian astronomy. Here, however, nothing about rising times was known, but only the rules by which the length of the days was calculated during the seasons. Each of the two systems mentioned above has a scheme of its own. The older one gives (expressed here in degrees) as the lengths the following list A, the younger one B.[15]

Neugebauer footnoted the last sentence with the remark "Unfortunately Kugler reversed the order of the two systems by calling the older one II, the younger I". In his 1945 paper on the treatment of eclipse magnitudes in Babylonian lunar theory Neugebauer made a similar remark:

> It need only be recalled that Kugler already recognized the existence of two different methods for the computation of lunar ephemerides: an older one, here called "System A," and a more recent "System B."[16]

Again, Neugebauer noted Kugler's names for the two systems in a footnote. It is interesting that Neugebauer was willing to state that System A was older than System B in such definite terms. He would later criticize Schnabel for attempting to date the two systems,[17] and by the time ACT was published he categorically denied that we could know which system was older:

> All that can be said with safety at present is that the methods for computing lunar and planetary ephemerides were in existence around 250 B.C. Their previous history is unknown to me.[18]

By 1945, Neugebauer had studied all of the previously published mathematical astronomical texts along with several additional texts in the British Museum and at the Oriental Institute in Chicago which he had been made aware of by Schaumberger. He later wrote that "work on this material was practically completed in 1945". But then he was sent photographs of more than one hundred astronomical texts from Uruk in the museum in Istanbul, and then a few years later he was sent copies of Strassmaier's notes on further astronomical tablets in the British Museum. Finally, Neugebauer's colleague Sachs obtained access both to original tablets at the British Museum and copies of over thirteen hundred astronomical tablets that had been made by the Assyriologist T. G. Pinches during the late nineteenth century.[19] From these and his own search of the British Museum's collection, Sachs identified more and more fragments, causing yet further delays and requiring additional re-writing of the manuscript of ACT until the point came where Neugebauer called a halt in

---

[15] Neugebauer 1941, 24. This passage is followed by a table giving the length of daylight for each sign of the zodiac according to System A and System B. Neugebauer later showed that the Babylonian schemes *were* based upon a rising time scheme (Neugebauer 1953).

[16] Neugebauer 1945, 10.

[17] Neugebauer 1953.

[18] Neugebauer 1955, 11.

[19] See further below.

1955 and published ACT. The work contained almost ten times the number of tablets he had initially expected to include.

It was not until the final build up to ACT that Neugebauer seems to have seriously turned his attention to the planetary texts. Whereas he had published several studies during the 1930s and 1940s as he worked on understanding the lunar theories, his first publication dealing with planetary theory appeared in 1951, followed by a second in 1954. This may have been in part because the planetary systems were better understood by Kugler, and their remained less to do in their analysis. An important development by Neugebauer to the study of the planetary systems was his extrapolation of the names "System A" and "System B" to planetary systems which calculated the longitude of the planet at one of its synodic phenomena by the same means as the longitude of the moon at syzygy was calculated in the two lunar systems.[20] Kugler, by contrast, had labeled the various planetary systems by consecutive letters of the alphabet, but the lunar systems by Roman numerals. For example, Kugler (1900, 208–209) discusses three systems for Jupiter. Kugler's "System A" corresponds to Neugebauer's "System A", Kugler's "System B" corresponds to Neugebauer's "System A′", and Kugler's "System C" corresponds to Neugebauer's "System B". Neugebauer's designations have the advantage of indicating the type of function (step or zigzag) employed in the system.

Neugebauer used one further convention in his discussion of the planetary texts: a series of Greek letters as a shorthand for the characteristic synodic phenomena of the planets such as first visibility, first station, etc. This shorthand is still regularly employed by scholars today, both in the study of Babylonian and Greek astronomy, and the synodic phenomena are often referred to as "Neugebauer's Greek Letter Phenomena". It is worth noting, however, that the Greek Letter designations first appear in a 1948 paper by Sachs.[21] Thus it is unclear whether the practice was originated by Sachs or Neugebauer; given their close collaboration during this period it is quite likely that the convention was developed by them together.

1955: Neugebauer's *Astronomical Cuneiform Texts* and Sachs's *Late Babylonian Astronomical and Related Texts*

In 1955, almost 20 years after it was announced, Neugebauer's edition of the Babylonian mathematical astronomical texts was finally published. The *Astronomical Cuneiform Texts* appeared in three hardback volumes. The book was published for the Institute for Advanced Study by Lund Humphries in London. Lund Humphries specializes in the publication of illustrated art books and were presumably entrusted with the publication because of the number of photographs of cuneiform tablets which needed to be printed in high quality. Presumably by oversight, the date of publication was omitted from the title page (Fig. 1).

[20] In Neugebauer's terminology, System A is used for systems that calculate longitudes using step functions and System B is used for systems that calculate longitudes using zigzag functions.

[21] Sachs 1948, 274.

ASTRONOMICAL CUNEIFORM TEXTS

BABYLONIAN EPHEMERIDES
OF THE SELEUCID PERIOD FOR THE MOTION OF THE SUN,
THE MOON, AND THE PLANETS

Edited by
O. NEUGEBAUER

Published for the
INSTITUTE FOR ADVANCED STUDY
PRINCETON, NEW JERSEY
BY LUND HUMPHRIES, 12 BEDFORD SQUARE, WC1
LONDON, ENGLAND

Fig. 1 Title page of Neugebauer's *Astronomical Cuneiform Texts*

Table 1 Division of tablets in ACT

Number Range	Content
1–99	System A lunar tables
100–199	System B lunar tables
200–299	Lunar procedure texts
300–399	Mercury tables
400–499	Venus tables
500–599	Mars tables
600–699	Jupiter tables
700–799	Saturn tables
800–899	Planetary procedure texts
1000–1100	Unidentified fragments

Astronomical Cuneiform Texts contains editions with commentary of around three hundred texts of mathematical astronomy. A large proportion of these texts, over two hundred, are tabular texts; the remainder are procedure texts. For each text Neugebauer gave a short summary of the tablet's provenance, previous publication (if any) and arrangement, followed by an edition of the text, critical apparatus and commentary. The tablets are numbered according to the scheme in Table 1. Within each number range the tablets are divided into groups (eg the range 600–608 is used for Jupiter System A ephemerides, 609–619 for System A′ ephemerides, 620–629 for System B ephemerides, and 650–659 for daily motion tables). Within each group, the tablets are presented chronologically with undated tablets at the end. Because the number of tablets to be included in ACT kept growing right up until publication, Neugebauer resorted to inserting tablets into his numbering system with letters after the number. For example, the first ten tablets (System A lunar ephemerides) are numbered 1, 2, 3, 3aa, 3a, 3b, 4, 4a, 5, 5a.

In addition to the discussion of the individual tablets, ACT contains a long introduction discussing the mathematics necessary for understanding Babylonian astronomy and for reconstructing and dating tabular texts, a study of the colophons found on the tablets, a detailed presentation of the different lunar and planetary systems, a glossary of Akkadian terminology, and an index.

The publication of ACT was intimately connected to the publication of an equally important work the same year: Neugebauer's Brown colleague Abraham Sachs's *Late Babylonian Astronomical and Related Texts* (LBAT). In 1948 Sachs had published a classification of the small number of the astronomical texts from the Hellenistic period that were known to him from the publications of Epping, Kugler and Schaumberger, plus a few unpublished tablets in the Oriental Institute in Chicago, at Yale and in Istanbul. Sachs's classification grouped the tablets into Astronomical Tables, Astronomical Diaries, Normal Star Almanacs, and Goal-Year Texts. In 1952 and 1953–1954 Sachs visited the British Museum to collate tablets for Neugebauer and to search for more astronomical texts. The Assyriologist Donald J. Wiseman later recalled what happened during those visits:

> Abe was already experienced in the history of mathematics and astronomy and their methodologies when he visited the British Museum to collate texts known to him and of which

> his classification was already widely accepted. My own work there included the provision of a list of unpublished tablets which bore historical data, so Sachs' advice was eagerly sought to help identify the references in astronomical diaries which I had noted. He was an able, friendly and sympathetic teacher and devoted some time to enable me to master the distinctive characteristics of Astronomical Tables, (Normal Star) Almanacs, Goal Year texts and especially the astronomical diaries to which my attention had been drawn. The Keeper, Cyril J. Gadd, wisely agreed that it would be far more expeditious if we made a joint exploration of the collections of tablets, for other urgent commitments meant that we could spare about one hour a day for this exercise. By the summer of 1952, with additions in the following years, some 1200 texts and fragments were listed and some joins made.[22]

In addition to the astronomical tablets Sachs identified himself he was given access to about 1350 copies of astronomical tablets made by T. G. Pinches at the end of the nineteenth century. Sachs shared these copies with Schaumberger and together they identified and dated a large proportion of the texts. As Sachs explained in the introduction to LBAT, the number of texts now known to him was drastically greater than only a few years earlier:

> The complete bibliography of the non-mathematical astronomical texts of the Hellenistic period could recently be presented on a page and a half; cf. JCS 2, 1948, p. 275f. The present volume contains more than 900 copies of the same type of texts from the identical period.[23]

Remarkably, Sachs's classification of 1948 proved valid for the vast majority of the more than twenty times greater number of tablets now known to him (as it still does today). In addition to the non-mathematical astronomical texts, Sachs identified among Pinches copies many texts of celestial divination and astrology, texts containing star lists and other miscellaneous material, and a significant number of texts of mathematical astronomy. These latter texts Sachs made available to Neugebauer, who included them in ACT with references to the publication of the copies in LBAT.

In contrast to ACT, LBAT does not contain editions of the texts it publishes. Instead, Sachs confined himself to publishing Pinches's copies (plus a few by Strassmaier) together with a "Descriptive Catalogue" of the texts which arranges them into different text groups and gives details of the date of the text. Some of the dates are preserved on the tablets, but in many cases the texts were dated by analysis of the astronomical data they contained. Sachs explained that lists of dated tablets were prepared independently by himself and Schaumberger and then compared. For the mathematical astronomical texts, only a reference to the number of the text in ACT is given.

LBAT was published by Brown University Press in 1955 (Fig. 2). Apart from the title page, the book is presented in Sachs's handwriting. The full title of the work was given as "Late Babylonian Astronomical and Related Texts Copied by T. G. Pinches and J. N. Strassmaier". The book is said to have been "Prepared for

[22] Wiseman 1988, 363.

[23] Sachs 1955, vi.

LATE BABYLONIAN ASTRONOMICAL
AND RELATED TEXTS

COPIED BY T. G. PINCHES AND J. N. STRASSMAIER

Prepared for publication by
A. J. SACHS
With the co-operation of J. Schaumberger

Brown University Press
Providence, Rhode Island
1955

Fig. 2 Title page of Sachs's *Late Babylonian Astronomical and Related Texts*

publication by A. J. Sachs With the co-operation of J. Schaumberger". Schaumberger died in the year of publication.

The publication of ACT by Neugebauer and LBAT by Sachs in 1955 can be seen as a partial fulfillment of the plan outlined by Neugebauer in 1938 to publish editions of all classes of cuneiform tablets related to astronomy. Neugebauer's ACT completed (at least temporarily) the publication and study of the mathematical astronomical texts: editions and translations of the texts with detailed commentaries and expositions of the various lunar and planetary theories. LBAT, however, was only the first step to the publication of the non-mathematical texts. Sachs had catalogued all the sources and published copies of the cuneiform texts, but it remained to edit, translated and analyze all of the texts. Sachs originally planned to publish the Astronomical Diaries in collaboration with Wiseman[24]—he even withheld some of Pinches copies from LBAT for inclusion in their joint publication—but the project was not completed during Sachs's lifetime, finally coming to fruition through the work of Hermann Hunger.

[24] Sachs 1955, vii, Wiseman 1988, 363.

The Reviews

In order to understand the reception of ACT it will be helpful to compare it with the reception of LBAT. Table 2 lists the reviews of ACT and LBAT that appeared in various journals. The journals are divided into the following groups: history of

Table 2 Reviews of ACT and LBAT

Journal type	Journal	ACT	LBAT
History of science	*Archive Internationales d'Histoire des Sciences*	A. Pannekoek 34 (1955), 281–283	–
	Isis	G. Abetti 49 (1958), 355–356	–
	Centaurus	–	–
Assyriology	*Zeitschrift für Assyriologie*	–	B. L. van der Waerden 52 (1956), 339–342
	Revue d'Assyriologie	–	M. Leibovici 53 (1959), 159–162
	Archiv für Orientforschung	–	–
Oriental studies	*Journal of the American Oriental Society*	G. Sarton 75 (1955), 166–172	–
	Orientalia, N. S.	J. De Kort, S. J. 25 (1956), 277–282	W. von Soden 26 (1957), 55–58
	Zeitschrift der Deutschen Morgenländischen Gesellschaft	B. L. van der Waerden 106 (1956), 371–372	–
	Bibliotheca Orientalis	I. J. Gelb 15 (1958), 36–38	P. Huber 13 (1956), 231–232
	Journal of Near Eastern Studies	A. L. Oppenheim 17 (1958), 157	A. L. Oppenheim 17 (1958), 157–158
	Journal of the Royal Asiatic Society	–	–
	Bulletin of the School of Oriental and African Studies	–	–
Science	*Nature*	R. W. Sloley 176 (1955), 569–570	–
	Science, N. S.	I. Bernard Cohen 123 (1956), 66–67	–
	Publications of the Astronomical Society of the Pacific	A. Pogo 67 (1955), 427–428	–

science journals, Assyriology journals, science journals, and oriental studies journals (many of which contain a significant Assyriological content). In addition, journals in the relevant fields field which regularly contain a substantial number of reviews and which might have been expected to review either ACT or LBAT but did not are also included. As might be expected, ACT was reviewed much more widely than LBAT. Reviews of the latter were restricted to Assyriology and oriental studies journals, hardly surprising when one considers that the book contained only a catalogue of tablets and drawings of those tablets which could only be appreciated by specialists. ACT, however, was reviewed in history of science journals and mainstream science publications as well as in oriental studies journals. Perhaps oddly, there were no reviews of ACT in the main Assyriology journals. The absence of a review of either ACT or LBAT in *Archiv für Orientforschung* is particularly surprising given that the journal was edited by Ernst Weidner, an Assyriologist who had published extensively on Babylonian astronomy, celestial divination and mathematics.

In his review of ACT published in the *Journal of Near Eastern Studies*, the Assyriologist A. Leo Oppenheim lamented that "Books of this type have the tragic fate that none but their author can be considered able and entitled to review them in an intelligent way".[25] Oppenheim's statement was not quite true: Antonie Pannekoek, Bartel van der Waerden and Johann Schaumberger had all published important studies on Babylonian mathematical astronomy, although only the latter could also read the texts.[26] Schaumberger, however, died within a few months of ACT's publication (and over 2 years before Oppenheim's review), depriving the book of its most qualified reader. Unsurprisingly, Pannekoek and van der Waerden were invited to review ACT, but beyond those two names journal editors had a problem in finding suitable reviewers. The solution adopted by most journals, including oriental studies journals, was to turn to either historians of science or to astronomers. For example, the journal *Orientalia*, published by the Pontificium Institutum Biblicum in Rome asked the Vatican Observatory astronomer J. De Kort S. J. to review the book. Kort wrote a strange review in which rather than discussing the book itself he gave a modern presentation of the accuracy of the synodic periods found in the ACT material. *Isis* sent the book to the director of the Osservatiorio Astrofisico di Arcetri, Giorgio Abetti, an astronomer and one of the editors of the collected works of Galileo. The two professional historians of astronomy who reviewed ACT, George Sarton and I. Bernard Cohen, were both big names in the history of science in the USA but were not specialists of ancient science. Only two Assyriologists reviewed ACT: A. Leo Oppenheim and I. J. Gelb, both of the Oriental Institute of the University of Chicago.

---

[25] Oppenheim 1958, 157.

[26] I exclude Sachs from this list as he can better be understood as a contributor, perhaps even an uncredited co-author, rather than a reader of ACT. For an overview of the scholarly background and publications of Schaumberger, Pannekoek and van der Waerden, see the chapter "Babylonian Astronomy 1880–1950: The Players and the Field" by Teije de Jong in this volume.

Most of the reviews of ACT were short, doing little more than summarizing the contents of the book and applauding Neugebauer for his work in producing it. In what follows I therefore focus on the five most important reviews: the reviews by Pannekoek and van der Waerden (the only reviewers with a background in Babylonian mathematical astronomy), the review by I. J. Gelb (the only detailed review by an Assyriologist), the review by R. W. Sloley in *Nature*, and the long review essay by the historian of science George Sarton. As would be expected, each reviewer focused on different aspects of the book and came to his own conclusion of its importance.

Of the two reviewers who had worked on Babylonian astronomy, the first to publish a review was Antonie Pannekoek. Pannekoek's review appeared already in the July–September 1955 issue of the *Archives Internationales d'Histoire des Sciences*, only a few months after ACT was published. The 2½-page review begins with a summary of Kugler's work and its extension by Schaumberger and Neugebauer in earlier publications, followed by a survey of the extant source material based upon the discussion in Neugebauer's preface to ACT. Pannekoek then outlined the contents of the book and the number of tablets included for the moon and each of the planets, pointing out that "It is not just this large number of texts that constitutes the value of Neugebauer's work, but chiefly the careful handling and thorough discussion of each of them".[27] Of particular importance, Pannekoek says, is that Neugebauer has generalized the methods of computations found in the ephemerides into a broad theory and Neugebauer's use of these methods to connect fragments. In the final paragraph, Pannekoek make some interesting remarks on the importance of ACT and what further avenues of research it may lead to, which are worth quoting in full:

> Thus Neugebauer's « Astronomical Cuneiform Texts » (quoted ACT) stands out as the most important work in the field of Babylonian astronomy since Kugler's books, and the author is to be congratulated on having been able to perform it in such a masterly way. The work ushers in a new phase of research; it will no doubt add a fresh incentive to further excavations—from the basements of museums as well as from Mesopotamian soil—, the results of which will give rise to later supplements. It is true that it can be no more than a collection of material needed as the basis of future science. In this respect it is to be compared to famous books on archaeology, describing excavations of ruined cities (e.g. Dörpfield's Troja); in both cases out of a jumble of *débris* an ancient civilization comes to light. There is, however, a difference: in the latter case it is mainly a material civilization that is revealed, while the civilization uncovered in the work under discussion is of a highly intellectual character. It is, however, wrapped in mystery. The origin of this mathematical astronomy is an enigma; it appears of a sudden, without any visible connection with earlier, more primitive, astronomical texts, some of which were also dealt with in Kugler's books. It seems reasonable to expect that upon a further increase of the texts available for study, simpler specimens may also be found, which will shed light on the origin of the class of texts here dealt with.[28]

[27] Pannekoek 1955, 282.

[28] Pannekoek 1955, 283.

Pannekoek highlights two issues that would be significant for future research on Babylonian mathematical astronomy: the discovery of new texts (something Sachs was already doing at the British Museum) and recovering the history of the origin and development of Babylonian mathematical astronomy. Neugebauer said nothing about the empirical basis of the Babylonian theoretical astronomy in ACT, nor to any great extent in any of his later publications, but this was a topic that interested Pannekoek. Pannekoek had criticized Neugebauer for an overly mathematical view of Babylonian astronomy at the end of his article on eclipse magnitudes published in *Eudemus* in 1941,[29] and himself published a proposal for the Babylonian methods that led to the discovery of the Saros in 1918.[30] As I shall discuss below, Pannekoek's prediction that further texts would be discovered which provided evidence for how the ACT theories had been constructed came true and became the major area of research into Babylonian astronomy during the 1960s and 1970s.

One point that Pannekoek did not raise in his review was the significance of the Babylonian material within the history of astronomy more broadly. Van der Waerden, by contrast, began his review with this very point. Van der Waerden's review appeared in the *Zeitschrift der Deutschen Morgenländischen Gesellschaft* in 1956. The review is short—about a single page in length—but longer than many other reviews published in this journal. Van der Waerden stressed the importance of understanding Babylonian astronomy because of its connection with Greek astronomy, and suggested that tracing the transmission of Babylonian astronomy to different cultures was a way of tracing cultural contact. After a brief overview of the history of research on Babylonian astronomy and an outline of the contents of the book, van der Waerden ended with the following remarks on the lasting value of Neugebauer's work:

> Twenty years of tireless work by the author and ten years of work by his Assyriological assistant A. Sachs have been put into these three volumes. In the interpretations the author has imposed the greatest possible restraint. He has described only those reconstructions that are absolutely necessary for the many fragmentary texts and for understanding the computational methods of the texts. Thus, the work has eternal value, so to say: it will perhaps be supplemented by more recent research and will be corrected in some points, but never be outdated.[31]

Van der Waerden himself would be one of the people whose further research supplemented the understanding of Babylonian mathematical astronomy presented in ACT, in particular through his work on the understanding of the underlying principles behind both the planetary theories and column Φ of the System A lunar theory.[32]

The only substantial review of ACT by an Assyriologist was written by Ignace J. Gelb of the University of Chicago and appeared in the January–March 1958 issue of *Bibliotheca Orientalis*. Gelb was the driving force behind the revival of the

---

[29] Pannekoek 1941.

[30] Pannekoek 1918.

[31] van der Waerden 1956, 372.

[32] van der Waerden 1957, 1966.

Chicago Assyrian Dictionary (CAD) project after the second world war, but had retired a editor-in-chief of the CAD at the end of 1954, "Tired of the administrative work and of the dissension" of the other editors, as he later wrote.[33] The "dissension" Gelb referred to was caused by a fundamental disagreement between Gelb and his colleagues Benno Lansberger and A. Leo Oppenheim about the structure of the CAD, indeed about the very nature and purpose of lexicography.[34] Gelb began his review by outlining the contents of ACT and its importance, describing it as a "splendid three-volume opus",[35] although noting that the texts themselves "represent the drabbest kind of material, consisting as they do of tables full of numbers and logograms for month names, zodiacal expressions, and for terms for other astronomical concepts",[36] and that he is not qualified to judge them. Instead "that which remains for an Assyriologist to evaluate is the system of writing and the conclusions that can be drawn from the colophons", and this is what he focuses on for most of the review. Gelb demonstrates that he had made a very careful study of Neugebauer's book, discussing issues such as the arrangement of tablets (eg how they turn from obverse to reverse) and the structure of the colophons (Gelb makes the perceptive remark that "the distinction between the scribe (indicated by *qât* PN) and the owner (indicated by *tuppi* PN) of the tablet cannot be established",[37] a question that has been the subject of recent study by Ossendrijver[38] and others). As might be expected, however, Gelb devotes most of his attention to ACT's glossary. He applauds Neugebauer's structuring of the glossary entries, although he notes a few inconsistencies and corrects some of Neugebauer's Akkadian normalizations or logographical readings.

The reviews by Pannekoek, van der Waerden and Gelb were all written by experts for experts, whether that be historians of science or Assyriologists. But Neugebauer's book was also picked up by the major interdisciplinary science journals *Nature* and *Science*. The latter had a short review by I. Bernard Cohen, a historian of Newton and the scientific revolution and at the time editor of *Isis*, the main history of science journal in the USA. *Nature* published a longer review written by Robert W. Sloley, a retired engineer with an interest in Egyptian astronomy and timekeeping. Sloley's review probably reached a larger audience than any of the other reviews of ACT and is important both for this reason but also because Sloley made some interesting observations about the nature of Babylonian astronomy.

Sloley began his review with the now familiar praise of Neugebauer for his work making these texts available and by a summary of their content. After describing some of the basic techniques of Babylonian astronomy such as the linear zigzag function, Sloley remarks that

---

33 Gelb 1964, xviii.

34 Reiner 2002.

35 Gelb 1958, 36.

36 Gelb 1958, 37.

37 Gelb 1958, 37.

38 Ossendrijver 2011.

> The Babylonian procedure shows a remarkably abstract attitude and unhesitatingly introduces quantities for purely mathematical convenience on the same principle as complex numbers are employed in modern mechanics.[39]

This is an important point about the mathematical sophistication and scientific nature of Babylonian astronomy that was taken as self-evident by Neugebauer and as such not explicitly articulated in ACT. Sloley obviously felt that this was a point that needed to be stressed for the readers of *Nature*—scientists who probably knew little about ancient astronomy, and who would likely have dismissed Babylonian astronomy as crude and unphysical because of it was not geometrical. Sloley highlights the scientific nature of Babylonian astronomy later in his review:

> The main incentive for the study of astronomy seems to have been the attempt to introduce some measure of regularity in the intercalations of the lunar calendar. Astronomy did not originate in astrology as has so often been stated; but the very widespread belief in astrology as the one science which gave insight into the causes of events on earth influenced the transmission of astronomical knowledge from one nation to another. The technique of Hellenistic astrology demanded knowledge of the position of the Sun, Moon and planets in the zodiacal signs at the moment under consideration, and this information was not immediately available from the Babylonian ephemerides, of which the elements played no part whatever in the practice of astrology. Magic, number-mysticism and astrology can no longer be regarded as the guiding forces in Babylonian science.[40]

Sloley seems intent here on establishing the scientific legitimacy of Babylonian astronomy to contemporary scientists (if anything, suggesting that it is Greek science, not Babylonian which is linked to astrology).

Sloley ended his review with more praise for the book and, echoing Pannekoek, the hope that further discoveries will lead to an understanding of the development of Babylonian astronomy:

> The texts considered in this new publication represent but a small fraction of the total available, most of which still remain unexamined. Few scholars are as competent as Prof. Neugebauer to deal with these, and perhaps in time he may be able to throw some light on the interesting problem of how the methods of computation were arrived at—quite possibly in a relatively short period, by some Newton or Einstein of the day.[41]

Throughout his career Neugebauer resisted speculating about the creation of Babylonian astronomy, and especially the idea that it was the creation of a single Newton- or Einstein-like genius.

I have reserved discussion of the longest and most important review of ACT until last: George Sarton's essay review "Chaldaean Astronomy of the Last Three Centuries B.C." which appeared in the 1955 volume of the *Journal of the American Oriental Society*. Sarton was a Belgian historian of science who emigrated to the USA during the First World War, founding editor of the journal *Isis*,[42] and, in

---

[39] Sloley 1955, 570.

[40] Sloley 1955, 570.

[41] Sloley 1955, 570.

[42] Pyenson 1995.

Neugebauer's sarcastic phrase, "the recognized dean of the History of Science",[43] an image Sarton would probably have felt comfortable with. Sarton was a generalist, dedicated to the production of bibliographies and synthetic works, very different to Neugebauer's detailed technical studies and text editions, and his writing carried weight among a broad community of scholars. Sarton's eight-page review, written the year before he died, follows his general approach to the history of science, asking broad questions about the role, nature and significance of Babylonian astronomy but saying very little about either the texts themselves or the details of the astronomy they contain.

Sarton begins by stressing the importance of Neugebauer's work ("The latest work of Professor Neugebauer on the astronomical cuneiform texts of the Seleucid period is so important and the field which it covers is relatively so new that it is worth while to devote a special article to it rather than a review"[44]) and gives a brief overview of the contents of the book. On the whole Sarton's overview is clear and his treatment of Neugebauer fair—Sarton's remark that Neugebauer's explanations of the mathematical method of Babylonian astronomy "are clear but terse"[45] will find no disagreement from anyone who has spent time working with ACT. Sarton does make one error, possibly a significant one, by claiming that ACT "is a corpus of all the astronomical tablets written in cuneiform during the last three centuries before Christ".[46] In fact, ACT contains only the texts of mathematical astronomy from this period; examples of other classes of astronomical texts could be found in the works of Epping, Kugler and Schaumberger, or in Sachs's LBAT (although in this last case only in cuneiform copies). However, as I shall discuss in the next section, for many historians of science in the mid to late twentieth century, ACT *was* Babylonian astronomy; Sarton's comment may have inadvertently lent support to that view.

After outlining the content of ACT, Sarton turned to the main theme of his review, the importance of the Babylonian achievements for teaching of the history of science. This theme was one that Sarton often brought up when discussing ancient science. For example, in 1940 Sarton wrote a short paper in *Isis* which outlined the debates between Neugebauer and Thureau-Dangin on the interpretation of Babylonian mathematics. Sarton wrote that this disagreement

> is very hard on historians of science and particularly on historians of mathematics, who realize the fundamental importance of Babylonian (as well as Egyptian) mathematics and would like to give their students as good an account as possible of pre-Hellenic mathematics yet cannot undertake the formidable task of interpreting themselves the original documents. When the experts disagree the non-experts are in an awful quandary.[47]

---

[43] Neugebauer 1951.

[44] Sarton 1955, 166.

[45] Sarton 1955, 167.

[46] Sarton 1955, 166.

[47] Sarton 1940, 398.

Sarton offered the pages of *Isis* as a place to debate the interpretation of Babylonian mathematics in the hope that harmony could be reached, and also urged the writing of a "primer of Babylonian mathematics", which would take the reader step-by-step through a few examples of Babylonian mathematical texts.

Sarton returned to the role of Babylonian and Egyptian science in the teaching of the history of science in a review of Neugebauer's *The Exact Sciences in Antiquity*:

> The teaching of the history of science is being gradually introduced into the curriculum of our universities, but instructors are not given much time. If they be permitted to give a short course on "ancient science," they may be tempted to speak mainly of Greek science and leave oriental science out. That would be a great mistake; they should devote at least a couple of lectures to Egyptian and Babylonian origins, not so much for the sake of the Egyptian and Babylonian knowledge which has been gradually eliminated from the main stream, but as a superb illustration of the complexity of scientific progress.[48]

Sarton made a very similar argument in the essay review of ACT. He stresses the time pressures in teaching the history of science but is adamant that "three or four" must be devoted to pre-Hellenic science, in other words Babylonian and Egyptian mathematics. Indeed, Sarton goes so far in stressing the importance of Babylonian science as to conclude that "the influence of ancient Babylonian science upon us is immense; instead of saying as is often done that the roots of our culture are Hebrew and Greek we should say Sumerian, Hebrew and Greek".[49] So what should the teacher of the history of science teach of Babylonian astronomy? After the importance Sarton placed on teaching Babylonian mathematics his answer comes as a shock: "I do not hesitate to say that the best that he could do would be to leave it out, or to refer to it incidentally in a lecture on Hipparchos".[50] He knew, of course, that this would be a controversial statement, to say that the subject matter of the book under review, a book which has taken its author 20 years to complete, should not be taught. He continues:

> Please do not misunderstand me and do not misquote me. All I mean is that the teacher who is asked to cover the whole of ancient science in 35 lectures must restrict himself to the essential, to the main story; he cannot afford to be sidetracked by aberrant developments such as Chaldaean astronomy, ancient Chinese astronomy, or Maya astronomy. I do not say that those developments are not important or not interesting.[51]

Sarton explained further in a footnote: "To put it otherwise, he must restrict himself to the main stations on the road to modern science and avoid the intriguing bypaths leading nowhere".[52] In Sarton's view, the history of science should be

---

[48] Sarton 1952, 72. Sarton continued by remaking that "I could not advise them, however, to build their course on the basis of Neugebauer's book alone" since the book did not contain enough material on Greek science. Sarton was very critical of the title of *The Exact Sciences in Antiquity* because of its focus on Babylonian and Egyptian science and lack of discussion of Greek science: "That is too much like the play *Hamlet* with Hamlet left out".

[49] Sarton 1955, 169.

[50] Sarton 1955, 171.

[51] Sarton 1955, 171–172.

[52] Sarton 1955, 172.

taught as the history of the development of science from its earliest beginnings up to today, tracing a direct line back from modern science. Babylonian mathematics should be taught because it is earlier than Greek science and could be seen as its predecessor because of the discoveries of texts dealing with what had been interpreted as Pythagoras's theorem and the like. Babylonian astronomy, however, was contemporary with Greek astronomy—Sarton makes a big point of this, and it is the reason for his instance on the use of the name Chaldaean rather than Babylonian for the astronomy—and had no influence on Greek developments. The reason for this, Sarton argues, is because of an essential failing of the Babylonians:

> The Chaldaean priests did not study Greek astronomy; not because they could not (it was relatively easy for them to master the Greek language and they were fully aware of astronomical problems) but because they would not. Their astronomy was the scientific armature of their folklore, the palladium of their folkways. ... It is only in such a manner that the development of Chaldaean astronomy can be understood; it could only develop in an isolated environment, and the isolation could be accomplished only because of their own religion and chauvinism.[53]

Sarton does not address why the Greeks did not learn Babylonian astronomy. For Sarton, the reason was probably self-evident: the Greeks would have felt they had nothing to learn from the Babylonians. Indeed, Babylonian astronomy should not even be termed science:

> To put it briefly (too briefly) the Greeks were philosophers as well as geometers, the Chaldaens were empiricists as well as sophisticated calculators. Their ephemerides were partly empirical and largely *a priori*; they suggest a complicated form of divination rather than a new branch of science.[54]

Neugebauer, of course, would have strongly disagreed with Sarton's arguments. Neugebauer knew that Babylonian astronomy was scientific, that it had nothing to do with numerology, that no model of oriental despotism was responsible for its creation, and, crucially, that Babylonian astronomy *did* influence Greek astronomy, as had been shown by Kugler already in 1900.[55] For Neugebauer, Babylonian astronomy was not an "aberrant development" but an essential part of the development of western astronomy, and therefore deserving of attention alongside Greek astronomy.[56]

---

[53] Sarton 1955, 171.

[54] Sarton 1955, 170.

[55] Kugler 1900; see also Aaboe 1955.

[56] Nevertheless, Neugebauer may well have had some sympathy with Sarton's basic point that what was worthwhile teaching are the steps along the road to modern science. In the introduction to *A History of Ancient Mathematical Astronomy*, Neugebauer explained that he had omitted any discussion of astronomy in China because of his lack of knowledge of the subject. He remarked, however, that Chinese astronomy's "influence upon the Islamic and Western development is probably not visible earlier than the creation of the Mongol states in Western Asia. Thus the damage done by omitting China is perhaps not to great and at any rate alleviated by ignorantia." (Neugebauer 1975, 2). Neugebauer's line here is not dissimilar to Sarton's remark that although study of the "bypaths" of astronomy is interesting, they are not part of the main story of the history of astronomy, which should be the main business of a historian of science.

Overall, ACT was widely reviewed in history of science, oriental studies and mainstream science journals and both the book and Neugebauer's work on Babylonian astronomy more generally were roundly praised by the reviewers. But it is noticeable that with the exception of Gelb's review which critically appraised the linguistic aspects of Neugebauer's work, offering several small corrections and clarifications, none of the reviews engage in detail with the book itself. In particular, none of the reviewers discussed any of the individual texts, suggested alternate interpretations to those of Neugebauer, attempted to understand any of the unidentified fragments, or challenged any of Neugebauer's views. One might have expected either Pannekoek or van der Waerden to have made some critical comments—both had previously disagreed with Neugebauer about the interpretation of certain aspects of Babylonian astronomy, and van der Waerden would again in the future. Sarton in his long essay review concentrated on discussing the nature of Babylonian astronomy and its place in the history of astronomy, topics Neugebauer discussed elsewhere but deliberately excluded from his presentation in ACT, but had nothing to say about the texts themselves or the astronomy they contain (which, in any case, he would have been quite unqualified to say anything about). It is informative in this regard to contrast the reviews of ACT with the reviews of Sachs's LBAT. This latter book was reviewed less extensively, and by and large in more specialist journals (and only one of the reviews, a short note by Oppenheim, was written in English), but several of the reviews engage deeply with the texts, offering interpretations of their contents that can only have been obtained by a careful reading of the cuneiform copies themselves. In general, it appears that although ACT was reviewed more widely, the reviewers of LBAT studied the book more carefully.

Other Readers of ACT

The impression left from a reading of the reviews of ACT is of a work that everyone thought was very important, that stood as a testament to the genius and tenacity of Neugebauer (nearly all the reviews mention that he worked on the project for nearly 20 years), that they were glad to know that the work had been done, but that they didn't really want to have to read about it. The question is, therefore, did *anyone* read ACT in detail? We can assume that Pannekoek and van der Waerden did study the work carefully, even though their reviews do not demonstrate this. But aside from these two, and of course Neugebauer's collaborator Sachs, did anyone else read the work? I suspect that only two others could claim to have read ACT in detail in the mid 1950s: Asger Aaboe, then a doctoral student working on Babylonian planetary theory under Neugebauer at Brown, and Peter Huber, a mathematics student in Zurich.

Aaboe had trained as a mathematician at the University of Copenhagen and was teaching mathematics at Tufts University when he entered Brown's Department of the History of Mathematics as their first graduate student.[57] Aaboe was granted leave from his position at Tufts for the academic year 1955–1956 and was awarded a President's Fellowship at Brown supplemented by an additional grant from the Danish Science Foundation. Aaboe's official arrival at Brown was only a few months after ACT was published. Aaboe's dissertation concerned Babylonian planetary theories and he was awarded the PhD in 1957. The dissertation was published as a long paper in the journal *Centaurus* the following year.[58]

In the introduction to his dissertation Aaboe explained that in ACT Neugebauer's aim was to uncover and explain the internal mathematical structures of the various planetary systems. Aaboe's study would attempt to understand that various techniques used by the Babylonian astronomers in these systems in the context of the astronomy of planetary motion and to estimate the accuracy of the Babylonian schemes. Unsurprisingly, Aaboe's work demonstrates that he read ACT carefully and thoroughly (we can also assume that he discussed Babylonian astronomy in detail with Neugebauer). Aaboe returned to Tufts after his year at Brown where he went back to teaching mathematics. At Tufts, and subsequently at Yale where he was appointed in 1962 to a joint appointment in the Department of the History of Science and Medicine and the Department of Mathematics, Aaboe continued his research on Babylonian mathematical astronomy (see next section).

Peter Huber was a student in mathematics at the ETH in Zurich when ACT was published. Earlier, as a student at the Gymnasium, Huber had learnt to read cuneiform as a respite from his studies of mathematics, physics and astronomy. In an interview with Andreas Buja and Hans R. Künsch which took place in 2005 Huber explained that he had read everything he could find on physics, astronomy and mathematics, but

> Then I suddenly had it up to here. I knew I would be going into mathematics or physics later, but I just couldn't continue right now. I had to do something completely different. Somehow I ended up learning cuneiform.[59]

In the spring of 1954 Huber attended an algebra class given by van der Waerden at the University of Zurich and subsequently discovered van der Waerden's interest in the history of mathematics. Huber read van der Waerden's *Ontwakende Wetenschap* (the original Dutch version of *Science Awakening*; the German and English translations had not yet been published), and through that Neugebauer's work on Babylonian mathematics. Huber disagreed with Neugebauer's algebraic treatment of VAT 8512 and, with the encouragement of van der Waerden, published a short note proposing an alternate geometrical approach in *Isis* in 1955. When he became a student at the ETH, Huber regularly participated in van der Waerden's historical seminars across the street at the University of Zurich. Huber told me what happened next in an email exchange on 3 November 2010:

---

[57] Neugebauer's first graduate student at Brown, Olaf Schmidt, was granted his PhD through the Mathematics department.

[58] Aaboe 1958.

[59] Buja and Künsch 2008, 14.

> After a while, after having worked through most of MKT and MCT, I lost interest in Babylonian mathematics, since I felt that only epsilon improvements beyond Neugebauer were possible. Clearly, Babylonian astronomy (in particular Kugler), was much messier, and therefore more interesting. Then ACT came out, and I bought it in the summer of 1955. Again, I attacked some texts where Neugebauer had left open ends, and again vdW encouraged me actively to publish the stuff.[60]

The "open ends" Huber attacked were a small fragment of a Mars ephemeris which he identified as having been computed according to a System B scheme and various texts having to do with the daily motion of Jupiter. Huber sent his reconstruction of the Mars ephemeris to Neugebauer and it was included (with Huber's permission) in Aaboe's dissertation.[61] The work on Jupiter, which for the first time provided a detailed understanding of the daily motion schemes, was published by Huber in the *Zeitschrift für Assyriologie* in 1957.[62]

In the spring of 1956 Huber obtained Sachs's LBAT and turned his attention to the non-mathematical astronomical texts. Huber wrote a detailed review of LBAT for *Bibliotheca Orientalis* in which he explained the structure of the eclipse texts LBAT 1413ff (Huber later contributed a further discussion of the eclipse texts to van der Waerden's *Die Anfänge der Astronomie*).[63] He also examined the Almanacs and Normal Star Almanacs in order to determine the zero-point of the Babylonian zodiac.[64] However, for Huber this was just a sideline to his PhD research in topology and once his work for van der Waerden was finished in 1960, he ceased work on Babylonian astronomy until the 1970s. Huber explains:

> For me, all this was an amusing hobby. I felt that history of science was not a serious profession, especially not for young people (and apart from that, the job situation predictably would be difficult).[65]

The Impact of ACT on Assyriology and the History of Science During the Second Half of the Twentieth Century

The publication of ACT was met with almost uniform praise by its reviewers but, as I have discussed in the previous section, the number of people who read the work in detail was very small. Nevertheless, ACT did have a considerable impact on both Assyriology and the History of Science during the second half of the twentieth

[60] Email Peter J. Huber to the author sent 3 November 2010. "MCT" refers to Neugebauer and Sachs's *Mathematical Cuneiform Texts* (1945) and "vdW" to van der Waerden.

[61] Aaboe 1958, 245.

[62] Huber 1957.

[63] Huber 1956.

[64] Huber 1958.

[65] Email Peter J. Huber to the author sent 3 November 2010.

century, although in many cases this impact was based upon reading only parts of the work.

Although Neugebauer published more than three hundred cuneiform texts in ACT—more than are contained in most other publications of collections of cuneiform texts—the individual tablets were by and large of little interest to Assyriologists. For Assyriology, ACT was of use primarily for two other parts of the work: the glossary and the study of the colophons. As we have already seen, Gelb discussed ACT's glossary in detail in his review of the book. In 1956, 1 year after ACT appeared, the first part (volume 6, the letter Ḫ) of the *Chicago Assyrian Dictionary* (CAD) was published, after a gestation period even longer than the 20 years Neugebauer spent preparing ACT. ACT appears already in the list of abbreviations at the beginning of the volume, and several entries contained citations of ACT texts. In the early 1930s Neugebauer, along with about forty other scholars, had been approached to write entries for the CAD and, although he initially accepted the assignment he was not able to fulfill it.[66] Neugebauer was presumably asked to write entries about some of the words found in mathematical texts. In the end, the dictionary staff simply cut up the glossaries found in Neugebauer and Sachs's *Mathematical Cuneiform Texts* and Thureau-Dangin's *Textes mathématiques babyloniens* to create filecards for these entries.[67] It seems reasonable to suppose that the ACT glossary was used in a similar way to produce the starting point for the entries concerning astronomical terminology.

Neugebauer's study of the ACT colophons initially received less attention. Before the mid 1960s, colophons were generally considered of little interest in Assyriology: the focus was on what the texts said not on who wrote or copied them. Indeed, some editions of cuneiform texts simply omitted the colophons found on the tablets. This situation changed in 1964 with the publication of a seminal article by Erle Leichty simply titled "The Colophon" in a volume in honour of A. Leo Oppenheim.[68] Leichty's article was followed 4 years later by the publication of Hermann Hunger's PhD dissertation on Babylonian and Assyrian colophons, which made full use of Neugebauer's work.[69] Over the past 20 years increasing attention has been paid to colophons, particularly in studies of the social and intellectual history of cuneiform scribes. Neugebauer's work on colophons in ACT was well ahead of its time.

For most historians of science, ACT probably made little impact. Instead, what historians of science knew of Babylonian mathematics and astronomy came from Neugebauer's *The Exact Sciences in Antiquity*, which gave a general presentation of the material, stressing the concepts and techniques used in these Babylonian sciences without discussing individual texts. Almost all treatments of Babylonian astronomy found in general histories of science or astronomy during the second half of the twentieth century describe only ACT-type astronomy: for these books the

---

[66] Gelb 1964, xiii.

[67] Gelb 1964, xvi.

[68] Leichty 1964.

[69] Hunger 1968.

early texts and the (mostly unpublished) non-mathematical texts such as the Astronomical Dairies did not exist. ACT became Babylonian astronomy, and Neugebauer the only source for learning about it.[70] This may in part have been an unintended consequence of Neugebauer's decision to change the name of what became ACT from "Mathematical Astronomical Cuneiform Texts" to simply "Astronomical Cuneiform Texts" when he started writing in English. The former name makes it clear that the subject of ACT is only part of Babylonian astronomy—the part concerned with mathematical methods—while the latter suggests that the book contains the whole of Babylonian astronomy.

The biggest impact of ACT was on the study of Babylonian astronomy itself. Before 1955, research on Babylonian astronomy was undertaken by a variety of scholars with backgrounds (and training) in Assyriology, mathematics and astronomy. After ACT, however, the study of Babylonian astronomy for the next 20 years or so—at least as far as published work—became the study of Babylonian mathematical astronomy and was carried out exclusively by mathematically trained scholars. Partly this was simply because the death of Schaumberger and Huber's move back into mainstream mathematics took away two of the three scholars who had shown an interest in the non-mathematical texts. Sachs, of course, continued to work on the Diaries and related texts until his death in 1983 but published almost nothing after LBAT. The scholars who were left to work on Babylonian astronomy were therefore van der Waerden, Aaboe and Neugebauer himself.

Perhaps surprisingly—or understandably given the years he had spent preparing ACT—Neugebauer published very little on Babylonian astronomy after 1955: other than a few general overview papers published from invited lectures, Neugebauer wrote a total of five papers on Babylonian astronomy between 1955 and 1975, three of which were jointly written with Sachs, and most of which dealt with non-standard ACT-type texts, plus a detailed mathematical treatment of ACT astronomy in his *History of Ancient Mathematical Astronomy*. Van der Waerden's work on Babylonian astronomy also slowed down during the late 1950s as he turned his attention to astronomy and mathematics in the Greek world. This left Aaboe more or less single-handed to continue to study of Babylonian astronomy.

During his visits to the British Museum in 1953–1954 Sachs had compiled a catalogue of astronomical fragments which were not included in either ACT or LBAT. Sachs passed this typewritten list to Aaboe (it is often referred to at the "Sachs-Aaboe list"—see Fig. 3). During the 1960s and 1970s Aaboe worked through the list identifying many tables and procedure texts that were related to ACT but were not normal ephemerides. Several of these texts allowed Aaboe to reconstruct parts of the System A lunar theory that were unknown to Neugebauer and also to understand something of the system's development.

[70] The only exception was van der Waerden's *Die Anfänge der Astronomie*, later published in English as *The Birth of Astronomy*, which deals with the whole of Babylonian astronomy. However, the work is marred by some unfounded speculations of Babylonian astral religion, which turned some readers away from this work.

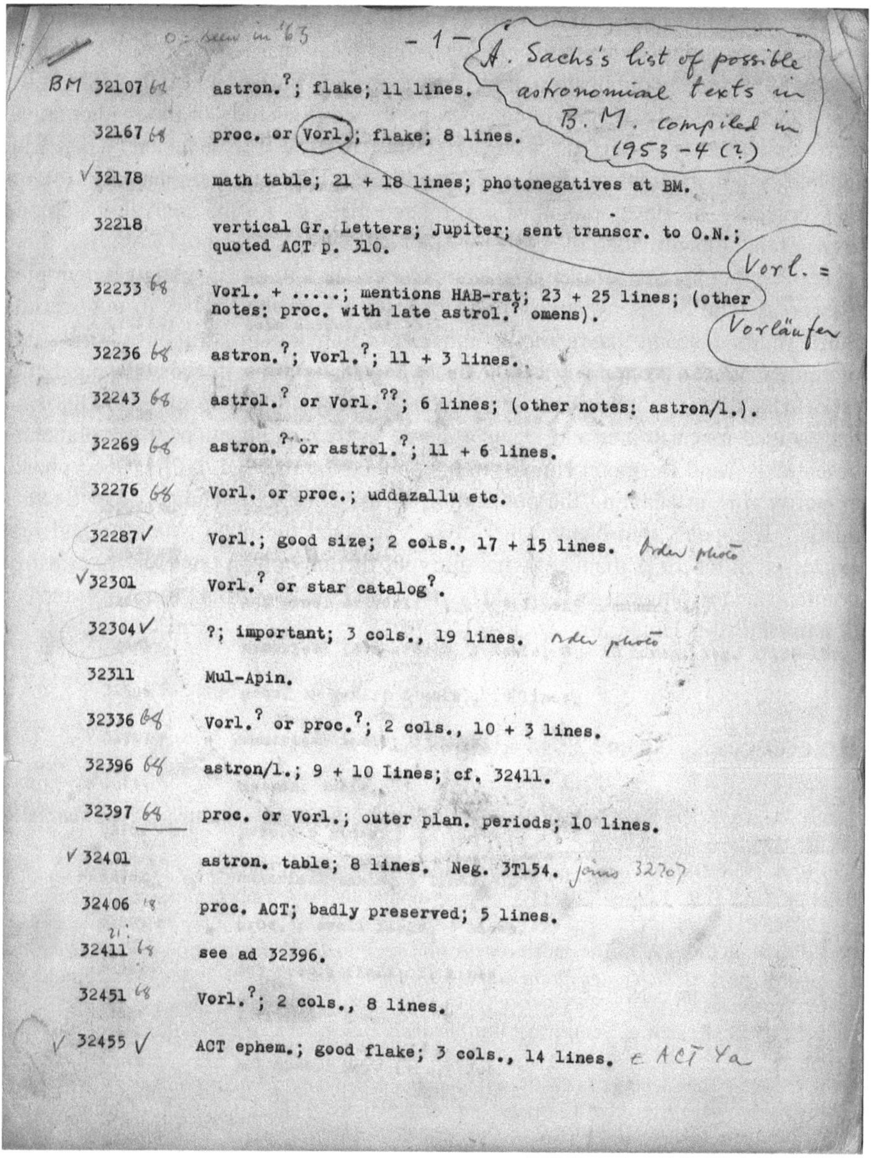

o = seen in '63

– 1 –

A. Sachs's list of possible astronomical texts in B.M. compiled in 1953–4 (?)

Vorl. = Vorläufer

BM 32107 68	astron.?; flake; 11 lines.
32167 68	proc. or Vorl.; flake; 8 lines.
√32178	math table; 21 + 18 lines; photonegatives at BM.
32218	vertical Gr. Letters; Jupiter; sent transcr. to O.N.; quoted ACT p. 310.
32233 68	Vorl. + ...; mentions HAB-rat; 23 + 25 lines; (other notes: proc. with late astrol.? omens).
32236 68	astron.?; Vorl.?; 11 + 3 lines.
32243 68	astrol.? or Vorl.??; 6 lines; (other notes: astron/l.).
32269 68	astron.? or astrol.?; 11 + 6 lines.
32276 68	Vorl. or proc.; uddazallu etc.
32287√	Vorl.; good size; 2 cols., 17 + 15 lines. order photo
√32301	Vorl.? or star catalog?.
32304√	?; important; 3 cols., 19 lines. order photo
32311	Mul-Apin.
32336 68	Vorl.? or proc.?; 2 cols., 10 + 3 lines.
32396 68	astron/l.; 9 + 10 lines; cf. 32411.
32397 68	proc. or Vorl.; outer plan. periods; 10 lines.
√32401	astron. table; 8 lines. Neg. 3T154. joins 32707?
32406	proc. ACT; badly preserved; 5 lines.
32411	see ad 32396.
32451 68	Vorl.?; 2 cols., 8 lines.
√32455 √	ACT ephem.; good flake; 3 cols., 14 lines. = ACT Ya

Fig. 3 The first page of the "Sachs-Aaboe list". This typed catalogue of astronomical fragments in the British Museum was prepared Abraham Sachs during his visit to the British Museum in 1953–1954. The list formed the basis for almost all of Aaboe's work on Babylonian mathematical astronomy. The handwritten notes are by Aaboe

The immediate impact of ACT, then, was to spur further work Babylonian mathematical astronomy. Indeed, it is not too strong to say that between ACT's publication in 1955 and the publication of the Astronomical Diaries and MUL.APIN in the late 1980s, the study of Babylonian astronomy was the study of the mathematical astronomical texts. In part this is due to the success of ACT: Neugebauer's penetrating and systematic mathematical analysis of the various lunar and planetary systems and, crucially, his development of generalized methods for the analyzing step and zigzag functions provided the tools for other scholars to extend his work. But there were other factors: Schaumberger's death, Huber's lack of time to pursue Babylonian astronomy, and Sachs's inability to bring his work on the Diaries to publication. Furthermore, it seems that restricting the field of Babylonian astronomy to the study of mathematical astronomy, was not what Neugebauer himself wished. Neugebauer repeatedly stressed the importance of Sachs's work on the Astronomical Diaries,[71] encouraged Erica Reiner and David Pingree in their publication of the planetary omen texts[72] and Hermann Hunger in his publication of MUL.APIN,[73] and played an active role in enabling the publication of the Diaries by Hunger after Sachs's death.[74] It appears that Neugebauer always retained the hope which he had first expressed in his 1938 manifesto that all types of Babylonian (and Assyrian) astronomical and astrological texts would be published: ACT was only his own contribution towards this larger aim.

References

Aaboe, A. 1955. On the Babylonian origin of some Hipparchian parameters. *Centaurus* 4: 122–125.

Aaboe, A. 1958. On Babylonian planetary theories. *Centaurus* 5: 209–277.

Buja, A., and H.R. Künsch. 2008. A conversation with Peter Huber. *Statistical Science* 23: 120–135.

Fotheringham, J.K. 1933. The indebtedness of Greek to Chaldaean Astronomy. *Quellen und Studien zur Geschichte der Mathematik, Astronomie und Physik* B 2: 28–44 [with notes by O. Neugebauer].

Gelb, I.J. 1958. [Review of Neugebauer, *Astronomical cuneiform texts*]. *Bibliotheca Orientalis* 15: 36–38.

Gelb, I.J. 1964. Introduction. In *The Assyrian dictionary of the oriental institute of the University of Chicago. Part I: A*, ed. A.L. Oppenheim, et al., vii–xxiii. Glückstadt: J. J. Augustin.

[71] For example in the preface to Tuckerman's tables of planetary, lunar and solar positions, Neugebauer wrote of the Astronomical Diaries: "one can hardly expect that sources will become available of greater significance than the almost continuous records of the cuneiform texts for the last centuries before our era. … The exacting work of slowly restoring a huge archive of well over a thousand texts to its full usefulness for the astronomer as well as for the historian of astronomy and the historian of the Hellenistic age is being carried out in all its aspects, philologically, historically, and astronomically, by Professor A. Sachs." (Tuckerman 1962, v).

[72] Reiner and Pingree 1975, 3.

[73] Hunger and Pingree 1989, 1.

[74] Sachs and Hunger 1988, 8.

Huber, P.J. 1956. [Review of Sachs, *Late Babylonian astronomical and related texts*]. *Bibliotheca Orientalis* 13: 231–232.
Huber, P.J. 1957. Zur täglichen Bewegung des Jupiter nach babylonischen Texten. *Zeitschrift fur Assyriologie* 18: 265–303.
Huber, P.J. 1958. Ueber den Nullpunkt der babylonischen Ekliptik. *Centaurus* 5: 192–208.
Hunger, H. 1968. *Babylonische und assyrische Kolophone*. Neukirchen-Vluyn: Butzon & Bercker Kevelaer.
Hunger, H., and D. Pingree. 1989. *MUL.APIN: An astronomical compendium in cuneiform*. Horn: Berger & Söhne.
Kugler, F.X. 1900. *Die Babylonische Mondrechnung*. Frieburg: Herder'sche.
Leichty, E. 1964. The Colophon. In *Studies presented to A. Leo Oppenheim*, 147–154. Chicago: Oriental Institute.
Neuenschwander, E. 1993. Der Nachlass von Erich Bessel-Hagen im Archiv der Universität Bonn. *Historia Matjematica* 20: 382–414.
Neugebauer, O. 1934. *Vorlesungen über Geschichte der antiken mathematischen Wissenschaften. Erster Band: Vorgreichische Mathematik*. Berlin: Springer.
Neugebauer, O. 1936a. [Review of Schaumberger *Sternkunde und Sterndienst in Babel: 3. Ergänzungsheft zum ersten und zweiten Buch*]. *Quellen und Studien zur Geschichte der Mathematik, Astronomie und Physik* B 3: 271–286.
Neugebauer, O. 1936b. Über eine Untersuchungsmethode astronomischer Keilschrifttexte. *Zietschrift der Deutschen Morgenländischen Gesellschaft* 90: 121–134.
Neugebauer, O. 1936c. Jahreszeiten und Tageslängen in der babylonischen Astronomie. *Osiris* 2: 517–550.
Neugebauer, O. 1938. Untersuchungen zur antiken Astronomie I. *Quellen und Studien zur Geschichte der Mathematik, Astronomie und Physik* B 4: 29–33.
Neugebauer, O. 1941. Some fundamental concepts in ancient astronomy. In *Studies in the history of science*, 13–29. Philadelphia: University of Pennsylvania Press.
Neugebauer, O. 1945. Studies in ancient astronomy. VII. Magnitudes of lunar eclipses in Babylonian mathematical astronomy. *Isis* 36: 10–15.
Neugebauer, O. 1951. The study of wretched subjects. *Isis* 42: 111.
Neugebauer, O. 1953. The rising times in Babylonian astronomy. *Journal of Cuneiform Studies* 7: 100–102.
Neugebauer, O. 1955. *Astronomical cuneiform texts*. London: Lund Humphries.
Neugebauer, O. 1975. *A history of ancient mathematical astronomy*. Berlin: Springer.
Oppenheim, A.L. 1958. [Review of Neugebauer, *Astronomical cuneiform texts*]. *Journal of Near Eastern Studies* 17: 157.
Ossendrijver, M. 2011. Exzellente Netzwerke: die Astronomen von Uruk. In *The empirical dimension of ancient near eastern studies*, ed. G.J. Selz and K. Wagensonner, 631–644. Vienna: LIT-Verlag.
Pannekoek, A. 1918. The origin of the Saros. *Koninklije Akademie van Wetenschappen te Amsterdam* 20: 943–955.
Pannekoek, A. 1941. Some remarks on the Moon's diameter and the eclipse tables in Babylonian astronomy. *Eudemus* 1: 9–22.
Pannekoek, A. 1955. [Review of Neugebauer, *Astronomical cuneiform texts*]. *Archive Internationales d'Histoire des Sciences* 35: 281–283.
Pyenson, L. 1995. Inventory and a route to understanding: Sarton, Neugebauer, and sources. *History of Science* 33: 253–282.
Reiner, E. 2002. *An adventure of great dimension: The launching of the Chicago Assyrian dictionary*. Philadelphia: American Philosophical Society.
Reiner, E., and D. Pingree. 1975. *Babylonian planetary Omens 1: The Venus tablet of Ammiṣaduqa*. Malibu: Undena.
Sachs, A. 1948. A classification of the Babylonian astronomical tablets of the Seleucid period. *Journal of Cuneiform Studies* 2: 271–290.

Sachs, A. 1955. *Late Babylonian astronomical and related texts*. Providence: Brown University Press.
Sachs, A., and H. Hunger. 1988. *Astronomical diaries and related texts from Babylonia. Volume I: Diaries from 652 B.C. to 262 B.C.* Vienna: Verlag der Österreichichen Akademie der Wissenschaften.
Sarton, G. 1940. Remarks on the study of Babylonian mathematics. *Isis* 31: 398–404.
Sarton, G. 1952. [Review of Neugebauer]. *The Exact Sciences in Antiquity* 43: 69–72.
Sarton, G. 1955. Chaldaean astronomy of the last three centuries B.C. [Essay review of Neugebauer, *Astronomical cuneiform texts*. *Journal of the American Oriental Society* 75: 166–173.
Schaumberger, J. 1935. *Sternkunde und Sterndienst in Babel: 3. Ergänzungsheft zum ersten und zweiten Buch*. Münster: Aschendorffschen.
Schott, A., and J. Schaumberger. 1941. Vier Briefe Mâr-Ištars an Asarhaddon über Himmelserscheinungen der Jahre –670/688. *Zeitschrift für Assyriologie* 47: 89–130.
Sloley, R.W. 1955. Babylonian astronomy [Review of Neugebauer, *Astronomical cuneiform texts*]. *Nature* 176: 569–570.
Tuckerman, B. 1962. *Planetary, lunar and solar positions 601 B.C. to A.D. 1 at five-day and ten-day intervals*. Philadelphia: American Philosophical Society.
van der Waerden, B.L. 1956. [Review of Neugebauer, *Astronomical cuneiform texts*]. *Zeitschrift der Deutschen Morgenländischen Gesellschaft* 106: 371–372.
van der Waerden, B.L. 1957. Babylonische Planetenrechnung. *Vierteljahrsschrift der Naturforschenden Gesellschaft in Zürich* 102: 39–60.
van der Waerden, B.L. 1966. *Die Anfänge der Astronomie: Erwachende Wissenschaft II*. Groningen: Noordhoff.
Wiseman, D.J. 1988. A note on some prices in late Babylonian astronomical diaries. In *A scientific humanist: Studies in memory of Abraham Sachs*, ed. E. Leichty, M. deJ. Ellis, and P. Gerardi, 363–373. Philadelphia: The University Museum.

Translating Babylonian Mathematical Astronomy: Neugebauer and Beyond

Mathieu Ossendrijver

Otto Neugebauer's involvement with Babylonian mathematical astronomy, one of the central topics of his research, can be traced through more than 30 publications stretching over the period from 1936 to 1991.[1] Foremost among these are *Astronomical Cuneiform Texts* (1955), a critical edition of the complete corpus known at the time, and *History of Ancient Mathematical Astronomy* (1975), a comprehensive survey of the astronomical algorithms from Babylonia and other ancient cultures. In this paper I aim to discuss Neugebauer's approach to the translation of Babylonian mathematical astronomy and assess it in the light of subsequent research. Apart from the critical editions contained in ACT and elsewhere we can hope to learn something about this topic from his other works, since he often displays a profound interest in methodological issues. However, Neugebauer rarely discussed his method of translation, focussing instead on the mathematical methods developed by him for analysing astronomical tables and reconstructing the underlying algorithms and empirical data. Nevertheless, these aspects of Neugebauer's methodology turn out to be relevant for understanding his approach to translation. I therefore begin by exploring the broader methodological framework underlying Neugebauer's research on Babylonian mathematical astronomy.

A few introductory remarks about this corpus should suffice to provide some necessary background for the discussion. Babylonian mathematical astronomy comprises about 440 cuneiform tablets and fragments from Babylon and Uruk, all written within the period 450–50 BC. This makes Babylonian mathematical astronomy the earliest known form of mathematical astronomy of the ancient world. A distinctive feature of these texts, which represent only a small fraction of the Babylonian astronomical corpus, is that astronomical quantities are computed with mathematical algorithms. The corpus can be divided into tabular texts, currently

[1] Aaboe et al. 1991.

M. Ossendrijver (✉)
Excellence Cluster TOPOI/Institute of Philosophy, Humboldt University, Berlin, Germany
e-mail: mathieu.ossendrijver@hu-berlin.de

A. Jones et al. (eds.), *A Mathematician's Journeys*, Archimedes 45,
DOI 10.1007/978-3-319-25865-2_11

numbering about 330, and procedure texts, numbering about 110. The former contain computed data for the Moon and the planets, e.g. positions and times, arranged into columns and rows; the latter contain detailed instructions for computing and verifying these tables. All computations are based on the sexagesimal place-value system which, along with some of the mathematical terminology, were handed down from Old Babylonian mathematics, which had flourished more than a 1000 years earlier.

When Neugebauer directed his attention to Babylonian mathematical astronomy in the mid 1930s he had just finished *Mathematische Keilschrift-Texte* (MKT), a critical edition of the Babylonian mathematical corpus. Since these texts have so much in common with Babylonian mathematical astronomy he was able to draw for his new project upon expertise acquired while writing MKT. As I will argue later, his approach to the translation of Babylonian mathematical astronomy can be partly traced back to MKT, but there are significant differences as well. Neugebauer's role in each of these fields was also different. Unlike Babylonian mathematics, where he had himself been the main pioneer, mathematical astronomy was already a well-established discipline in the mid 1930s. During the preceding 55 years a lot of groundbreaking research had been done by Joseph Epping, Franz Xaver Kugler, Johann Schaumberger and a few other scholars. Kugler had written two comprehensive standard works: *Die babylonische Mondrechung* (1900) and *Sternkunde und Sterndienst in Babel* (1907–1924), to which Schaumberger (1935) had added an important supplement. Using only a modest number of tablets, mostly copied in London by the Assyriologist J. N. Strassmaier, these pioneers had succeeded in reconstructing many of the lunar and planetary algorithms and establishing their astronomical significance. Neugebauer repeatedly expressed his admiration for their achievements, for instance in a review of Schaumberger's supplement (Neugebauer 1936a). Elsewhere he acknowledged that "the period of uncovering has now essentially ended" (Neugebauer 1936c), at the same time affirming that much remained to be done, namely (the emphasis is that of Neugebauer):

> to carry out the task [...] of discovering the *methods* of this discipline so accurately that one is able (in principle) to solve their problems *with their own means*. [...] In spite of the large number of texts already edited this source material is still full of lacunae and in itself so many-faceted, that it is presently not possible at all to proceed in a truly, strictly Babylonian manner, without having to resort to modern tools at every step.

Neugebauer saw it as his goal to uncover the Babylonian methods, but for the time being the fragmentary nature of the source material forced him to pursue this with modern tools. What kind of tools he meant is revealed in the paper "Über eine Untersuchungsmethode astronomischer Keilschrifttexte" written that same year (Neugebauer 1936b). It is here that we encounter for the first time some of Neugebauer's characteristic mathematical and graphical methods: diophantine equations for analysing astronomical tables and graphs of zigzag functions and step functions. In these years he also introduced several new notations that are still widely used today, e.g. for transliterating sexagesimal numbers (Neugebauer 1932), for the nomenclature of the astronomical functions (Neugebauer 1936c) and Greek capitals for the synodic phenomena of the planets (Neugebauer 1952). Methodological issues again feature prominently in a long series of papers titled "Untersuchungen zur antiken Astronomie" (UAA), five of which appeared alone in 1938 in *Quellen und Studien zur Geschichte der Mathematik, Astronomie und Physik*, a new journal cofounded by Neugebauer (Neugebauer 1938a, b, c, d). Their purpose was to provide auxiliary investigations that need to be carried out before the critical edition can be produced. The edition itself, which was to become ACT, was announced as follows in UAA I:

> I will in the first part ("Texts") of this edition of the mathematical-astronomical cuneiform texts (*Mathematisch-Astronomische Keilschrift-Texte)* which is in progress (I will henceforth cite it as MAKT) publish the entire textual material as completely as possible.

In a second volume to be called "Explanations" he intended to lay out his mathematical methods for dating and reconstructing the astronomical tables. That Neugebauer remained skeptical about the possibility to reconstruct the methods of the Babylonians is clear from the final volume, to be called "Lectures", in which he hoped to "sketch only in rough strokes the foundations (Grundgedanken) of Babylonian astronomy". In UAA II (Neugebauer 1938b) he returned to his auxiliary investigations by presenting new mathematical tools for dating and reconstructing tables of lunar system A, an endeavor which he characterised in the introduction with typical sharpness:

> The method of the present paper has nothing to do with astronomy, nor with history. It only fulfills a task for a certain group of astronomical cuneiform texts which would otherwise fall on the custodian of a museum [namely joining tablet fragments] [...] It is essential to emphasise that the solution of this task becomes possible here without any hypothesis about the content of the texts, since nothing else is used but the generative laws of the series of numbers that are empirically derived from the fragments.

Neugebauer thus viewed these mathematical tools as constituting a stage of interpretation wholly separate from astronomical issues. In UAA III (Neugebauer 1938c) he set out to "discover the methods" as promised earlier (Neugebauer 1936c). Hence the aim was now not to date or reconstruct tables as in UAA II, but:

> The present paper confronts the exactly opposite aim: to recover the lines of thought and the empirical data that led to the computation of these series of numbers. Hence this centers on the theory contained in certain text passages as opposed to a purely formal treatment connected with the bare problem of reconstruction and dating.

The two interpretative steps envisioned by Neugebauer can be summarised as follows: one ("the formal treatment") is aimed at dating and reconstructing tables, the other at reconstructing the underlying astronomical theories and empirical data. They were to remain the pillars of his research on Babylonian astronomy. Common to both is that they are achieved through mathematical analysis. At this stage of the project, issues of translation are notably absent—the entire focus of his discussions is on the tabular texts with their sequences of numbers. Neugebauer's strongly mathematical perspective may be attributed to his background in mathematics. Only later, when he begins to tackle the procedure texts, does translation become an issue.

This contrasts with the more eclectic approach of his predecessor Kugler, who from the very beginning combined the analysis of tables with efforts to translate procedure texts (notably so in Kugler 1900). His mathematical outlook also permeated his view on the nature of Babylonian mathematical astronomy, as expressed e.g. in UAA II (p. 196):

> Here [in Babylonia], for the first time in the history of mankind, one has succeeded in controlling (beherrschen) the laws of a very complicated natural phenomenon through purely mathematical methods.

Hence Babylonian mathematical astronomy is, in its core, perceived as a form of pure mathematics. Elsewhere he expressed the primacy of mathematics in ancient astronomy using even more drastic terms (Neugebauer 1946):

> For methodological reasons it is obvious that a drastic restriction in terminology must be made. We shall here call "astronomy" only those parts of human interest in celestial phenomena which are amenable to mathematical treatment.

However, Neugebauer's views on Babylonian astronomy are certainly more complex than what is suggested by these sharply formulated passages. He repeatedly displayed flexibility by changing a previous view, for example on the relation between astronomy and astrology, a contentious issue for many historians of astronomy of his age. While he flatly denied the possibility of a beneficial relationship between the two in his early years[2] he later came close to the opposite conclusion (Neugebauer 1975, p. 475). As will be argued later, it is less obvious that his views on the relation between astronomy and mathematics changed significantly. His work on the edition of Babylonian mathematical astronomy was interrupted in 1938 by the events in Germany that eventually led to his relocation to Providence (USA) in 1939. Further delays resulted from discoveries, made between 1945 and 1952, of significant numbers of unpublished texts in Chicago, Istanbul and London.[3] In 1941 he was able to enlist the Assyriologist Abraham Sachs, who became his main collaborator on Mesopotamian subjects. Sachs was important for the project not only because he discovered numerous new fragments with mathematical astronomy in the British Museum, but also because his profound expertise in Assyriology left its traces in Neugebauer's critical editions, including his translations. When ACT finally appeared in 1955, almost 20 years after having been announced, it is essentially the first publication in which Neugebauer is concerned with the translation of procedure texts.[4] A tremendous amount of research on the technical terminology underlying these texts is compiled in the glossary (Vol. II) and summarised in the Introduction (Vol. I). However, we look in vain for a substantial discussion of the method of translation. One passage in which this topic is addressed, if not directly, is found in the Introduction (pp. 3–4), where he explains the conventions underlying his transliterations:

---

[2] E.g. Neugebauer 1938c, 196, n. 2.

[3] Cf. the Preface in ACT (pp. xi–xii).

[4] On one prior occasion Neugebauer (1953) published a few passages from a procedure text (ACT 200).

> Serious problems arise, however, in the transcription of the procedure texts. Here we meet a great number of technical terms of wholly unknown reading, if not unknown meaning. We are far from being able to give the Akkadian correspondences for many words, not to mention details such as determining the special verbal forms, etc. to be used.

The only direct reference to his method of translation occurs at the beginning of the Glossary on p. 467:

> The translations given are not intended to be strictly literal but rather try to convey the general meaning, especially for technical terminology.

From these passages we can nevertheless glean the justifiable reason why Neugebauer did not want to spend much time on translation issues. The procedure texts are full of technical terms written as (pseudo-) Sumerian logograms or phonetic abbreviations that are unique to the corpus and whose Akkadian reading was not always known in Neugebauer's time. Rather than speculate about which Akkadian words may underlie the badly known logograms he preferred to infer their technical mathematical or astronomical meaning from the context, thus skipping the step of translating the Akkadian terminology. He was therefore not aiming for semantic accuracy as required by modern standards of translation, but for pragmatic adequacy. As he admits, the resulting translations are not literal but, on closer inspection, especially regarding the technical terminology, modernising representations of the assumed pragmatic meanings of the cuneiform signs. This approach is consistent with Neugebauer's strongly mathematical perspective on ancient astronomy. From the outset he aimed to reconstruct the methods of Babylonian astronomy primarily through mathematical analysis of the astronomical tables. The translation and interpretation of procedure texts thus proceeded within a conceptual framework rooted in modern mathematics in which the precise Babylonian formulation, even if it could be established, appeared less relevant.

What kind of English terms did Neugebauer use in his translations of the astronomical procedures and what can we say about the pragmatic meanings that they represent? In order to answer these questions we have to turn to the translations and the glossary. Not surprisingly, the technical terminology of his translations is borrowed from two sources: modern mathematics and modern astronomy. The former is evident in the arithmetical terminology. Different logograms or Akkadian terms representing apparently synonymous arithmetical terms are translated by Neugebauer with one and the same English term. A similar approach had been identified by J. Høyrup (1996) in Neugebauer's translations of Old Babylonian mathematical problem texts, where this led to certain distortions. In particular it had the effect of obscuring the geometrical nature of the arithmetical operations in these texts (Høyrup 2002). The same modernising and homogenising tendencies can be observed regarding the arithmetical terms in the astronomical procedure texts.[5] For instance, several apparently synonymous words for additive and subtractive operations are translated with the same word "to add" or "to subtract". However, the consequences are less severe than in Old Babylonian mathematics. The geometrical

[5] For a detailed discussion of the terminology of the procedure texts cf. Ossendrijver 2012.

operations that are characteristic of Old Babylonian mathematics are virtually absent from the astronomical procedures, and the arithmetical terms that replace them, mostly innovations of the Late Babylonian period, do not have significant geometrical connotations that could be lost in Neugebauer's translation. The only semantic differentiation that can be recognised among some of these terms concerns the identity of the involved quantities and the symmetry of the operation. Both notions manifest themselves most clearly in the additive operations. For instance, addition by means of "appending" results in the sum inheriting the identity of the quantity to which something is "appended", i.e. the identity of one quantity is conserved. This occurs in situations where an increment is added to a named quantity—for instance a displacement to a position, resulting in an updated position. When the identity of the sum is different from that of any summand, i.e. there is a loss of identity, the expressions "to append together with" or "to accumulate" are used. On the other hand, the four different subtractive operations that occur in the astronomical procedures ("to subtract", "to diminish", "to deduct", "to tear out")[6] do not reflect a clear semantic differentiation with regard to the identity of quantities. The notion of symmetry applies when the quantities involved in an operation can be exchanged without a change of meaning. As in Old Babylonian mathematics, a rather strict semantic distinction is maintained between symmetric addition, for which the verb "to accumulate" is used, and asymmetric addition, which is expressed by the verb "to append". These rather subtle semantic features may have helped the Babylonian astronomers in keeping track of the astronomical meaning of different quantities during the execution of algorithms, some of which involve numerous steps and auxiliary quantities. Although they are almost completely lost in Neugebauer's translations, his interpretations of the algorithms are not really affected by this.

The second source from which Neugebauer borrowed terminology for his translations is modern astronomy. Perhaps the most revealing example concerns the quantity represented in cuneiform by the logograms **NIM** and **SIG**, which Neugebauer translated as "positive" and "negative" (latitude) (cf. the corresponding entries on p. 485 in the Glossary to ACT). At this point Neugebauer's approach to translation turns out to cause considerable confusion. The modern astronomical concept of latitude introduced by him into the Babylonian procedures allows for positive and negative values corresponding to positions below and above the ecliptic, respectively. However, the literal translation of **NIM** and **SIG** is "height" and "depth", both of which denote (positive) distances with respect to the ecliptic. In fact a concept of negative numbers exists neither in Babylonian mathematical astronomy nor in Babylonian mathematics (Høyrup 1993). Neugebauer was certainly aware of this since he avoided using the word "negative" in translations of Old Babylonian mathematical problem texts. For instance, (positive) numbers that are subtracted from other numbers are carefully referred to as "subtractive" rather than "negative" numbers. For some reason he was less hesitant to introduce negative numbers into his translations of Babylonian mathematical astronomy, not only for

[6] Cf. Ossendrijver 2012, 22–24.

the concept of latitude, but also for subtractive quantities, which have now become "negative".[7] When translating **SIG** as "negative latitude" he presumably kept in mind that this is actually a positive quantity, expecting us to do the same. Obviously things can go wrong here. If one adds something to, or subtract something from, the supposedly negative latitude the result will coincide with the Babylonian computation only if one engages in some mental gymnastics. Examples of that can be found in the lunar procedures.

Figure 1 shows two configurations of the Moon moving towards the ecliptic and the point diametrically opposite to the Sun before, during and after Full Moon.[8] They illustrate step 6 of the so-called Lunar Six module, a complex, 13-step algorithm for computing six time intervals between the rising or setting of the Moon and that of the Sun near New Moon or Full Moon.[9] In step 6 the Moon's distance to the ecliptic is computed for the sunrise or sunset immediately before or after Full Moon from its value at Full Moon by adding or subtracting a correction. In Fig. 1 the Moon is moving towards the ecliptic; analogous graphs can be made for the case when the Moon moves away from the ecliptic. These drawings cannot be found in Neugebauer's works, but he did make similar ones for other steps of the Lunar Six module. Furthermore, his astronomical commentary leaves no doubt that they rep-

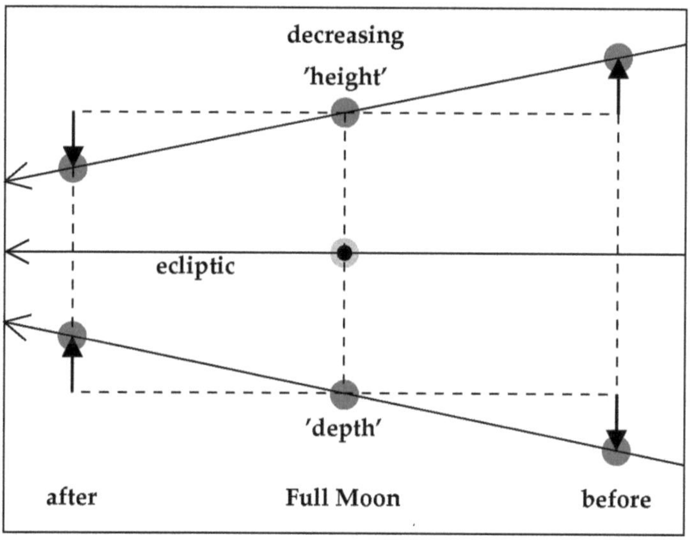

Fig. 1 Schematic view of the sky showing the Moon before, during and after Full Moon and the point on the ecliptic diametrically opposite to the Sun. *Arrows* indicate the direction of motion. *Top*: Moon above the ecliptic and descending. *Bottom*: Moon below the ecliptic and ascending

[7] Cf. for instance the entry 'lal (≈ *maṭû*) "negative"' in the Glossary to ACT (p. 481).

[8] The same arguments presented here apply to the alternative situation when the Moon moves away from the ecliptic.

[9] Ossendrijver 2012, 161–178.

resent his interpretation of the algorithm, which remains valid today. Since step 6 is not represented by a column in the lunar tables, its reconstruction is entirely based on several passages in the procedure texts, which I translate as follows[10]:

> You put down the Moon's "height and depth".

This refers to the Moon's latitude at Full Moon, which is considered known, so that it can be "put down". If the Lunar Six interval occurs after Full Moon (case 1) this is followed by:

> You multiply the Moon's displacement by 0;4 and you add it with the Moon's "height and depth" if it is increasing, subtract if it is decreasing.

Here the Moon's displacement times 0;4 represents the correction to be applied to the Moon's latitude at Full Moon. If the Lunar Six interval occurs before Full Moon (case 2) it is followed by:

> You multiply the Moon's displacement by 0;4 and you subtract it from the Moon's "height and depth" if it is increasing, add if it is decreasing.

In ACT things go wrong here. Neugebauer assumes a scribal error in the procedure for case 2 (p. 234), because he thinks that it should be identical with case 1. He thus arrives at the following identical translation for both cases:

> The positive or negative latitude of the Moon (at opposition) you shall put down.
> Multiply the distance traveled by the Moon (since opposition) by 0;4.
> With the positive or negative latitude of the Moon you shall add in case of increasing values, you shall subtract in case of decreasing values.

By doing so he overlooked that the additive or subtractive sense of the correction to be applied to the Moon's latitude at Full Moon is reversed in case 2, because one is here going back in time. A careful analysis of the Babylonian formulation reveals that there is no scribal error, so that Neugebauer's emendation of the text is erroneous. His mistake may be a consequence of the mental gymnastics required by his use of "negative latitude". He must have noticed that here, more than in any other procedure text, his modernising translation was causing confusion. However, my guess is that Neugebauer did not spend much time on a detailed interpretation of the Babylonian formulation of step 6. Already a casual reading of the procedures was probably sufficient for him to conclude what their mathematical and astronomical interpretation had to be. As soon as he established this, the actual formulation used by the Babylonian astronomers was no longer very interesting to him. In other words, even though his translation paradigm broke down, it did not prevent him from establishing the correct mathematical and astronomical interpretation of the procedure. Neugebauer's mathematical and astronomical skills enabled him to "discover the methods" of the Babylonian astronomers, which for him meant their algorithms construed in modern terms. The fact that his mathematical and astronomical interpretations in ACT and HAMA remain largely valid, with mostly minor changes here and there, may justify that approach. Subsequent research on Babylonian

---

[10] Ossendrijver 2012, 395–409. Case 1 corresponds to NA and GI_6, case 2 to $ŠU_2$ and ME.

mathematical astronomy has proceeded from the advanced stage where Neugebauer left it, by redefining what it means to "discover the methods". In particular, the aim has shifted from understanding Babylonian mathematical astronomy in our own terms to interpreting its algorithms and the underlying mathematical and astronomical concepts in Babylonian terms. In the same spirit the internalist approach with its focus on the reconstruction of algorithms and the underlying empirical data is making way for a more holistic one that aims to explain Babylonian astronomy in its institutional, political, religious and social contexts. The following statement by Neugebauer (1946) may give some hope that he would approve of these developments:

> Science must work with methods and must consider its problems from viewpoints which correspond to the methods and standards of other branches of historical research. The idea must definitely be abandoned that the history of science must adapt its level to the alleged requirements of the teaching of the modern fields of science.

References

Aaboe, A., J.P. Britton, J.A. Henderson, O.N. Neugebauer, and A.J. Sachs. 1991. Saros cycle dates and related Babylonian astronomical texts. *Transactions of the American Philosophical Society* 81(6): 1–75.

Høyrup, J. 1993. On subtractive operations, subtractive numbers, and purportedly negative numbers in old Babylonian mathematics. *Zeitschrift für Assyriologie* 83: 42–60.

Høyrup, J. 1996. Changing trends in the historiography of mesopotamian mathematics: An insider's view. *History of Science* 34: 1–32.

Høyrup, J. 2002. *Lengths, widths, surfaces. A portrait of old Babylonian algebra and its kin*, Sources and studies in the history of mathematics and physical sciences. New York: Springer.

Kugler, F.X. 1900. *Die babylonische Mondrechnung. Zwei Systeme der Chaldäer über den Lauf des Mondes und der Sonne*. Freiburg i.Br: Herder Verlag.

Kugler, F.X. 1907–1924. *Sternkunde und Sterndienst in Babel*, Buch I-II. Münster: Aschendorffsche Verlagsbuchhandlung.

Neugebauer. 1932. Zur Transkription mathematischer und astronomischer Keilschrifttexte. *Archiv für Orientforschung* 8:221–223.

Neugebauer, O.N. 1936a. *Quellen und Studien zur Geschichte der Mathematik, Astronomie und Physik* B3, 271–286 [review of J. Schaumberger, *3. Ergänzungsheft zu F.X. Kugler, S.J., Sternkunde und Sterndienst in Babel*].

Neugebauer, O.N. 1936b. Über eine Untersuchungsmethode astronomischer Keilschrifttexte. *Zeitschrift der Deutschen Morgenländischen Gesellschaft* 90: 121–134.

Neugebauer, O.N. 1936c. Jahreszeiten und Tageslängen in der babylonischen Astronomie. *Osiris* 2: 517–550.

Neugebauer, O.N. 1938a. Untersuchungen zur antiken Astronomie I. *Quellen und Studien zur Geschichte der Mathematik, Astronomie und Physik B* 4: 29–33.

Neugebauer, O.N. 1938b. Untersuchungen zur antiken Astronomie II. Datierung und Rekonstruktion von Texten des Systems II der Mondtheorie. *Quellen und Studien zur Geschichte der Mathematik, Astronomie und Physik B* 4: 34–91.

Neugebauer, O.N. 1938c. Untersuchungen zur antiken Astronomie III. Die babylonische Theorie der Breitenbewegung des Mondes. *Quellen und Studien zur Geschichte der Mathematik, Astronomie und Physik B* 4: 193–346.

Neugebauer, O.N. 1938d. Untersuchungen zur antiken Astronomie V. Der Halleysche "Saros" und andere Ergänzungen zu UAA III. *Quellen und Studien zur Geschichte der Mathematik, Astronomie und Physik B* 4: 407–411.
Neugebauer, O.N. 1946. History of ancient astronomy: Problems and methods. *Publications of the Astronomical Society of the Pacific* 58(17–43): 104–142.
Neugebauer, O.N. 1952. *The exact sciences in antiquity*. Princeton: Princeton University Press [second edition: 1957, 1969. New York: Dover Publications].
Neugebauer, O.N. 1953. The rising times in Babylonian astronomy. *Journal of Cuneiform Studies* 7: 100–102.
Neugebauer, O.N. 1955. *Astronomical cuneiform texts*. London: Lund Humphries.
Neugebauer, O.N. 1975. *A history of ancient mathematical astronomy*. New York: Springer.
Ossendrijver, M. 2012. *Babylonian mathematical astronomy: Procedure texts*, Sources in the history of mathematics and physical sciences. New York: Springer.
Schaumberger, J. 1935. *Sternkunde und Sterndienst in Babel. Assyriologische, astronomische und astralmythologische Untersuchungen. 3. Ergänzungsheft*. Münster: Aschendorffsche Verlagsbuchhandlung.

MIX
Papier aus verantwortungsvollen Quellen
Paper from responsible sources
FSC® C105338

If you have any concerns about our products,
you can contact us on
ProductSafety@springernature.com

In case Publisher is established outside the EU,
the EU authorized representative is:
Springer Nature Customer Service Center GmbH
Europaplatz 3, 69115 Heidelberg, Germany

Printed by Libri Plureos GmbH
in Hamburg, Germany